Latest Developments in Reality-Based 3D Surveying and Modelling

Fabio Remondino, Andreas Georgopoulos,
Diego González-Aguilera, Panagiotis Agrafiotis
(Eds.)

Latest Developments in Reality-Based 3D Surveying and Modelling

MDPI

Editors

Fabio Remondino
Bruno Kessler Foundation (FBK)
Italy

Andreas Georgopoulos
National Technical University of Athens
Greece

Diego Gonzalez-Aguilera
University of Salamanca
Spain

Panagiotis Agrafiotis
Cyprus University of Technology
and National Technical University of Athens
Greece

Editorial Office
MDPI AG
St. Alban-Anlage 66
Basel, Switzerland

Publisher
Shu-Kun Lin

Production Editor
Elvis Wang

First Edition 2018

MDPI • Basel • Beijing • Wuhan • Barcelona • Belgrade

ISBN 978-3-03842-684-4 (Pbk)

ISBN 978-3-03842-685-1 (PDF)

Cover photo: Rendering of the 3D reconstruction of the Maritime Theatre at Hadrian's Villa, Italy.

Table of Contents

Chapter 1. Latest Developments in Data Acquisition and Processing

Chapter 2. Sensor Fusion and Data Integration

Chapter 3. 3D Modelling and VR/AR

Chapter 4. Underwater 3D Surveying and Modelling

Chapter 5. BIM and HBIM

About the Editors

Fabio Remondino received his PhD in Photogrammetry in 2006 from ETH Zurich, Switzerland and he now leads the 3D Optical Metrology unit (http://3dom.fbk.eu) of the Bruno Kessler Foundation (http://www.fbk.eu), a public research center in Trento, Italy. His research interests include geospatial data collection and processing, heritage documentation, 3D modelling, sensor and data integration. He is author of over 200 scientific publications in journals and at international conferences, 7 books, 8 book chapters, 8 special issues in journals and 15 conference proceedings. He has received 10 best paper awards at conferences and organized more than 25 scientific events, 20 summer schools and 10 tutorials. He is serving as President of ISPRS Technical Commission II, Vice-President of CIPA Heritage Documentation and Vice-President of EuroSDR.

Andreas Georgopoulos is Professor of Photogrammetry and Director of the Lab of Photogrammetry of the School of Rural & Surveying Engineering of NTUA. He holds a Diploma of Surveying (NTUA 1976) and an MSc (1977) and a PhD (1981) in Photogrammetry from University College London. He has been teaching Photogrammetry and Documentation of Monuments since 1980 in UCL, NTUA and as visiting professor in KU Leuven (RLICC), CUT (Dept. of Civil Eng.) and University of Aegean. He has been Vice-Head (1998–2002) and Head (2002–2006) of the School of Rural & Surveying Eng. and member of the Research Committee of NTUA since 1999. Since 2006, he is a member of the Executive Board of CIPA-Heritage Documentation and has served as Secretary general (2010–2014), while currently he is President. He has published more than 250 scientific papers in international journals and conference proceedings. His research interests focus on 3D modelling of cultural heritage, photogrammetric automation and digital contemporary techniques.

Diego González-Aguilera received his B.Sc. degree in Surveying Engineering and M.Sc. degree in Geodesy and Cartography Engineering from Salamanca University, Spain, in 1999 and 2001, respectively. He was a Research Assistant at INRIA, Grenoble, France, in the Institute of Computer Vision and Robotics, where he also conducted his Ph.D. research on 3D reconstruction from a single view, graduating in 2005. He has authored over 150 research articles in international journals and conference proceedings. He is co-inventor of 11 patents and 21 intellectual properties (software). Based on his PhD results, he has received seven international awards from the International Societies of Photogrammetry and Remote Sensing (ISPRS/ASPRS) and 11 national awards, one of them awarded by the Royal Academy of Engineering. In 2005, he founded the research group TIDOP (http://tidop.usal.es) (Geotechnologies for the 3D digitalization and modelling of complex objects) at the University of Salamanca. He is Full Professor and Head of the Cartographic and Land Engineering Department at the University of Salamanca, co-founder and CEO of the start-up ITOS3D (Image TO Smart 3D).

Panagiotis Agrafiotis is a PhD student in the Lab of Photogrammetry of National Technical University of Athens (https://www.ntua.gr) and a researcher in the area of underwater 3D modelling and mapping in the Photogrammetric Vision Lab of Cyprus University of Technology (https://www.cut.ac.cy). He holds a Diploma in Rural and Surveying Engineering and a M.Sc. in Geoinformatics. In recent years, he has worked on various EU research projects with the responsibility of applied Computer Vision algorithms design and 3D modelling and mapping. He was a member of the Local Organizing Committee of 3D ARCH 2017 international workshop while he is serving as secretary of the ISPRS WG II/9: Underwater Data Acquisition and Processing (http://www2.isprs.org/commissions/comm2/wg9.html).

Preface

Reality-based 3D surveying and modelling are playing an important role in various domains and applications. They are the protagonist in the documentation, preservation and valorisation of Cultural Heritage but also for city planning, energy assessment and audit, territorial monitoring, hazard recording, etc.

Recent technological advances supported by the development of fast and efficient algorithms and computer power have enabled the gradual adoption of the above by almost all scientific communities including Cultural Heritage experts. Consequently, the members of this community follow and study reports about this progress with great interest. Specialized fora on these very interesting and attractive topics, such as conferences, workshops and summer schools, are always well attended by multi-disciplinary audiences.

This book originates from the ISPRS/CIPA Workshop "3D-ARCH 2017—3D Virtual Reconstruction and Visualization of Complex Architectures", which was held in March 2017 in Nafplio, Greece. The workshop's main scope was to bring together scientists, developers and advanced users in 3D surveying and data processing and to encourage cooperation and practice sharing in the various fields where 3D technologies are nowadays used. The workshop focused primarily on multi-source and multi-sensor approaches; low-cost sensors and open-source algorithms for terrestrial 3D modelling; automation in data registration; image matching and 3D reconstruction; point cloud analysis; 4D modelling; Building Information Modelling/Heritage Building Information Modelling (BIM/HBIM) and procedural modelling; accuracy requirement and assessment in 3D reconstructions; 3D applications in terrestrial and underwater environments; Virtual and Augmented Reality (VR/AR) applied to the visualization and conservation of complex architectures and heritage. The most exciting and innovative papers presented at the workshop were selected to be extended and included in this cornerstone collection.

In this book, composed of 16 peer-reviewed articles structured in five chapters, different viewpoints and experiences are presented, each related to a different aspect of 3D surveying and modelling of terrestrial scenes.

CHAPTER 1—Latest developments in data acquisition and processing

New sensors and algorithm developments are described in complex Cultural Heritage applications.

This chapter reports the latest developments in photogrammetry and laser scanning, with particular attention to fish-eye photogrammetry, SLAM methods, point cloud generation and processing.

CHAPTER 2—Sensor fusion and data integration

For the benefit of the final 3D results, new developments both in hardware and software integration are taking place. Hence, newly developed sensors and their data are combined in order to produce enhanced results.

This chapter includes interesting results in the field of data and sensor fusion. Experiences range from small artefacts to large heritage sites or hazarded areas.

CHAPTER 3—3D modelling and VR/AR

Virtual and Augmented Reality have become more accessible in the last 10 to 15 years. It was inevitable that they would find a niche in Cultural heritage documentation.

This chapter shows open challenges and developments for heritage digitization and 3D data access through VR/AR solutions.

CHAPTER 4—Underwater 3D surveying and modelling

Almost half of mankind's Cultural Heritage lies underwater, a hostile environment for humans posing special challenges for 3D surveying and modelling.

This chapter reports the latest developments in underwater 3D documentation, including image enhancement, visual odometry and 3D reconstruction.

CHAPTER 5—BIM and HBIM

BIM and HBIM are slowly becoming useful tools for professionals, although open issues with regard to geometries, semantics, databases and formats are still present.

This chapter reports the latest developments in BIM and HBIM realizations using commercial and open source tools.

We would like to thank all authors who have submitted their extended manuscripts to be compiled in this MDPI book and all reviewers for their valuable work during the reviewing process.

Enjoy the reading!

Fabio Remondino, Andreas Georgopoulos,
Diego González-Aguilera and Panagiotis Agrafiotis
Editors

Chapter 1

Latest Developments in Data Acquisition and Processing

Fisheye Photogrammetry to Survey Narrow Spaces in Architecture and a Hypogea Environment

Luca Perfetti [a], **Carlo Polari** [a], **Francesco Fassi** [a], **Salvatore Troisi** [b], **Valerio Baiocchi** [c], **Silvio Del Pizzo** [b], **Francesca Giannone** [d], **Luigi Barazzetti** [e], **Mattia Previtali** [e] and **Fabio Roncoroni** [f]

[a] 3D Survey Group, Politecnico di Milano, Architecture, Built environment and Construction engineering (ABC) Department, Milan, Italy; (luca.perfetti, carlo.polari, francesco.fassi)@polimi.it

[b] Centro Direzionale Isola C4, Parthenope University of Naples, Naples, Italy; (salvatore.troisi, silvio.delpizzo)@uniparthenope.it

[c] DICEA, Sapienza University of Rome, Rome, Italy; valerio.baiocchi@uniroma1.it

[d] Niccolò Cusano University, Rome, Italy; francesca.giannone@unicusano.it

[e] Gicarus, Politecnico di Milano, Architecture, Built environment and Construction engineering (ABC) Department, Milan, Italy; (luigi.barazzetti, mattia.previtali)@polimi.it

[f] Gicarus, Polo Territoriale di Lecco, Lecco, Italy; fabio.roncoroni@polimi.it

Abstract: Nowadays, the increasing computation power of commercial grade processors has actively led to a vast spreading of image-based reconstruction software as well as its application in different disciplines. As a result, new frontiers regarding the use of photogrammetry in a vast range of investigation activities are being explored. This paper investigates the implementation of fisheye lenses in non-classical survey activities along with the related problematics. Fisheye lenses are outstanding because of their large field of view. This characteristic alone can be a game changer in reducing the amount of data required, thus speeding up the photogrammetric process when needed. Although they come at a cost, field of view (FOV), speed and manoeuvrability are key to the success of those optics as shown by two of the presented case studies: the survey of a very narrow spiral staircase located in the Duomo di Milano and the survey of a very narrow hypogea structure in Rome. A third case study, which deals with low-cost sensors, shows the metric evaluation of a commercial spherical camera equipped with fisheye lenses.

Keywords: fisheye; photogrammetry; videogrammetry; narrow spaces; 3D modelling

1. Introduction

Fisheye camera models for photogrammetric applications were extensively studied, tested and validated in the first decade of the 2000s. Calibration procedures were presented by Abraham and Förstner (2005), Schwalbe (2005), Van den Heuvel et al. (2006) and Schneider et al. (2009), among others.

The recent introduction of the fisheye camera model in some commercial packages for automated image-based 3D modelling (such as Agisoft PhotoScan, Pix4D and Bentley ContextCapture) has allowed both professional and "less expert" users to generate 3D models in a fully automated way, starting from a set of digital images. Results presented in Strecha et al. (2015) confirm the new level of automation achievable for the different steps of the image modelling workflow: camera calibration, dense matching and surface generation.

Such a level of automation for fisheye cameras is quite similar to the automation already achievable in projects based on central perspective cameras (pinhole cameras). However, the risk of unreliable and "crude" digital reconstructions because of the lack of expertise in basic surveying concepts has already been described in Nocerino et al. (2014), in which the authors presented inaccurate reconstructions obtained from pin-hole (rectilinear projection) images.

In the case of a fisheye lens, the short focal length coupled with an extreme distortion makes automated 3D modelling more complicated. This could provide inaccurate 3D models without metric integrity.

The incorporation of the fisheye camera model in commercial software is a clear indicator of how users are becoming more familiar with such distorted projections, not only for photographic purposes but also for metric applications. Nowadays, automated fisheye image processing is possible without turning them into pinhole images.

This paper presents three different case studies with the aim of testing the current state-of-the-art regarding the usability of fisheye optics into different productive workflows.

The manuscript is divided as follows:

- The introduction opens with a small summary of the differences that exist among the available optical projections;
- in Section 2, the authors explain the motivations behind the three case studies presented here;
- Sections 3,4,5 deepen the topic by trying to evaluate whether it is currently possible, or not, to implement a functional workflow based on fisheye lenses, in particular: for 3D architectonic modelling, the localisation of narrow spaces and the extraction of fast shapes and volumes.
- Finally, Section 6 draws conclusions from the results.

1.1. Difference between Fisheye Lenses and Rectilinear Lenses

The main issue with regard to using fisheye lenses concerns the high probability of obtaining incomplete, weak and inconsistent results—a consequence of the considerable radial distortion. The main drawback appears to be the difficulty to take control of the variables that could invalidate the photogrammetry process: first of all, the questionable approximation of the radial distortion by the radial distortion coefficients (K1,K2,K3,...), and secondly, the unfamiliar correlation between the fisheye projection and the design of the capturing phase.

This happens when one tries to use the same consolidated pipeline designed for the rectilinear projection lenses with fisheyes. A fisheye lens is not a rectilinear lens: it is critical to understand the difference between the different optical projections in order to avoid incorrect survey planning in the first place.

A very wide FOV and a very short focal lens do not make a lens a fisheye. It is the particular interaction between the two, focal length and FOV, which makes the difference. The relation between the two parameters, which consists of the optical function, defines the characteristics of the lens. Each optical function maintains its own relation between focal length and FOV. For the same focal length, a different FOV can correspond to each available optical projection. The main advantage of fisheye lenses is that the incoming light beam converges on a circumference of a shorter radius on the sensor than a rectilinear lens at a given focal length.

The main type of available optical projections are the following (Ray, 2002; Kannala, 2006):

$$\text{Rectilinear:} \quad r = f \tan(\theta) \tag{1}$$

$$\text{Equidistant:} \quad r = f\,\theta \tag{2}$$

$$\text{Equisolid:} \quad r = 2f \sin\left(\frac{\theta}{2}\right) \tag{3}$$

$$\text{Stereographic:} \quad r = 2f \tan\left(\frac{\theta}{2}\right) \tag{4}$$

$$\text{Orthographic:} \quad r = f \sin(\theta) \tag{5}$$

where r = distance from the centre of the sensor

f = focal length
θ = angle of incidence of the light beams

The first one is the classical "pin-hole" projection also known as perspective projection (when the medium in which the light beams travel remains the same both outside and inside of the dark camera). The others are all different types of fisheye projections.

It must be noted that a FOV of 180° is impossible with the perspective projection, whereas with the fisheye projections a 180° FOV angle is always possible. The equidistant and equisolid angle projections can theoretically reach 360° (depending both on the focal length and the sensor size). The widest fisheye lens ever designed, though never pushed to mass production, is an equidistant-based projection which can reach a 270° field of view: Nikon 5.4 mm f/5.6 (U.S. Patent 3,524,697).

The advantage of a wider FOV alone can be crucial to obtain manageable data, where a rectilinear lenses approach would be prohibitive. This advantage comes with a price: the issues regarding the use of fisheye optics are numerous and linked to the fact that fisheyes follow a different type of optical projection.

1.2. The Ground Sampling Distance Degradation

The Ground Sampling Distance (GSD) is a fundamental concept to start planning the survey. It expresses the resolution, in object space, of the acquired images and, as a consequence, the potential accuracy of the 3D reconstruction.

The rectilinear optical projection (1) can be schematized in the pinhole projection scheme, where it is easy to recognise a similarity between triangles, and therefore the following ratio is instantly deduced:

f (focal length) : D (capturing distance) = pixel size : GSD, this straightforward and standard tool gives control over the results of the survey itself to the survey operator. The concept of GSD and the possibility of calculating/imposing it, provides the connection to the precision of the two- or three-dimensional representation to be extracted from the survey.

Though fisheye lenses follow a different optical projection, the abovementioned simple tool can no longer be employed in the survey designing phase. The GSD is indeed variable across the image frame; it goes from the minimum value, in the principal point, where the value is the same as the rectilinear projection at the same focal length, to a maximum value in the frame corner. The GSD can also reach infinity when the field of view is 180°.

Since the accuracy of the 3D reconstruction depends on the inferior resolution, the variable GSD hampers the operator's ability in designing the survey and, as a consequence, the control over the result. The risk of measurement's failure in the case of incorrect photogrammetric planning is very high and it leads to a warped model (bent, stretched, compressed, etc.) (Nocerino, 2014).

Figure 1. Two circular fisheye images of a straight wall; the same area was rendered at a very different resolution when placed in the centre or on the border of the frame.

2. Why Use Fisheye Lenses for Photogrammetry

2.1. Fisheye Lenses to Survey Narrow Spaces in a Complex Architectonic Environment

Nowadays, the use of Building Information Modelling (BIM) technologies and 3D representations in the architectural field has increased the demand for 3D survey techniques. For this reason, in the last few years, the world of Cultural Heritage has seen a strong and constant development of survey technology and practices aimed at acquiring complete, dense, and high-precision 3D information.

Dealing with a full 3D reconstruction of complex architectures, secondary spaces (such as staircase, corridors, passages and tunnels) play a major role. If the goal of the survey is to produce a BIM model of the building, they cannot be forgotten just because of their reduced accessibility.

Due to the lack of the "right instrument", surveying those areas results oddly in the most time-consuming and challenging part of the measurement campaign. The fisheye lenses approach could be the most promising solution to speed up the acquisition phase by reducing the total amount of pictures to be acquired, without compromising the achievable accuracy.

2.2. Fisheye Lenses for Quick Surveys in Critical Underground Environments

Very similar difficulties are encountered when surveying complex structures of subterranean spaces present in modern and historic city centres. Most of them are used to provide an efficient traffic-free transit system (i.e., subway, underpass, etc.), while others host a series of infrastructures such as several cable ducts or a

7

drainage system. Every underground space should be accurately mapped and georeferenced in order to simplify both the ordinary and extraordinary maintenance.

Actually, most parts of big cities do not have an accurate map of their underground spaces, although new evolutions of GIS (Geographic Information System) technology are starting to allow the management of such data. Moreover, while new infrastructures are accurately mapped, old ones often remain unsurveyed.

Generally, whether the metric survey is carried out in indoor or in hypogea environments, it is usually expressed in a Local Reference System (LSR), which in turn is linked to a global reference system, such as the geocentric coordinate system ETRS-2000. The relationship between the LRS and the global one is not always well known. The transformation parameters could be determined by a classical topographic survey performed with the use of a total station, where both the accuracy and the precision achievable are strictly related to the survey adopted techniques (i.e., open and/or close traverses). Furthermore, the specific technique adopted is conditioned by several factors: the morphology and position of the hypogea site, the extension of the site as well as the number of breaches that allow open spaces to be reached. In specific critical cases, when the extension of the hypogea site is huge, and the quality of the air can change quite rapidly even becoming unbreathable, a different kind of survey could be suitable: quicker than the topographic one, even if less rigorous. The solution suggested is to walk through the tunnel taking pictures (video frames) from a single camera, equipped with a fisheye lens to ensure a wide FOV, pointed forward while advancing in the tunnel. The reconstruction of camera movement as well as the 3D model of the hypogea environment will be successively obtained by the use of Structure from Motion (SfM) algorithms. Finally, in order to geolocate and to scale the obtained 3D model, several GCPs (Ground Control Points) have to be placed where possible; furthermore, several scale constraints have to be imposed in order to limit the model deformations.

2.3. Fisheye Lenses and Panoramic Cameras for Speeding up the Survey Phase of Closed Environments

Consumer-grade cameras able to capture 360 photos and videos are becoming more popular due to the opportunity to look in any direction, exploiting immersive visualisation with virtual reality headsets. Different cameras are available on the market. Some examples are the Ricoh Theta S, 360fly 4K, LG 360 CAM, Kodak PIXPRO SP360 4K, Insta360, Kodak PIXPRO SP360, and the Samsung Gear 360. More professional (and expensive) cameras are the GoPro Odyssey, Sphericam V2, Nokia OZO, and GOPRO OMNI. Typically, these cameras are equipped with at least two fisheye lenses and provide a spherical 360°

panorama (Fangi, 2010). Due to the stable geometry of a cylindrical panorama, photogrammetric bundle adjustment can be performed with very few object points. Once each panorama is oriented in the global coordinate system, photogrammetric object reconstruction procedures such as space intersection can be applied (Luhmann, 2010). The idea behind the use of this kind of camera is to reduce the acquisition time using popular commercial instrumentation as well as simplifying the capture geometry approach so that even the less experienced operator has the possibility of performing a 3D reconstruction of close and complex environments. The camera's constructive and photographic qualities, the low-resolution acquisitions and the internal panorama stitching can negatively influence the quality of the results. Chapter 5 will investigate this aspect.

3. Fisheye Lenses for Cultural Heritage Survey

3.1. General Model to Calculate the GSD for Fisheye Lenses

The main drawback of fisheye lenses for photogrammetric use can be identified in the degradation of the GSD across the image frame. The GSD is the link to the chosen representation scale. It is commonly used as the primary and only reference on which to base the capturing phase design; losing the ability to calculate it a priori led to us losing control over the final results of the matching process, giving rise to the topic of fisheye lenses not being reliable enough for metric purposes.

Although an accurate calculation of the GSD is not required in the design phase, an approximate calculation is mandatory to have control over the image resolution.

A general mathematical model for describing the GSD is presented here; this theoretical approach allows one to evaluate different lenses and different optical projections a priori by deriving a specific GSD formula from the general one for each of them. One can say that the commonly used GSD definition is a specific formulation suitable only for rectilinear lenses, but not for fisheye lenses.

For all fisheye projections, unlike the pinhole scheme, it is true that the GSD varies also depending on the radial distance from the centre of the frame. Therefore, the new model has to express the GSD as a function of focal length, f, capturing distance, D, and pixel size, pix, as well as the parameter r.

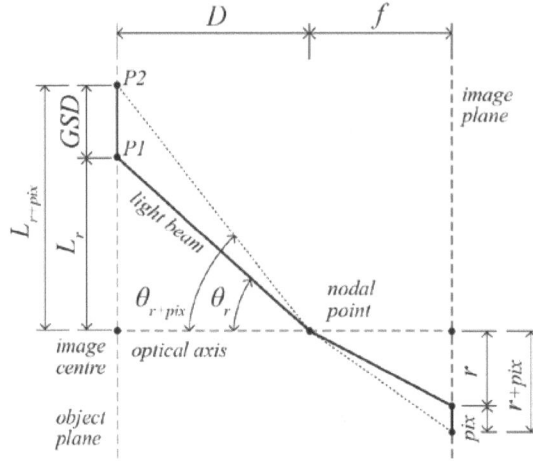

Figure 2. The relation between pixel size and GSD for each projection function.

Equation (6) expresses the GSD considering both the classical parameters usually considered in photogrammetry, D, f and pixel size with the addition of r to take into account the GSD degradation through the frame.

$$GSD = D \cdot \left\{ \tan \left[2 \arcsen \left(\frac{r + pix}{2f} \right) \right] - \tan \left[2 \arcsen \left(\frac{r}{2f} \right) \right] \right\} \tag{6}$$

In Figure 2, θ_r and θ_{r+pix} depend on the optical projection of the lens, where theta is always related to r. It is possible to reverse the mapping function of each lens to make theta explicit in order to obtain the GSD formula for the chosen one. Below, a list of the formula for the equidistant (7), equisolid (8), stereographic (9) and orthographic (10) projection function is outlined.

$$GSD = D \cdot \left[\tan \left(\frac{r + pix}{f} \right) - \tan \left(\frac{r}{f} \right) \right] \tag{7}$$

$$GSD = D \cdot \left\{ \tan \left[2 \arcsen \left(\frac{r + pix}{2f} \right) \right] - \tan \left[2 \arcsen \left(\frac{r}{2f} \right) \right] \right\} \tag{8}$$

$$GSD = D \cdot \left\{ \tan \left[2 \arctan \left(\frac{r + pix}{2f} \right) \right] - \tan \left[2 \arctan \left(\frac{r}{2f} \right) \right] \right\} \tag{9}$$

$$GSD = D \cdot \left\{ \tan \left[\arcsen \left(\frac{r+pix}{f} \right) \right] - \tan \left[\arcsen \left(\frac{r}{f} \right) \right] \right\} \tag{10}$$

The presented system allows the operator to calculate the resolution distribution across the images beforehand, and, as a consequence, to monitor the minimum value of GSD for the various optical projections; since the lower resolution will influence the outcome of the photogrammetric process, it is important to understand which parts of the images can be used in relation to the chosen representation scale. This method makes it possible to calculate the radius r

of the maximum circumference within which the resolution is compatible with the chosen scale; the remaining part, expressing a resolution lower than the acceptable minimum, can be discarded (Figure 3).

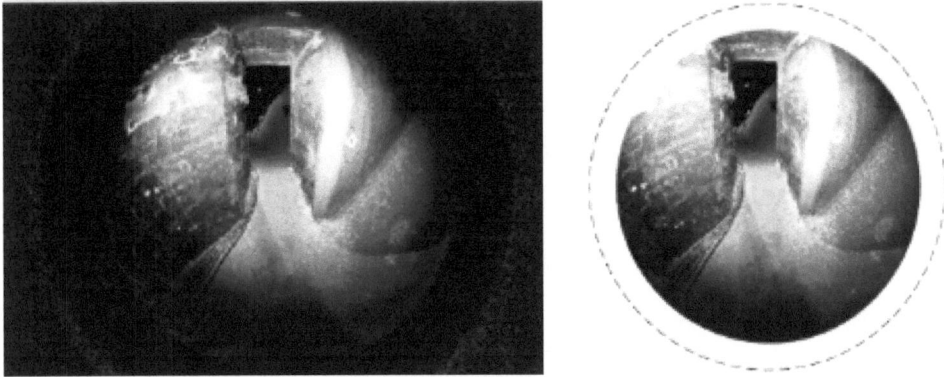

Figure 3. Original picture taken using a fisheye lens and the same image cropped to comply with the desired scale.

By drawing a parallel between the two charts below (Figure 4), the operator can predict in advance, for each case study, the behaviour of fisheye lenses as well as rectilinear lenses, managing to check, always in advance, if there is a significant advantage to using one over the other.

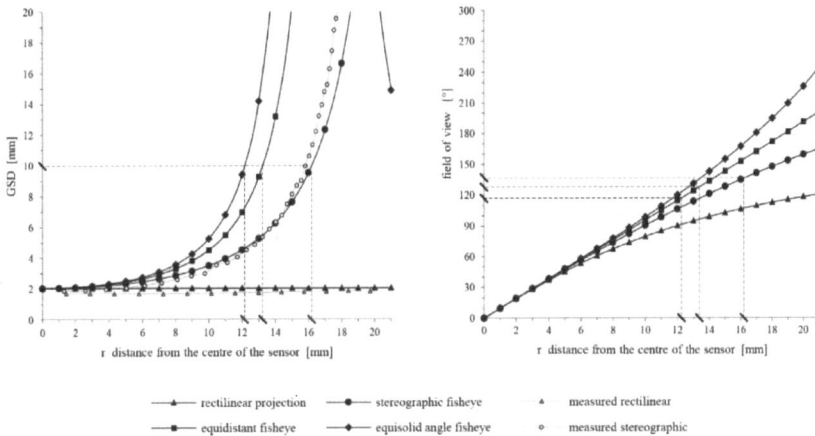

Figure 4. Comparison graph among different optical projection behaviours (left) and FOV that can be reached with each of them at a given radius distance (right). The simulation is done for a camera with a pixel size of circa 4.9 microns, a focal length of 12 mm and a taking distance from the object of 2.5 m.

In the scale 1:50, for instance, typically used in the survey of cultural heritage, the minimum resolution that the photographs must have is expressed by the value of the GSD, maximum 10 mm. At this point, according to the optical projection in use, it is possible to calculate the value of the radius r which defines the circumference that places the photographic area with sufficient resolution on the inside and the one with insufficient resolution on the outside. This information can be used to crop the images, discarding the marginal areas to keep only the portion on the inside of the aforementioned circumference.

3.2 Experimental and Validating Tests

3.2.1. Experimental Test on GSD Distribution

The graph in Figure 4 reveals the behaviour of the different optical projections that have to be taken into consideration in order to correctly design the survey and obtain the desired results. However, the actual distortion of any physical lens is characterised by some small differences when compared to their theoretical mathematical model. Therefore, in order to verify the theoretical model, a test was conducted, making it possible to measure the decrease in the GSD with two fisheye lenses: Samyang 12 mm fisheye stereographic and Sigma 8 mm fisheye equisolid.

The aim of this test was to obtain a sort of "scale factor" able to parameterize the fisheye lens behaviour, in terms of lens distortion and in relation to the rectilinear theoretical parameters. This will then take two aspects of the distortion into consideration at the same time: the difference between the two optical projections in question (rectilinear and fisheye), as well as manufacturing defects of the lens.

To obtain this "scale factor", a target was designed and attached horizontally to a wall, while the camera was positioned tilted, ensuring that the target was lying exactly on the diagonal of the sensor passing exactly into the centre of the frame (Figure 5). At this stage, a great deal of attention was paid to ensure the maximum parallelism and centring of the sensor with the target. Using this configuration, it was possible to measure the compression of the lengths on the metric scale of the target, and therefore, to compare it with the theoretical model.

Although the behaviour of the real lens (fisheye) differs slightly from the theoretical one (Figure 4), the variation is sufficiently limited to allow the theoretical model to be used for the design of the survey with real optics (Perfetti et al, 2017).

Figure 5. Photo of the metric target obtained with a diagonal fisheye, the Samyang 12 mm.

3.2.2. Experimental Tests on Ideal Capture Geometry: Base Distance/Capturing Distance Ratio

The next step is to define an optimal capturing geometry on which the photogrammetric survey can be based. Although the capturing geometry varies from case to case depending on the volumetric features of the survey object, it is important to define the *base distance/capturing distance* ratio which underlies the capturing geometry design.

While with rectilinear projection lenses the calculation of the *base distance/capturing distance* ratio comes from the percentage of overlap between adjacent images (about 70-80%), with fisheye lenses this approach fails. Due to the wider field of view of fisheye lenses, the overlap of two adjacent images lying on the same plane would be bigger, being even equal to infinity when the FOV is equal to 180° or more. It is clear that the percentage of overlap is not a useful parameter on which to base the capturing geometry. Many empirical tests were held in order to overcome this issue: for instance, a photogrammetric survey of a straight wall was performed using a high number of photographs precisely spaced; little by little, some images were removed with the aim of obtaining a correct survey with a minimum number images.

After these tests, a 1:1 *base distance/capturing distance* ratio turned out to be the best ratio to be used with fisheye lenses. This information makes it possible to design a suitable capturing geometry when the working distance is known.

3.3. The "Fisheye-Grammetric" Survey of the Minguzzi Spiral Staircase

The Minguzzi staircase is a marble stone spiral staircase located inside the right pylon of the main facade of the Milan cathedral. Along its extension of about 25 m in height, it connects three different levels of the building: at the upper end, the lower level of the roofs; in the middle, the central balcony of the façade and, at the base, the floor of the church.

The staircase is extremely dark. The artificial illumination is of poor quality, and there are only a few openings towards the outside placed at regular intervals. From the inside, these openings are relatively large (85 cm) and deep (circa 2 m) but due to the considerable thickness of the wall, they narrow down noticeably towards the outside and end up as small vertical embrasures (30 cm width). At the centre of the staircase, there is a stone column with a diameter of about 40 cm around which the ramp rolls up; the space left for the passage is extremely narrow, about 70 cm. This extremely narrow space was complex enough to represent a serious test of our research topic. In this situation, there is simply no space to use the regular terrestrial laser scanner instrumentation. Moreover, many other factors would nullify the whole scan, making the job unnecessarily time-consuming and burdensome in terms of the amount of data.

Before moving on to the field, it was imperative to fully understand the geometry of the object in order to properly design the capturing geometry and to completely cover all the spaces useful for the re-design of the staircase.

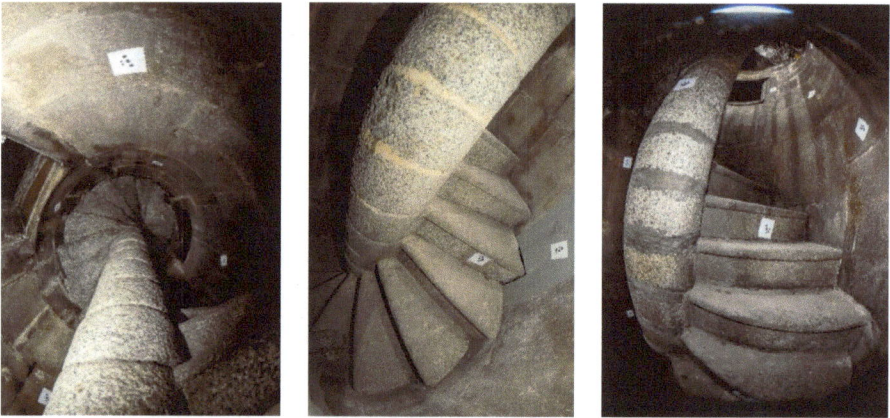

Figure 6. Fisheye views of the internal winding of the staircase.

3.3.1. Acquisition Phase

The adopted shooting geometry was defined throughout some experimental tests aimed at finding out the right *base distance/capturing distance* ratio to be used with these optics (Perfetti et al, 2017).

The image acquisition phase was divided into five main placements of the camera. In addition, a series of integration acquisitions were taken to complete some complex areas.

The survey was carried out with two different cameras and three different lenses: Canon 5D mark III coupled with Sigma 8 mm circular fisheye and Nikon D810

14

coupled with Samyang 12 mm diagonal fisheye and a Sigma 12–24 mm rectilinear lens. The whole image acquisition process was accomplished with the aid of three portable synchronised photographic flashes, which made it possible to light up the otherwise too dark area of the staircase.

The different field of view and the GSD calculated value have determined the number of shots needed for each configuration.

Some scans using TLS Leica C10 were performed at the base and the top of the staircase. These scans were georeferenced in the topographic network built for the cathedral survey activity; some photogrammetric targets were measured at this stage as well. These measurements were performed in order to check the accuracy of the survey and in particular if any bending, stretching or compression distortion occurred to the final model. In this way, we had two check-stations at the base and the top of the staircase. In the interior of the staircase, where no additional instrumental check was possible, many reference distances among markers were taken in order to monitor the alignment manually.

3.3.2. Data Elaboration and Results

First, before proceeding with the software elaboration, it was important to apply the discussed methodology to crop the marginal areas of the pictures when not suitable for the desired scale.

For all the three tests, but in particular, for those with the fisheye lenses, it was necessary to manually intervene on the alignment after the automatic matching process, to provide additional manually selected constraints by manually picking the targets (Photoscan coded targets) that the software could not recognise due to fisheye distortion. This operation was necessary in order to optimise the calculation of the camera alignment.

Table 1. Residual errors RMSE on Ground Control Points (GCPs) for the different elaborations during the photogrammetric process.

		CANON 5D 8mm fisheye	NIKON D810 12mm fisheye	NIKON D810 12mm rectilinear
Process	Allignment	0.032m	/	0.055m
	Marker optimization	0.031m	0.013m	0.051m
	Scale-bars optimization	0.010m	0.014m	0.051m
	Topography optimization	0.009m	0.014m	0.051m

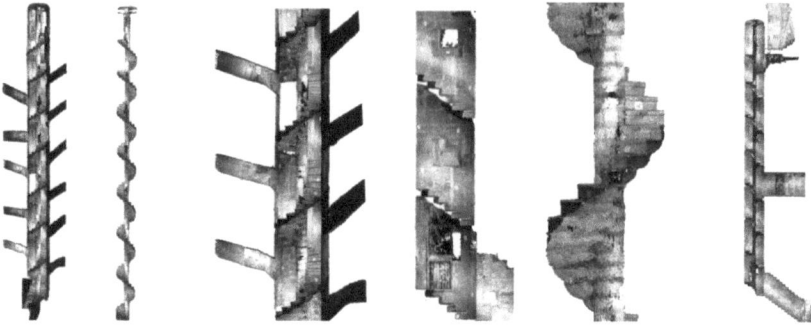

Figure 7. Dense point cloud obtained with Nikon D810 coupled with 12 mm stereographic fisheye; elaboration was done using Agisoft Photoscan Pro.

The results show (Figure 7) that using the GSD calculation to hold the resolution under control leads to a significant improvement of the alignment and matching process that produces more reliable results and less noisy dense point cloud; it could also help the camera calibration model by removing the most distorted portion of the frame. The best results were obtained using the Sigma 8 mm equisolid fisheye; in this case, the use of the cropping method was very successful (Table 1); the final Root Mean Square Error (RMSE) of 9 mm corresponds to the expected/designed accuracy.

The remaining issue is the very complex capturing geometry that has to be arranged case by case. No standards can be followed a priori, and the use of a large number of markers and a lot of manual refining work represented the key to success in the process: a complete autonomous elaboration when using fisheye lenses can only lead to a failure. Topographic data are necessary to improve the final accuracy.

4. Fisheye Lens Survey to Georeference Complex Hypogea Environments

4.1. Site Description

The ancient Romans mined the soft rock from underground to erect buildings (i.e., "pozzolana"); actually, this activity was carried out until the middle of the twentieth century. Such action has involved the appearance of a series of tunnels, about 10–15 m deep, realising an intricate labyrinth in Rome underground. There is no topographic map of this intricate labyrinth excavated more than 2000 years ago; furthermore, the total extension is still unknown. In this application, an accurate mapping of part of this complex system was needed to find the planimetric position of a specific point of one of these galleries located in the southern part of the city.

16

Several researchers focused their efforts on the survey of underground structure, such as necropolis (Remondino et al., 2011) or catacomb (Bonacini et al., 2012). Very low luminosity is a typical feature of a hypogea environment; therefore, the laser scanner is usually employed as the main instrument, while photogrammetry is generally used to obtain a high-quality texture. The results are very noteworthy, and the quality of the survey is very high, but these techniques are very time-consuming. For this specific application, a very agile methodology is required, because in every moment, due to a change of air current direction, some dumps present in the galleries itself could yield toxic air. Furthermore, for heritage conservation reasons, one is not allowed to leave permanent and invasive markers.

4.2. Photogrammetry Setup

The survey was carried out using a full frame DSLR camera Nikon D800E with a pre-calibrated Nikkor 16 mm fish-eye lens. Fisheye lenses allow a wide diagonal FOV of up to 180° to be achieved. Furthermore, the choice of fisheye lenses allows a substantial overlap between two consecutive frames to be achieved. The camera was set in video mode, and recording was performed in standard HD (High Definition) 1080 px on 30 fps (frame per second). Such camera settings allow to obtain a pixel size of 18.7 microns.

4.2.1. Camera Calibration

The camera calibration procedure is a fundamental task in the photogrammetric workflow. The well-known self-calibration method (Fraser, 1997) is generally used to determine the camera calibration parameters.

Specifically, this technique employs analytical calibration methods to derive the calibration parameters indirectly from photogrammetric image coordinate observations. The mathematical model is the classical Brown model (Brown, 1971) that is composed of classical internal orientation parameters (principal distance, coordinates of the principal point, pixel size) extended by the inclusion of additional parameters that model the image distortion effects. Fisheye lenses utilise a different optical design that departs from the central perspective imaging model to produce image circles of up to 180°. If the image format sensor is larger than the resultant image circle, the camera is termed as a fisheye system. Conversely, if the format is smaller than the image circle, such that the image diagonal covers about 180° of the field of view, a quasi-fisheye system is attained. It is well known that under planar perspective projection, images of straight lines in space have to be mapped into straight lines in the planar perspective image. However, such assertion, for fisheye cameras, is not true: the central perspective mapping is replaced by another model such as stereographic, equidistant and orthographic projection.

When modelling the distortions in a fisheye lens, conventional radial lens distortion corrections are mathematically unstable. In the peripheral region of the image sensor, the gradient of the distortion curve describing the departure from the central perspective case is high. In such a case, it is necessary to apply the appropriate fisheye lens model before using the conventional radial distortion model (Luhmann et al., 2014).

In this work, a quasi-fisheye system was realised; therefore, the peripheral portion of the image circle is not recorded on the sensor, and the fisheye distortions can be reasonably modelled using the conventional perspective camera model and its classical radial distortion coefficients. Such a model is more flexible than the fisheye model because it can be imported/exported in/from any photogrammetric software as well as integrated into any SfM algorithm.

The camera was preliminary calibrated in a controlled environment before starting any survey operations. Two types of calibration camera models were tested:

1   *Fisheye camera*: it is the more rigorous camera model; it combines two types of distortion: the fisheye distortion that the lens provides as well as the conventional pin-hole ones.

2   *Perspective camera*: it is the classical model used in photogrammetry, i.e., the Brown one. It uses only the conventional distortion model.

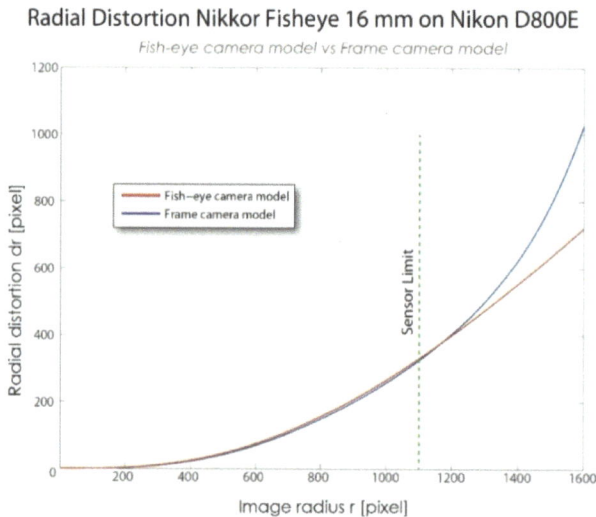

Figure 8. Radial distortion curves: in blue are the distortions obtained using the frame camera model, while in red are the distortions determined with the fisheye camera model.

18

In our tests, the two models provided the same values of image radial distortions in the central part of the sensor (close to the principal point), whereas on the peripheral parts the gap between two models increases progressively.

In order to quantify the differences between the two models, the components of computed radial distortion were plotted in Figure 8. Basically, the two curves are identical; just on the image periphery, the difference becomes approximately a few pixels.

In this case, study of the gap between the two camera models is primarily restricted because video recording automatically crops the frame and the image sensor is entirely included in the image circle (where the FOV is 180°) as shown clearly in Figure 9.

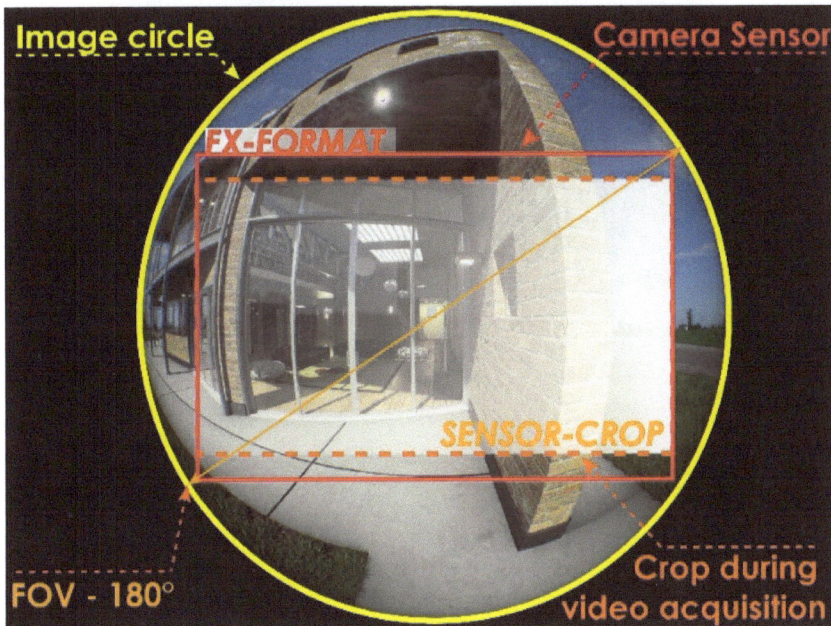

Figure 9. Representation of the full frame fisheye lens. In yellow is the image circle, in red is the coverage of the FX sensor; the orange dotted line shows the sensor active area during the acquisition.

4.2.2. Acquisition Phase

The acquisition phase started outside with the aim of acquiring sufficient targets in order to georeference the entire survey. The camera operator slowly brought it downhill to the tunnel labyrinth. The transition between the very bright environment (open-air) and the dark location (tunnels) was gradually performed. The focus camera was set to infinite, maximum aperture, while the sensitivity of the digital image system ISO automatically changed, to adjust the great bright gradient. The data collection was performed holding the camera optical axis aligned with the path axis (Figure 10). The acquisition phase lasted less than two hours to cover about 1 km and it consisted of a simple video tape acquisition recorded along the path. For the duration of the acquisition, some caution was taken: the light source is never located in front of the camera and within the scene; no person or moving object was detected, except for the shadow of the camera operator projected by the light source located behind him.

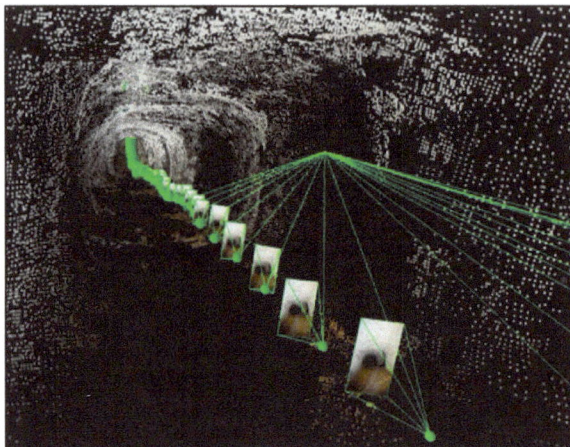

Figure 10. Camera geometry carried out during the acquisition phase, moving forward along the optical axis.

4.3. Elaboration and Results

4.3.1. Relative Image Orientation

The acquired video was processed using a standard procedure employed in 3D image-based modelling by sampling, on average, a frame every 1–3 seconds, for a total of 4425 camera stations.

In order to correctly position the light source, the video tape was not recorded continually; indeed, 33 video clips were acquired. Each clip was processed as a different sub-project and was oriented independently using internal camera parameters computed during the calibration task. Two following sub-projects have at least 20 images in common. Such overlap permitted the sub-projects to be joined among them. Figure 11 shows the number of extracted tie-points in the joint model. Such a parameter is very high, both for the great overlap assured by the fisheye lens and for the feature extractor employed during the orientation phase. A 3D sparse point cloud has been obtained by about 4200 images correctly oriented; indeed, due to imposed overlapping camera stations, the number of unique images was reduced. The bundle result obtained with such camera stations provided about 2.5 million tie-points.

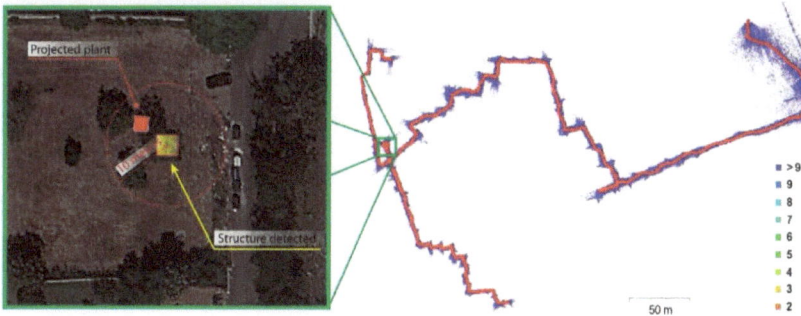

Figure 11. On the right: an orthographic projection of 3D tie-points, the colour shows the associated multiplicity whereas in red is the track of the moving camera. On the left: a detailed comparison between the structure inspected in the underground environment (in red) and the corresponding aboveground environment.

4.3.2. Scaling and Georeferencing

The photogrammetric model needed to be geo-referred using at least three GCPs (Ground Control Points). This step was performed at the beginning of the video recording. Specifically, the acquisition started outside the tunnel where three GCPs were signalised by targets. The GNSS observations, collected in fast-static mode, were processed and adjusted; the final 3D global accuracy is about 2–3 cm. Such achieved accuracy is more than enough for the goal of this work, where an accuracy of about 10 m is required. Further scale constraints were added to the project; indeed, as described in the previous section and during the acquisition phase, several measurements of some distances were carried out. The 3D model was scaled and georeferenced by three GCPs and two scale constraints. Initially, the bundle adjustment was conducted in free network mode; afterwards, adding GCP measurement, it was performed in minimal constraint mode. The total error reported on the GCP is about 1 cm. In order to limit these scale deformations, the bundle adjustment was carried out again, enhancing the constraints. The total error in the final three-dimensional reconstruction has grown up to about 6 m.

4.3.3. Obtained Results

The result is a geo-referred 3D point cloud of the explored tunnels. Finally, such a geo-referred sparse point cloud was further processed to obtain an orthoimage map. The hypogea site does not provide any reference point to control the solution, and the tunnels network is very complex; in any case, on the path, several structures—probably well casing or building foundations—are present. Unfortunately, such structures are not easily recognisable on the surface. In order to verify the accuracy of the survey, every structure encountered along the underground pathway was inspected afterwards. By projecting the plant on Google Earth, a rectangular structure on the surface was discovered (Figure 11). Such a discovery verifies the 3D model obtained (only in planimetry). Comparing the positions of the barycentre, an estimated accuracy within 7 m was achieved.

4.4. First Conclusions on Complex Hypogea Environments Test

The complexity of this specific experimentation was further increased both by the environmental conditions (such as the low brightness, no electric energy, low level of oxygen, etc.) and by the morphology of the site. A generic camera HD and a simple measuring tape, further to a GNSS receiver, composed the equipment. Initially, no reference point was detected to check the solution or to add additional constraints. This type of approach allows one to conduct a prompt survey in a short time; furthermore, the essential equipment is easily reachable, and it does not require high profile skill to be managed. On the other hand, expertise and time are necessary for post-processing to obtain a reliable solution.

5. Photogrammetry with Fisheye Images from a Low-Cost Spherical Camera

The camera considered in this work is the Samsung Gear 360, which has a dual 15 MP CMOS sensor with integrated f/2.0 fisheye lenses, dual cam video resolution of 2840 × 1920 pixels, and dual cam photo resolution of 7776 × 3888 pixels. The camera requires a Bluetooth connection with a Samsung mobile phone (such as the Samsung S6 or S7) to obtain real-time visualisation and control acquisition parameters. The images can be downloaded from the camera as circular fisheye images or equirectangular projections for 360° visualisations.

The aim of this work was to try out the Samsung Gear 360 for 3D modelling, considering front and rear-facing fisheye images. The interest in this kind of acquisition tool is motivated by the wide field of view available, which makes spherical cameras very attractive for interior scenes that usually require a large number of pinhole images.

5.1. Evaluation of Metric Accuracy

The aim of this experiment was to test the metric accuracy of the Samsung 360 with a long image sequence. A set of 60 images of a wall was acquired with the Samsung Gear 360 placed on a tripod. Images have the typical configuration for camera calibration, i.e., several convergent images with roll variations. The aim was to run a markerless calibration procedure as described in Barazzetti et al. (2011) and Stamatopoulos and Fraser (2014), in which calibration parameters are estimated from a block of target-less images. The used software is ContextCapture, which allows fisheye images to be processed with a mathematical formulation based on the asymmetric camera model.

The estimated calibration parameters were then assumed as constant values for a 3D reconstruction project of a straight wall, on which a set of targets was installed and measured with a total station. The sequence was acquired only with the front-facing camera (Figure 13). Images were oriented with ContextCapture, using 12 targets as ground control points and seven targets as check points. The sequence is 42 m long, and the camera object distance is 1.2 m. Statistics are shown in Table 2 and reveal an error of about 5 mm, that confirms a good metric accuracy for the project carried out with the front-facing camera. Such results confirm the good metric quality of the Samsung Gear 360 when the original fisheye images are used for photogrammetric applications.

5.2. 3D Modelling from Fisheye Images Acquired with the Samsung 360

The results described in the previous section revealed a good metric accuracy when the single fisheye images are used; a reconstruction based on two fisheye images (front and rear) seems feasible. Figure 14 shows the image orientation results inside the oratory of Lentate sul Seveso in the case of fisheye image processing (30 images, i.e., 15 + 15 image pairs). No constraint was used to fix the relative position of the images, which were instead processed as independent images. The used software is ContextCapture, in which camera calibration parameters were assumed as fixed for both front- and rear-facing images.

The achieved mesh is of better quality than that generated by equirectangular image processing presented in Barazzetti et al. (2017). On the other hand, the reconstruction is partially incomplete, especially the area of the vault, which was instead modelled in the case of equirectangular projections.

Figure 12. Example of front- and rear-facing fisheye images acquired with the Samsung 360.

Table 2. Accuracy achieved with the front-facing camera. The good metric accuracy is achieved by the short camera–object distance, as well as the short baseline between the images, i.e., the same point is visible on a large number of images providing multiple intersections.

Number of GCPs	RMS of reprojection errors [pixels]	RMS of horizontal errors [m]	RMS of vertical errors [m]
12	0.8	0.001	0.001
Number of Check Points	RMS of reprojection errors [pixels]	RMS of horizontal errors [m]	RMS of vertical errors [m]
7	2.1	0.004	0.001

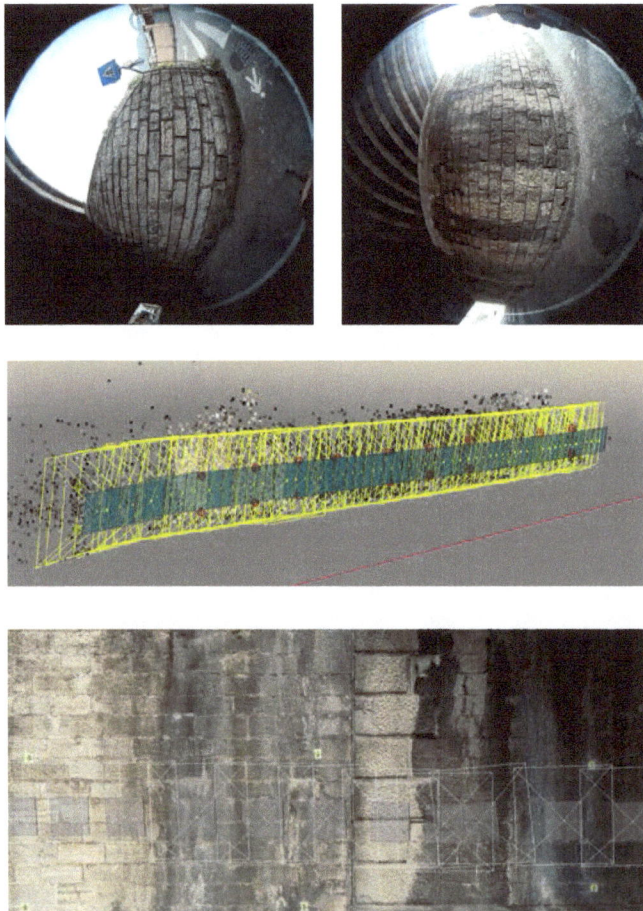

Figure 13. Two images of the wall captured with the front-facing camera (top), the sequence of images, control points and checkpoints (middle), and a detail of the extracted mesh for a portion of the wall (bottom).

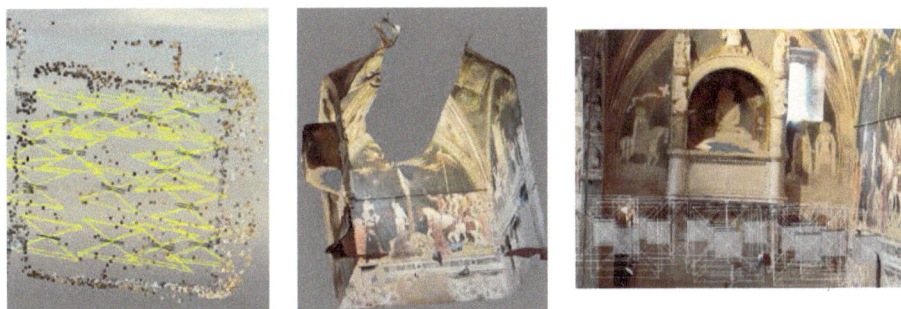

Figure 14. The reconstruction from front- and rear-facing fisheye images is quite detailed but incomplete, especially the area above the camera; notwithstanding, the area is visible (with a very narrow angle) in the original fisheye images.

6. Conclusions

The three presented case studies aim to briefly describe the possible use of fisheye photogrammetry to solve some common environmental and architectural measurement problems. First, the conducted research tried to address the problem of designing a complete photogrammetric survey of narrow spaces, characterised by scarce illumination and intricate geometry. The second goal was to speed up the photogrammetric acquisition in this type of areas, especially when they are complicated and also huge in their extension, by reducing the number of images to be acquired. This can be obtained either by using extremely wide-angle lenses, or by reducing the number of images using low-cost commercial panoramic cameras. The idea for solving both problems was to use fisheye lenses and find rules that could standardise this type of survey and the ensuing elaboration process.

In the first case study, the architectonic reconstruction of narrow internal spaces, the goal was to obtain an accurate, high-resolution 3D reconstruction of the building. The theory of fisheye distortion and some preliminary tests suggest concentrating the attention on the non-uniform distribution of the image resolution across the frame. The idea was to use only the portion of the image that provides the GSD within the maximum acceptable limit for the desired restitution scale. The test applied to the Minguzzi staircase case study demonstrates the effectiveness of the approach and the possibility to obtain complete and accurate results at the architectonic representation scale of 1:50.

The second case study describes the use of alternative methods for a rapid reconstruction of underground environments. The aim was not to obtain a complete, high-resolution 3D reconstruction, but to quickly derive the position on the outside of an underground point. The research focused on testing the "videogrammetry approach" to speed up the acquisition and check the capabilities of automatic photogrammetric algorithms in the arduous task of orienting a high number of images with the purpose of substituting the classical topographic open

25

traverse. The tests showed the real possibility to conduct a prompt survey in a short period.

The third case study showed some preliminary tests of a ready-made instrument based on multiple fisheye lenses, able to speed up the acquisition phase and regularise the capturing geometry. The first results of the experiments revealed that the Samsung Gear 360 (the camera used in the test) is suitable for metric reconstruction, although the achieved metric accuracy is not comparable with a traditional photogrammetric approach based on pinhole images. The achieved accuracy is within the tolerance of the 1:300 representation scale, which could be sufficient for applications aimed at determining the overall size or volume of a room. The proposed setup is surely less expensive than a laser scanner and also allows rapid data acquisition where, instead, a large number of pinhole images would be needed. On the other hand, there is limited control on camera parameters and images are acquired in an almost entirely automated way. This makes the camera a low-cost photogrammetric tool for people that have limited experience in 3D modelling from images.

The presented case studies show how fisheye lenses can actually solve the problem of speeding up the acquisition and therefore, how their employment allows the reconstruction of complex narrow environments. Fisheye lenses are therefore a valid tool in practical applications where traditional lenses (central perspectives) would require an enormous number of images. The outcomes of the previous experiments allowed us to define a set of good practices for "fisheye photogrammetry":

- fisheye lenses mounted on SLR cameras are already a valid tool for metric applications;
- the large field of view of fisheye lenses allows a reduction of the number of images, but the use of the whole field of view produces outputs with a very variable resolution (very low close to the image edges); this is a significant limitation for some steps of the production workflow (e.g., camera orientation and texture mapping);
- camera calibration plays a vital role because the extreme image distortion could result in poor metric accuracy;
- good practices for traditional photogrammetry (e.g., inclusion of control points as pseudo-observations in the bundle adjustment instead of basic absolute orientation techniques) are still valid for fisheye photogrammetry;
- low-cost sensors (such as the Samsung 360°) are not ready for productive work, except for the case of the simplified model with limited metric integrity. On the other hand, significant technological improvement is expected (in terms of resolution and sensor quality), so they have remarkable potential;

The use of a multiple fisheye lenses instrument can significantly reduce the acquisition time but, in the future, it could be necessary to use higher-level instrumentation to improve the quality and the accuracy of the reconstruction in order to fit the architectonic scale requirements. Future works could develop a hand-held fisheye-based instrument capable of solving all the problems described here as well as increasing the resolution and the quality of the results.

References

1. Barazzetti, L.; Mussio, L.; Remondino, F.; Scaioni, M. Targetless Camera Calibration. *Int. Arch. Photogramm. Remote Sens. Spat. Inf. Sci.* **2011**, *XXXVIII-5/W16*, 335–342.
2. Barazzetti, L.; Previtali, M.; Roncoroni, F. 3D modelling with the Samsung Gear 360. *Int. Arch. Photogramm. Remote Sens. Spat. Inf. Sci.* **2017**, *XLII-2-W3*, 79–84, doi:10.5194/isprs-archives-XLII-2-W3-79-2017.
3. Bonacini, E.; D'Agostino, G.; Galizia, M.; Santagati, C.; Sgarlata, M. The catacombs of San Giovanni in Syracuse: Surveying, digital enhancement and revitalization of an archaeological landmark. In *Euro-Mediterranean Conference*; Springer: Berlin/Heidelberg, 2012; pp. 396–403.
4. Brown, D.C. Close-range camera calibration. *PE&RS* **1971**, *37*, 855–866.
5. Covas, J.; Ferreira, V.; Mateus, L. 3D reconstruction with fisheye images strategies to survey complex heritage buildings. *Digit. Herit.* **2015**, *1*, 123–126, doi:10.1109/DigitalHeritage.2015.7413850.
6. Fangi, G. Multi scale, multi resolution spherical photogrammetry with long focal lenses for architectural survey. In Proceedings of the ISPRS Midterm Symposium, Newcastle, UK, 22–24 June 2010; pp. 228–233.
7. Fassi, F.; Achille, C.; Mandelli, A.; Rechichi, F.; Parri, S. A new idea of BIM system for visualization, web sharing and using huge complex 3d models for facility management. *Int. Arch. Photogramm. Remote Sens. Spat. Inf. Sci.* **2015**, *XL-5/W4*, 359–366, doi:10.5194/isprsarchives-XL-5-W4-359-2015.
8. Fraser, C.S. Digital camera self-calibration. *ISPRS J. Photogramm. Remote Sens.* **1997**, *52*, 149–159.
9. Kannala, J.; Brandt, S.S. A generic camera model and calibration method for conventional, wide-angle, and fish-eye lenses. *IEEE Trans. Pattern Anal. Mach. Intell.* **2006**, *28*, 1335–1340.
10. Luhmann T. Panorama Photogrammetry for Architectural Applications. *Mapping* **2010**, *139*, 40–45.
11. Luhmann, T.; Robson, S.; Kyle, S.; Boehm, J. *Close-Range Photogrammetry and 3D Imaging*; Walter de Gruyter: Berlin, Germany, 2014.
12. Marčiš, M.; Barták, P.; Valaška, D.; Fraštia, M.; Trhan, O. 2016. Use of image based modelling for documentation of intricately shaped objects. *Int. Arch. Photogramm. Remote Sens. Spat. Inf. Sci.* **2016**, *XLI-B5*, 327–334, doi:10.5194/isprs-archives-XLI-B5–327-2016.

13. Nocerino, E.; Menna, F.; Remondino, F. Accuracy of typical photogrammetric networks in cultural heritage 3d modeling projects. *Int. Arch. Photogramm. Remote Sens. Spat. Inf. Sci.* **2014**, *XL-5*, 465–472, doi:10.5194/isprsarchives-XL-5-465-2014.

14. Perfetti, L.; Polari, C.; Fassi, F. Fisheye photogrammetry: tests and methodologies for the survey of narrow spaces. *Int. Arch. Photogramm. Remote Sens. Spat. Inf. Sci.* **2017**, *XLII-2/W3*, 573–580, doi:10.5194/isprs-archives-XLII-2-W3-573-2017.

15. Pierrot-Deseilligny, M.; Clery, I. APERO, an open source bundle adjustment software for automatic calibration and orientation of set of images. *Int. Arch. Photogramm. Remote Sens. Spat. Inf. Sci.* **2011**, *5/W16*.

16. Ray, S.F. *Applied Photographic Optics: Lenses and Optical Systems for Photography, Film, Video and Electronic Imaging*; Focal Press: Oxford, UK, 2002.

17. Remondino, F.; Rizzi, A.; Jimenez, B.; Agugiaro, G.; Baratti, G.; De Amicis, R. The Etruscans in 3D: From space to underground. *Geoinformatics FCE CTU* **2011**, *6*, 283–290.

18. Stamatopoulos, C.; Fraser, C.S. Automated Target-Free Network Orientation and Camera Calibration. *Int. Arch. Photogramm. Remote Sens. Spat. Inf. Sci.* **2014**, *1*, 339–346.

Perfetti, L.; Polari, C.; Fassi, F.; *et al.* Fisheye Photogrammetry to Survey Narrow Spaces in Architecture and a Hypogea Environment. In *Latest Developments in Reality-Based 3D Surveying and Modelling*; Remondino, F., Georgopoulos, A., González-Aguilera, D., Agrafiotis, P., Eds.; MDPI: Basel, Switzerland, 2018; pp. 3–28.

Photogrammetric Tools for Restoration Purposes: Open-Air Bronze Surfaces of Sculptures

Fabrizio I. Apollonio [a], **Marco Gaiani** [a], **Wilma Basilissi** [b] and **Laura Rivaroli** [c]

[a] Dept. of Architecture - Alma Mater Studiorum University of Bologna, Italy;
     (fabrizio.apollonio, marco.gaiani)@unibo.it
[b] Istituto Superiore per la Conservazione ed il Restauro, Rome, Italy;
     vilma.basilissi@beniculturali.it
[c] Rome, Italy; laura_rivaroli@yahoo.it

Abstract: The restoration of an outdoor bronze artefact, among the various processes, provides the cleaning activity, a delicate task that requires a work of synthesis to define—checking step by step its peculiar irreversible process—an effective operating protocol on a limited patch area to be extended to the entire artwork's surface. The paper presents a user-friendly, semi-automated three-dimensional (3D) photogrammetry-based solution, developed and validated during the ongoing restoration of the Neptune Fountain in Bologna, that is able to support restorers in the open-air bronze artwork cleaning from corrosion and weathering decay. The purpose of the developed solution was to offer an operative and useful support tool for restorers, throughout the restoration work, to ascertain the 'surfaces' to be cleaned, for testing and gathering information (as direct feedback) of the achieved results. The proposed solution, in fact, thanks to a customized interface and using an OpenGL viewer, can reproduce with a high level of fidelity in colour and shape the bronze surface before, during, and after the clear-out treatment.

Keywords: photogrammetry; 3D models; colour management; physically-based rendering; bronze cleaning

1. Introduction

The restoration of an outdoor bronze artefact is a delicate activity since the operations to be performed are very complex, and for restorers it is very difficult to control the phases of the surface cleaning. In fact, the cleaning of outdoor bronze artefacts is a difficult and complex task due to the complex corrosion mechanisms, which trigger the interaction between the material and the surrounding environment.

The surface of a bronze object that is exposed to air presents corrosion products that are generated owing to electrochemical corrosion and chemical reactions with atmospheric pollutants. These chemical products form different layers of corrosion, typically called 'patina', which offer partial protection to the metal substrate. Outdoor bronze patina layers mainly consist of copper oxides, copper sulphides, and sulphates (green copper carbonates are extremely rare in atmospheric exposure). A healthy layer is thin, stable, and compact. A thick and dusty layer acts as a spongy wrap and keeps corrosive agents and moisture in direct and continuous contact with the bronze surface. The presence of chlorides readily causes the localized formation of copper chlorides that are known to have a weakening effect on copper alloy patinas. The main problem is the removal of the unstable corrosion products, encrusted and laminated, on the metal surface, as much as those substances that are applied in previous restoration or maintenance (i.e., coatings layers, paints, etc.), while keeping intact the original so-called 'patina'. Therefore, the restorers should consider the stratification system that is distributed unevenly over the artwork (Basilissi and Marabelli, 2008) in order to identify which is this 'original surface' of the artefact. The cleaning process is carried out like a stratigraphic excavation: throughout different cleaning steps restorers gradually advance and, analysing micro-areas, they slowly proceed to identify the original surface layer.

The evaluation and understanding of stratigraphy of the bronze's surface is currently tackled by a cautious and progressive operating procedure that is based primarily on the identification of representative sample areas that are indicative of the state of conservation; then, the action is gradually performed with a series of cleaning tests. When the supposed level is reached, the objective feedback is given by a chemical evaluation of the type of material removed. Different analytical investigation techniques exist to characterize the deterioration of the 'patina', such as Non-destructive Test or destructive ones: e.g., cross sections using reflected light microscopy and analytical techniques as infrared spectroscopy, Fourier-Transform Infrared Spectroscopy (FTIR) and FTIR microscopy, in conjunction with Attenuated Total Reflection (ATR) spectroscopy, also making possible the surface mapping of the identified corrosion products (Mazzeo and Joseph, 2004). However, these tools are not quickly available to restorers during the cleaning work and are employed mainly to control the end of the cleaning procedures.

A second solution consists in the surface's check using magnifiers and/or a digital microscope, which is able to ensure microscale accuracy. Nevertheless, it is very difficult to identify the same area in different cleaning phases and digital microscopes are not able to ensure a faithful colour reproduction (Figure 1).

In any case, to understand which is the proper layer, we not only need the experience of the restorer and specific tests, but also to interrelate heterogeneous data coming from different sources and fields of knowledge (restoration, chemistry, physics, art history, etc.).

To overcome this complex and long procedure we developed a new, quick, semi-automated three-dimensional (3D) photogrammetry-based solution allowing to assess in Real-Time Rendering (RTR), and with a high level of fidelity in colour and shape, the comparison of 'surfaces' before, during, and after the cleaning treatment. The application is designed to be used by unskilled operators (i.e., the restorers) and aims to be an effective and valuable 'verification tool' to support the search and identification of the original layer of the bronze surface. Overall, the solution consists in an OpenGL viewer allowing to compare side-by-side high quality RTR of the bronze surfaces in the different phases of cleaning, from the dirty surface to the final one and in a robust pipeline for data acquisition, management, and visualisation, using a 6-click software. The focus of the development was on the colour management and final accuracy. Metric accuracy of the solution was checked against laser scanner data (see paragraph 7.).

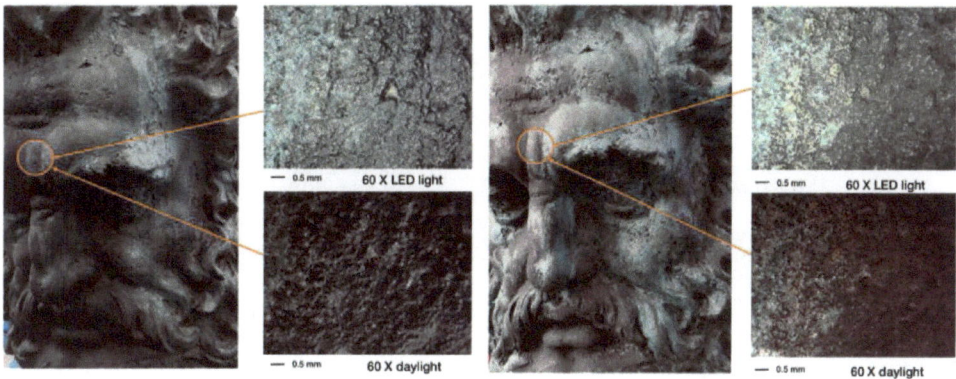

Figure 1. Optical evaluation of cleaning process: Neptune Head before (**left**) and after cleaning (**right**).

2. Solution Overview

The main problem faced in the developed solution is a correct visualisation using OpenGL graphics of real material aging. Aging phenomena could be classified as a process that deteriorates objects over time. On metals, there are several external factors of deterioration, generally gathered into two main classes: chemical and physical. Those changes are due to matter that is removed from the object's surface. That can occur at a variety of geometric scales. Some effects of aging could lead to the formation of corrosion products and introduce structural

31

damage of the material, such as destructive corrosion that would destroy a metal surface with the passage of time (e.g., archaeological metal artefacts). A real-life surface might not be affected by just a single class of events, but of any combination of different factors. Moreover, simultaneous processes could influence each other (Merillou and Ghazanfarpour, 2008). Various methods have been developed to simulate material aging, including patinas. A complete survey is given by Lu et al. (2007) and Rushmeier (2009). Methods include modelling the mechanics of adhesion and cohesion of paint and substrates to simulate the cracking and peeling and flow over objects, resulting in the deposition of dirt and corrosive agents. Julie Dorsey and Pat Hanrahan have developed a physically based method for the modelling and rendering of such patinas (Dorsey and Hanrahan, 1996).

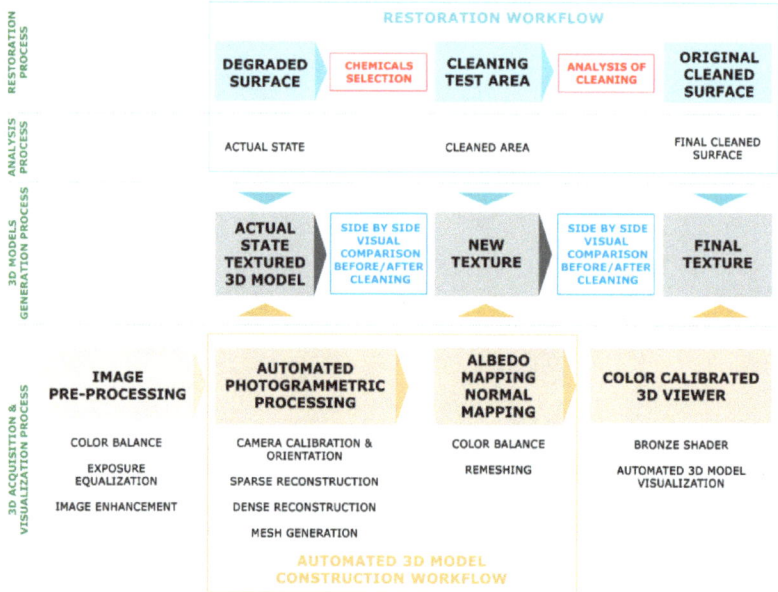

Figure 2. The proposed solution in parallel with the bronze cleaning phases.

Most simulation methods are computationally slow and can result in noticeable visual artefacts. Anyhow, simulations have only been validated by general comparisons of synthetic results with photographs of the same phenomena that show similarities, such as crack density and not being concerned with the faithful reproduction of real cases. Also, the basic approach, consisting in the acquisition of just the so-called apparent colour and mapping those samples to the 3D model, fails, giving a limited realism, as a diffuse texture cannot represent an object's true reflection properties. Summarizing, results are limited and do not concern the faithful reproduction of real cases, that is our case. Starting from studies demonstrating that open-air bronze patinas present average values of thickness between 80 and 130 µm before cleaning, which reduces to 50 to 110 µm afterwards (Letardi, 2004), we modelled the material surface behaviour using the types of light interactions described in Westin et al. (1992), where they are roughly grouped by the size of the geometric structures into three different levels:

- macroscopic scale: features on this scale are low-frequency details defining the shape and the geometry of a model and will be stored as polygonal meshes;
- mesoscopic scale: features on this scale are high-frequency details still just visible with the naked eye but are usually not considered to be part of the defining shape of an object (e.g., small bumps). These fine structures are usually captured using data-driven representation techniques that directly stores measured data in a form allowing for faithful depiction at low computational cost: i.e., multiple textures, mapped onto the 3D geometry; and,
- microscopic scale: on this level, the light is thought to interact with microscopically small facets on the surface of a material. The amount of reflected light can be influenced by the orientations of surface microfacets and the facets can occlude or cast shadows or inter-reflections on one another. A method that is usually adopted could model only the statistical distribution of the reflected light in dependence on the directions. In practice, the key issue to solve is to model correctly the Bidirectional Reflectance Distribution Function (BRDF) (Nicodemus, 1965), allowing to render the specular and diffuse reflectance components.

The characterization of these three levels in the case of outdoor bronze surfaces has been defined as follows:

- macroscopic scale: the shape of the surface has been reconstructed as a triangle mesh, with a 0.2 mm accuracy starting from a simple dataset of 9 images;
- mesostructure scale: today's state of the art method for modelling the mesostructure is certainly the Bidirectional Texture Function (BTF) (Filip and Haindl, 2009). To overtake the usual difficulties characterizing this method (i.e., huge amount of data and the need of skilled operators), a new solution was developed starting from Kautz (2005), who shows that in many cases, acquiring some portions without making perceptible errors is enough. The new solution reconstructs the mesostructure, basically using a simple normal map obtained from a new procedure. Normal maps are an excellent solution to capture bronze surface mesostructure also because they are easy to be stored in the modern graphics hardware (Akenine-Mo¨ller et al., 2008); and,
- microstructure scale: to represent it, we developed a new shader able to overcome the typical problem of the completely Lambertian reflexion of the

patinas and the presence of specular reflections in the case of the cleaned bronze using a unique BRDF representation.

The proposed solution is illustrated in Figure 2, and some results are demonstrated in Figure 3. From the point of view of the 3D model acquisition, construction, and visualisation, it presents three distinct phases:

- the delivering of a photogrammetric reality-based 3D model with a 0.2 mm accuracy to model the macrostructure of the bronze surface;
- the acquisition and processing of some dataset of images to faithfully represent mesostructure and microstructure, modelling albedo, and normals of the bronze surface; and,
- the side-by-side visualisation of the textured 3D model, allowing to compare in real-time inside the OpenGL viewer chromatic and microscale characteristics at different steps of the surface cleaning.

The solution is based on our experience in the field of colour acquisition and reproduction, 3D photogrammetric pipeline automation (Gaiani, 2015) and RTR (Gaiani et al., 2015) applied to Cultural Heritage artefacts.

From the point of view of the operator, it requires: to place a target X-Rite ColorChecker Classic (CC) in the scene; to take some images with a SLR camera, using a simple image sequence network geometry able to ensure correct camera calibration (Barazzetti et al., 2010) and orientation, and consistent data from dense stereo matching (Remondino et al., 2014) from automated process; and, to handle a button-click pipeline using a unique interface, specifying a measurement to scale the results.

The pipeline is a seven steps process:

1. images pre-processing: RAW images development, denoise, and colour balance;
2. automatic 3D model construction: photos alignment, dense cloud building, mesh construction;
3. 3D model scaling;
4. 3D model remeshing and parametrization;
5. texture map creation;
6. normal map creation; and,
7. 3D OpenGL based visualisation.

Phases 1., 4., 5., 6., 7., are handled with in-house software; phases 2., 3., and partially 5. and 6., are managed using Agisoft PhotoScan (Agisoft, 2017); phase 7. is managed by Autodesk Showcase (Autodesk, 2017).

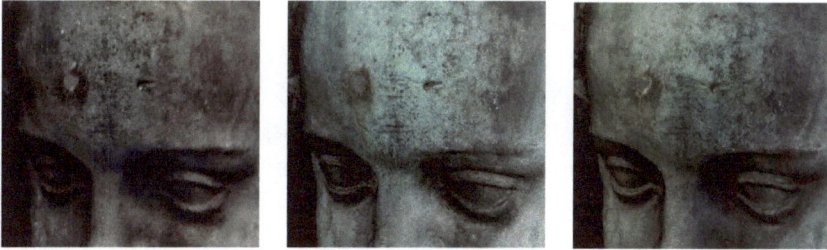

Figure 3. Female Putto. Comparison side-by-side on OpenGL viewer of bronze surfaces in the different phases of cleaning: before cleaning (**left**), phase 1 (**middle**), phase 2 (**right**).

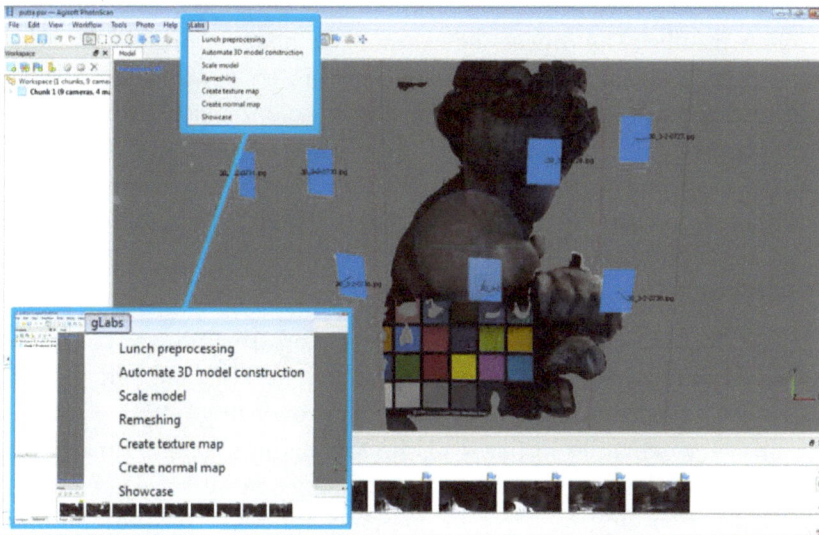

Figure 4. Customized interface based on PhotoScan to manage some steps of the pipeline.

Each phase inside PhotoScan was accurately checked and some improvements were done to overcome some emerged problems. Autodesk Showcase allowed to speed-up the development, but an in-house solution is under development. The current interface is based on a customized version of PhotoScan (Figure 4), but in the future, a web-based separate interface handling a new client server solution will be implemented, as well as an in-house solution to replace Autodesk Showcase.

Automated image pre-processing is rooted in the well-delineated context in which Stamatopoulos et al. (2012) demonstrated that a pre-processing approach for RAW imagery can yield significant photogrammetric accuracy improvements over those from JPEG images. With the aim of avoiding as far as possible modifications of RAW pixel values, only the basic in-camera processing was retained: black point subtraction; bad pixel removal; dark frame, bias subtraction and flat-field correction; green channel equilibrium correction; and, Bayer interpolation. We did not allow on-camera: denoising; colour scaling; image sharpening; colour space conversion; Gamma correction; and, format conversion. These pre-processing steps were done off-camera using an automated procedure described in (Ballabeni et al., 2015). This procedure consists in a calibrated and customized version of the on-camera processing, which consistently fits our purposes. For image denoising, a variant of the Color Block Matching 3D (CBM3D) filter (Dabov et al., 2007) was implemented (named CBM3D-new) with some customizations. For every image of a dataset, the method, reading initialization data from the EXIF, automatically selects the necessary parameters based on the type of camera, ISO sensitivity, and stored colour profiles. The processing selection of the latter one is based on image features and camera capabilities. Dealing with professional or prosumer setups, when source images are stored as RAW in a wide colour space, such as the Adobe-RGB 1998, then opponent colour spaces are chosen. When consumer cameras are used and source images are stored in JPG format and a narrower colour space, such as sRGB, in that case YCbCr colour space is chosen.

3. Photogrammetric Data Acquisition

The photogrammetric-based 3D model construction is based on the completely automated pre-processing procedure, as illustrated in Gaiani et al. (2016a; 2016b), allowing for a robust image matching starting from RAW photographs. The remaining phases of the automatic photogrammetric pipeline (Structure-from-Motion (SfM), Bundle Adjustment, dense stereo matching), the mesh reconstruction, and the texture mapping are based on the automation of PhotoScan, with some customizations and enhancements listed below. The software allows also a quick orientation of the new images captured in the different steps of the cleaning process to obtain the new maps representing the albedo. We paid attention to the problems that have arisen in automated tools that are used to produce a metrically precise 3D point cloud or textured 3D model, where apparently successful solutions are instead a local minimum and not the fully correct one. Typical examples are SfM results where, despite a message of successful image orientation and a very small re-projection error, there are some cameras wrongly oriented. A non-expert user could only spot such errors with difficultly and would proceed to the successive processing stages, negatively affecting the results. To minimize these problems in our automated solution, we:

- established a camera network, easy to use by restorers, appropriate for each bronze surface shooting setting and assuming areas of maximum size of 30 × 30 × 30 cm, corresponding to the typical surface area investigated. The dataset includes nine images, vertical and horizontal. Two images include a CC target acquired from a greater distance, the other seven are captured in an arc within a 45° angle from the perpendicular to the object;

- automated all of the steps inside PhotoScan to avoid any human error without introducing further software errors;
- tested repeatedly all of the steps performed in PhotoScan. Mainly we evaluated the cameras orientation using a different software solution consisting in our robust feature extraction based on a new implementation (Apollonio et al., 2014) of the ASIFT detector-descriptor (Morel and Yu, 2009) and a customized version of the software VisualSFM (Wu, 2013), slower but more accurate of the original software. We also tested the results made with the dense image matching against laser scanner data; and,
- defined a very simple technique to put in scale the 3D model. The user just need to place markers in two different photos in correspondence of the four crosses placed on the corners of the CC. Distances between crosses are known, constant across different targets and accurately measured in the laboratory.

4. Colour Workflow

Colour acquisition, management, and visualisation solution extends an existing efficient solution that is developed for an interior, light controlled environment (Gaiani et al., 2015) to the outside environment, with the limited condition, always satisfied in a restoration site of a statue, to have only indirect lighting. Implementation includes:

1. colour management and calibration throughout the acquisition and rendering pipeline, working in the context of colour space in which images are observed (i.e., sRGB); and,
2. development of a specific shader simulating the microstructure behaviour of the metallic material.

Essentially, we dealt with the issue of colour management for visualisation using OpenGL API, a very limited system when compared to the perceptive quality of the human eye and to that reproducible with analogical systems.

4.1. Colour Management and Calibration

The aim of our colour processing essentially consists in obtaining radiometric calibrated albedo maps that are able to ensure consistency and fidelity of surfaces colour reproduction at runtime in the OpenGL-based application. Albedo is the ratio of total reflected energy to incident energy as a function of incident angle. Diffuse albedo is the albedo of the diffuse component of the lighting, which is a constant in a Lambert diffuse model.

For digital images, colour characterization methods refer to the techniques of converting camera responses (e.g., RGB) to a device-independent colorimetric representation (e.g., CIEXYZ) (Green and MacDonald, 2002). The main problem consists in recovering a linear relationship between the irradiance values and the pixel encoding produced by the camera, typically non-linear. We need therefore to model the non-linearities introduced by in-camera processing for enhancing the visual quality of recorded images. Between the two general approaches to colour characterization—spectral sensitivities-based and colour targets-based (Hong et al., 2001)—the latter technique is adopted, which uses a set of differently coloured samples that are measured with a spectrophotometer. The colour characterization is performed using a CC target (McCamy et al., 1976) in two different phases:

- captured images development and pre-processing; and,
- albedo map pre-processing for rendering.

The colour characterization in the second phase was introduced to overcome a limitation of PhotoScan, which does not manage colour, producing unwanted artefact and even considerable colour differences in the image due to mismatches between the colour spaces. The solution is a two-steps process:

- a first colour balance was performed against the CC using a polynomial regression approach realized with a known illuminant (e.g., known values of the patches) using images in the camera's native linear colour space (Cheung et al., 2006). At the end of the process, before the rendered colour space assignment, we applied a standard gamma correction ($\gamma = 2.2$), improving the visualisation quality in the rendered colour space; and,
- a refinement that is based on Bruce Fraser's calibration procedure for successive approximations (RAGS, 2017). After colour balance, a conversion of image colorimetric coordinates in a scene-referred non-linear colour space was carried out. Colour correction settings are applied automatically using a developed software solution consisting in a RAW image processing implemented in MATLAB and supported by DCRaw, an open-source command-line program, easily coupled with MATLAB (DCRaw, 2017), allowing for image demosaicing, white balance, output file in a rendered colour space, gamma correction, brightness control, 8-bit/16-bit conversion.

Our automatic workflow is partially described in (Gaiani et al., 2016a) and is as follows:

1. RAW image 16-bit linearization and devignetting;
2. ColorChecker localisation;
3. polynomial fitting of observed values with expected values;
4. image correction using the fitting function found at point 3;
5. white balance of the correct image;

6. ΔE^*_{00} mean error evaluation on the CC;
7. iterative image correction using the fitting function found at point 3 increasing the degree of the polynomial at each step (iteration stops when ΔE^*00 stops decreasing);
8. image correction using the new fitting function;
9. ACR scripts calibration; and,
10. colour space assignment.

A crucial issue in the project concerned the use of an appropriate colour space. This is a major issue using PhotoScan since different colour spaces provide different results.

We used two different colour spaces. The first is the sRGB IEC 61966-2-1, a default colour space for multimedia application (IEC, 1999) with the white-point at 6500K temperature (D65) that we used as output colour space. The sRGB is consistent with the visualisation using different devices and is incorporated by the two dominant programming interfaces for 3D graphics, OpenGL and Microsoft Direct3D. The main drawback is the gamma built inside that cannot be expressed as a single numerical value. The overall gamma is approximately 2.2, but it consists of a linear (gamma 1.0) section near black, and a non-linear section elsewhere, involving a 2.4 exponent and a gamma changing from 1.0 through about 2.3. A second drawback is the range of colours, which is narrower than that of human perception (i.e., it does not display properly saturated colours, such as cadmium yellow and cobalt blue). This last disadvantage is not a problem in our case, because misrepresented colours are rarely found. The second one is the AdobeRGB (1998), a RGB colour space that is developed by Adobe Systems in 1998 that we used along the processing pipeline. This colour space is larger than that sRGB (mainly in cyan-green hues), well represents the colours that are acquired by the cameras and could be entirely (99%) displayed on a professional monitor. sRGB and AdobeRGB gamut are similar in mid-tone values (~50% luminance), but clear differences are evident in shadows (~25% luminance) and highlights (~75% luminance) as well. In fact, AdobeRGB expands its advantages to areas of intense orange, yellow, and magenta regions. We used the AdobeRGB colour space for images acquisition and processing, and only at the end of the workflow are the images compressed in the visualisation colour space sRGB for a minimum loss of information.

4.2. Shader Development

To faithfully represent microstructure, reflectance can be described in terms of the BRDF that defines in which way light is reflected off an object's surface. A detailed presentation of BRDF theory and applications can be found in Dorsey et al. (2007) and Weyrich et al. (2008). The basic problem for its use is that the

BRDF is not known a priori in analytical form and its measurement is very difficult. However, efficiently modelling surface reflectance behaviour needs to be coupled with the use of an adequate model, which should have as few parameters as possible to still enable an easy to use faithful depiction of the material in a synthetic image. Usually for metal and patinas simulations layered models are used, but this is not our case where:

- cleaned bronze surface presents a slight specular reflection;
- metallic patina results in a relatively rough and diffuse surface which completely covers the bronze before the cleaning; and,
- cleaning operations could change specular reflection of surface in the different steps of cleaning.

Recent measurements on open-air bronze artworks (Franceschi et al., 2006) demonstrated that specular contribution is limited and only appears when patinas cover a very limited area on the surface, which is at the end of cleaning. Following these observations, we defined a BRDF incorporating a Lambert diffuse term and a microfacet BRDF for the specular contribution (Torrance and Sparrow 1967; Cook and Torrance, 1981; Walter et al., 2007):

$$f(l,v) = \frac{c_{diff}}{\pi} + \frac{F(l,h)G(l,v,h)D(h)}{4(n \cdot l)(n \cdot v)} \tag{1}$$

where l light vector, v view vector, h is the halfway vector between the light direction vector and the viewing direction vector, c_{diff} diffuse albedo, $F(l, h)$ is the Fresnel reflectance, $G(l, v, h)$ is the proportion of microfacets, which are not shadowed or masked, and $D(h)$ is the microfacet normal distribution function, determining the size, brightness, and shape of the specular highlight.

A key step to define our BRDF is to have a Fresnel reflectance correct but practical to be used in RTR. The Fresnel reflectance term computes the fraction of light that is reflected from an optically flat surface. Its value depends on two things: the incoming angle (angle between light vector and surface normal) and the refractive index of the material. Since the refractive index may vary over the visible spectrum, the Fresnel reflectance is a spectral quantity. With a higher refractive index of the material, the surface will start to look equally reflective, no matter the viewing angle. A higher index of refraction 'weakens' the Fresnel effect. The full Fresnel equations are complex, require the material parameter (complex refractive index sampled densely over the visible spectrum), and are impractical to be used in RTR. Starting from Schlick's approximation for the Fresnel term (Schlick, 1994):

$$F = c_{spec} + (1 - c_{spec})(1 - h \cdot v)^5 \tag{2}$$

we modelled its behaviour starting from using two different values at 0° and 90° (in our case 0.8 to 1.0) and with an interpolating curve using the Nvidia formula for CG programming, which considers the three parameters of bias, scale, and power (Fernando and Kilgard, 2003):

$$F = bias + scale * (1 - h \cdot v)^{power} \tag{3}$$

$G(l, v, h)$ in our case was fixed to 0.2 starting from direct observations. For $D(h)$, we used the Beckmann distribution, which is physically-based and could deliver better results (Walter et al., 2007).

The Beckmann normal distribution of microfacets is as follows:

$$D = \frac{1}{\pi m^2 (n \cdot h)^4} e^{\left(\frac{(n \cdot h)^2 - 1}{m^2 (n \cdot h)^2}\right)} \tag{4}$$

where m is a user-defined variable controlling the surface roughness, placed to 0.5 according to Mihaylov (2013).

Shader implementation was done using the OpenGL Shading Language (GLSL), where components are computed separately per-pixel and then added together to get the pixel's final colour.

5. Albedo and Normal Mapping, Mesh Processing

Our albedo and normal reconstruction is a two-step process:

- firstly, the selected set of images is projected on the mesh, exploiting camera pose and camera calibration previously recovered; and,
- in the second step, a texture atlas is created.

Both of these functions are implemented in PhotoScan reliably starting from Lempitsky and Ivanov (2007), allowing for excellent colour blending of images dataset and a compact atlas as output.

We led enhancement in the parametrization to have less texture distortion, introducing an extra-step consisting in the remeshing of the geometry from dense stereo matching (typically noised and irregular) and exploiting the excellent blending to develop a new solution for normal mapping.

Our remeshing technique relies on the method described in Dunyach et al. (2013), and is extended from Botsch and Kobbelt (2004), which are creating a uniform isotropic mesh, to an approach for curvature-adaptive remeshing.

The normal mapping solution arises from the observation that we do not need a lower density surface to include finer detail, but accurately show these details. There are basically three methods to create normal maps that could to be used together: 3D modelling and baking from high-poly geometry to a low-poly geometry (Cohen et al., 1998; Cignoni et al., 1998); photometric stereo techniques (Horn, 1989); and, two-dimensional (2D) image processing, using filtering techniques over a heightmap, e.g., the Sobel filter (Pratt, 2007), which gives a greater weight to the directly adjacent neighbours (Ginsburg et al., 2001). In

41

practice, the Sobel operator is simply a pair of 3×3 convolution kernels. One highlights the horizontal gradient/edges, and the other one highlights the vertical gradient/edges. The normal map could be obtained from the gradient vectors, normalized, and displayed as an RGB-encoded normal map.

In our case, the first two methods are impractical and/or ineffective. Preserving attribute values on simplified meshes present the only advantage of a lighter polygonal model, but the quality of the mesostructure is that explicitly modelled. Shape-from-shading is an ill posed problem and constraints needed to give solution in the outside environment without allowing high accuracy detail. Normal maps created in 2D could be an excellent solution. Sobel operator is a simple way to approximate the gradient of the intensity in an image. However, it works best when tiled across 3D models that have a uniform direction in tangent space, like terrains or walls. On these models, the UVs are not rotated; they are all facing roughly in the same direction. To get seamless lighting, rotated UVs require specific gradients in the normal map, which can be created properly by baking a 3D model.

To fill this requirement, we developed a reconstruction pipeline that combines data from two different sources: the cameras orientations obtained from automated photogrammetry and a few 2D normal maps calculated using the Sobel operator, starting from our dataset of nine images. The solution is inspired, with some customization, by the work of Nehab et al. (2005), who proposed combining the low-frequency components of the geometry with the high-frequency details of the normal maps. However, that work uses the 3D geometry, instead of the simple camera orientation parameters. Also, their optimization for full 3D models is an approximation that forces the surface to be locally planar, resulting in the over-smoothing of high-frequency details. Our pipeline consists of three main steps:

- extraction of 2D normal maps applying the Sobel filter to the images used for the photogrammetric process converted into greyscale diffuse images;
- alignment of the normal maps to the mesh using existing camera orientation; and,
- blending of multiple normal maps to produce a seamless normal field using Lempitsky and Ivanov (2007).

The system is completely integrated in the photogrammetric pipeline and does not require further alignment, pre-processing, or resolution/precision compatibility between different data types. Results are presented in Figure 5.

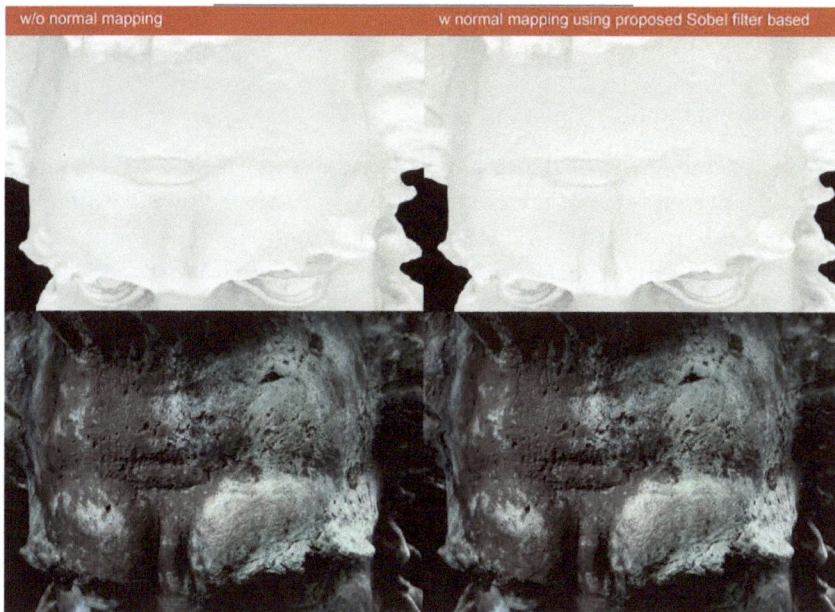

Figure 5. Normal mapping of the Neptune's Head, colour rendering: without normal mapping (**left**); using the proposed solution based on Sobel-filter (**right**).

6. Data Visualisation

The developed 3D model visualisation solution is based, at current state, on the RTR commercial software Autodesk Showcase. which is enabled by the PhotoScan interface. It consists basically of an OpenGL user-friendly viewer, allowing for quick prototyping of the visualisation system and the automatic variants change of the different maps representing the colour of the bronze surface without any intervention of the user. The viewer supports texture maps of up to 16,384 × 16,384 pixels (a strict requirement to ensure the quality of the visualised surface). In Photoscan, a Python language script (Python, 2017) enabled by a pop-up menu, allow to open a new file in Showcase, where geometry, texture mapping, and normal mapping come from Photoscan processing and shaders, lighting and environment are generated on-fly using the Showcase SDK environment (Autodesk, 2013), which is also enabled by the Python language programming.

To improve the visualisation quality, four solutions have been implemented on the top of our RTR system:

1. colour management: performed throughout the rendering pipeline by working within the context of the final colour space, in which images can be viewed (i.e., sRGB);

2. shading enhancement: a pre-computed ambient occlusion and a series of light maps were implemented;
3. colour balance; and,
4. image-based lighting (IBL) (Debevec, 2002).

In more detail, colour balance allows to solve typical RGB rendering failures that are also present in our case: starting the process with measured colours; simulating colour appearance under another illuminant.

We achieved colour balance by implementing the Ward and Eydelberg-Vileshin (2002) 'Picture Perfect RGB Rendering' technique in the sRGB version as follows:

1. identification of dominant illuminant spectrum by prefiltering of material spectra to obtain tristimulus colours for rendering with von Kries transformation; and appropriate adjustment of source colours;
2. perform tristimulus (RGB) rendering; and,
3. apply white balance transform and convert pixels to display colour space sRGB.

This technique presents the benefit that no modification at all is required for a conventional RGB rendering engine. The spectral pre-filtering is realized by performing colour-balance by a fixed approach against a CC placed inside the scene using the ACR calibrator tools and by applying the results of the calibration step through a series of tone mapping algorithms, implemented in our software, having the same behaviour as in Adobe Photoshop. As suggested in the original paper, for the rendering in sRGB space, white balancing is performed ahead of time, prior to rendering. Light sources matching the dominant illuminant spectrum were modelled as neutral, while spectrally distinct light sources were modelled as having their sRGB colour divided by that of the dominant illuminant. IBL instead to simulate the complete interaction between the virtual and real objects, simplifies the simulation by dividing the real scenes into two parts, a distant part, which is not affected by the virtual objects, and a local part, in which the reflectance properties and geometry are modelled. The distant scene is, in contrast to the local scene, only represented by its emitted radiance, or, equivalently, by the incident illumination on the virtual objects from the real scene. To compute the final, composed rendered image in IBL, a technique known as differential rendering (Debevec, 1998) has been adopted, using as latitude-longitude HDR map, a physically accurate model of the typical bronze surface observation scenario lighting, synthetically recreated for the application.

The main advantage of the system is a faithful visualisation that is able to provide the possibility to visualise the fine details of the surface from any angle of observation and under different types of lights. Restorers, as well as the scholars, can zoom in on high-resolution images to measure the characteristics of an image and compare side-by-side different cleaning states from the same point of view, observing not only the cleaning level of the surface, but also many extraordinary details, invisible so far.

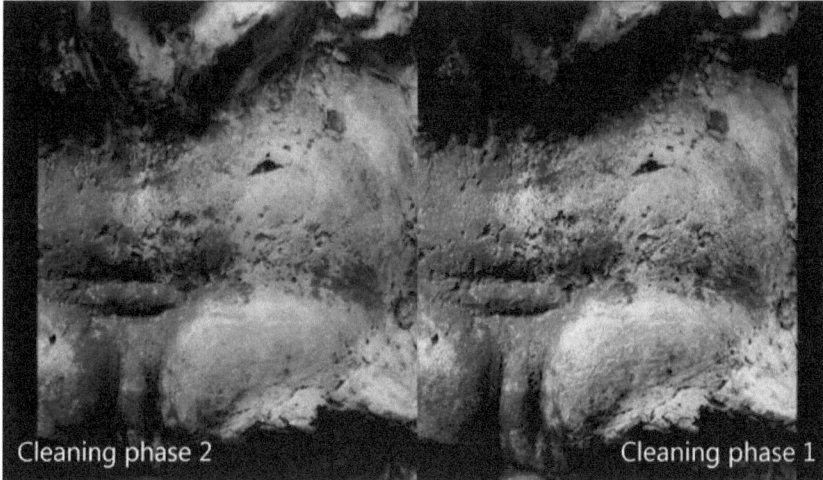

Figure 6. Side by side visualization in Showcase of two different cleaning phases of the Neptune Head.

7. Case Study and Assessment

The implemented workflow was evaluated exploiting the on-going restoration works of the Neptune Fountain in Bologna, undoubtedly the most emblematic monumental complex of the city, and one of most beautiful Renaissance fountain worldwide (Tuttle, 2015). The restoration promoted by the Municipality of Bologna has been carried out in collaboration with the University of Bologna and the *Istituto Superiore per la Conservazione ed il Restauro* (ISCR). One author of this paper is the coordinator for the bronze restoration.

We selected two case studies that were concerning bronze surfaces consisting in an area of a female Putto and in an area of the head of the Neptune statue. Each case study includes three datasets, concerning distinct cleaning stages: previous, during, and at end of the restoration activities. The datasets (acquired with a Nikon D5300, CMOS Image sensor 23.5 × 15.6 mm, 18 mm focal length) include nine images, two of which, containing the CC (Figures 6 and 7).

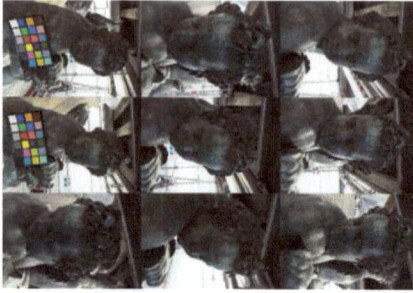

Figure 6. The female Putto datasets and camera networks.

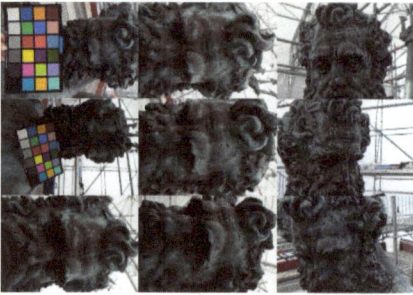

Figure 7. The Neptune head datasets and camera networks.

Table 1. ΔE^*_{00} and exposure error (in f-stops) evaluation.

File Name	ΔE^*_{00} Captured Images	Exposure Error Captured Images	ΔE^*_{00} Albedo	Exposure Error Albedo
Neptune before	2.33	−0.14	2.45	−0.13
Neptune phase 1	2.80	−0.10	2.71	−0.08
Neptune phase 2	2.53	−0.12	2.61	−0.08
Putto before	2.42	0.13	2.51	−0.05
Putto phase 1	2.56	0.12	2.64	−0.08
Putto phase 2	2.75	0.15	2.95	−0.04

We assessed both case studies in two different aspects:

Colour fidelity. We assessed the colour accuracy of our characterization process using the metric distance ΔE^*_{00}, on the CIExyz chromaticity diagram (Sharma et al., 2005), and using the Imatest Studio software (Imatest, 2017). As reference, we used the 8-bit sRGB values tabulated by Pascale (2006). For all of the case studies, colour accuracy ΔE^*_{00} was <3.0. Applying the ACR scripts, ΔE^*_{00} improved up to <2.3, both at the end of captured images development and pre-processing and of albedo map pre-processing for rendering. Exposure error in f-stops obtained were less than 0.05 f-stops. Results are shown in Table 1.

Geometric reconstruction. The implemented procedure was evaluated on the nine image networks. We evaluate the statistical output of the Bundle Adjustment (re-projection error) with two different software (besides PhotoScan, also with our customized version of Visual SFM); and, the accuracy of the dense matching results against the TLS mesh in the Cloud Compare software using Iterative Closest Point (ICP) alignment algorithm. Limits of the re-projection error measurement are known (results could be too optimistic), but no ground truth measurements are available to have an accurate evaluation. The accuracy evaluation of the dense cloud points results was done using Terrestrial Laser Scanning (TLS) data as reference, as acquired with an Artec Space Spider and a resolution of 0.5 mm. The two polygonal models obtained by TLS were compared with the photogrammetric ones. The average image GSD (Ground Sample Distance) in both of the datasets (reference and data comparison) is ca. 1 mm, obtained with a uniform resampling the TLS data. Results (see Figures 8 and 9) show an excellent consistency and completeness respect to the TLS ground truth data.

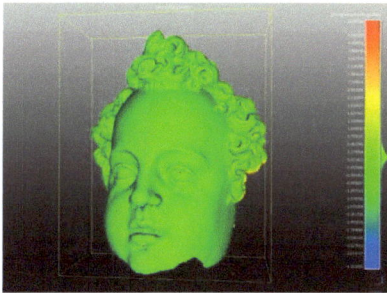

BA reprojection error (px)
- Visual SFM: 0.879
- PhotoScan: 0.937

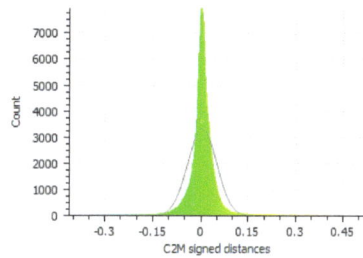

Photogrammetric Model
Numb. Pts. 160,461 / Trian. 313,801
Reference (TLS) Model
Numb. Pts. 409,681 / Trian. 818,403
Gauss mean (mm): 0.04528
Std. Dev. (mm): 0.44880
Average dist. (mm): 0.07052

Figure 8. Female Putto dataset. Bundle Adjustment results and dense cloud points compared with the ground truth TLS data using ICP in CloudCompare cloud to mesh without scale.

Gauss: mean = -0.008873 / std.dev. = 0.062789 [513 classes]

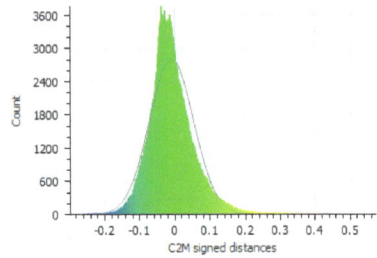

Photogrammetric Model
Numb. Pts. 262,235 / Trian. 521,226
Reference (LS) Model
Numb. Pts. 594,835 / Trian. 1,186,299
Gauss mean (mm): 0.08873
Std. Dev. (mm): 0.62789
Avg. dist. (mm): 0.08399

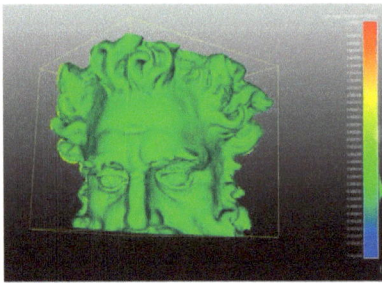

BA reprojection error (px)
-   Visual SFM: 0.961
-   PhotoScan: 0.831

Figure 9. Neptune head dataset. Bundle Adjustment results and dense cloud points compared with the ground truth Terrestrial Laser Scanning (TLS) data using ICP in CloudCompare cloud to mesh without scale.

8. Conclusions and Future Works

In this paper, we presented a global and easy-to-use solution allowing the restorer to have a useful control tool during the restoration work, especially for 3D artefacts. The software is a 6-click solution requiring the user just to check the camera settings before starting the imaging, to place in the scene a CC and to follow a simple image capture protocol. Model scaling is giving, using as reference points the distances between the cross at the CC corners, which are standard and well-known distances. The software can be useful during cleaning phases for testing and gathering information (as direct feedback) of the reached results, considering the irreversible nature of the operation. Furthermore, it becomes an objective tool for documenting the restoration during the work and to illustrate it, giving an easy and direct access to other professionals that are interested. In addition, the versatility of the system and its easy use will be useful to verify all of its potentialities in lab, without the difficulties in the data acquisition phase (stability of the scaffolding, distribution of light, etc.), as demonstrated during the experimentation that was carried out on a large bronze statue outdoors.

Acknowledgements: The authors are thankful to the Bologna City Council for their collaboration in the project. The authors acknowledge also M.C.M. srl for laser scanner data.

References

1. Agisoft. Available online: http://www.agisoft.com/.
2. Akenine-Möller, T.; Haines, E.; Hoffman, N. *Real-Time Rendering*, 3rd ed., A K Peters/CRC Press: Wellesley, MA, USA, 2008.
3. Apollonio, F.; Ballabeni, A.; Gaiani, M.; Remondino, F. Evaluation of feature-based methods for automated network orientation. *Int. Arch. Photogramm. Remote Sens. Spat. Inf. Sci.* **2014**, *40*, 47–54.
4. Autodesk. Available online: http://download.autodesk.com/global/docs/showcasesdk2013/en_us/.
5. Autodesk. Available online: http://www.autodesk.com/products/showcase/.
6. Ballabeni, A.; Apollonio, F.I.; Gaiani, M.; Remondino, F. Advances in image pre-processing to improve automated 3D reconstruction. *Int. Arch. Photogramm. Remote Sens. Spat. Inf. Sci.* **2015**, *50–5(W4)*, 315–323.
7. Barazzetti, L.; Scaioni, M.; Remondino, F. Orientation and 3D modelling from markerless terrestrial images: combining accuracy with automation. *Photogramm. Record* **2010**, *25*, 356–381.
8. Basilissi, V.; Marabelli, M. Le patine dei metalli: Implicazioni teoriche, pratiche, conservative. In *L'arte Fuori dal Museo: Problemi di Conservazione Dell'arte Contemporanea*; Rinaldi, S., Ed.; Gangemi: Rome, Italy, 2008; pp. 74–89.
9. Botsch, M.; Kobbelt L. A Remeshing Approach to Multiresolution Modeling. In Proceedings of the SGP '04, Nice, France, 8–10 July 2004; pp. 185–192.
10. Cheung, V.; Westland, S. Methods for Optimal Color Selection. *J. Imaging Sci. Technol.* **2006**, *50*, 481–488.
11. Cignoni, P.; Montani, C.; Scopigno, R.; Rocchini, C. A general method for preserving attribute values on simplified meshes. In Proceedings of the VIS '98, Los Alamitos, CA, USA, 18–23 October 1998; pp. 59–66.
12. Cohen, J.; Olano, M.; Manocha, D. Appearance-preserving simplification. In Proceedings of the SIGGRAPH '98, Orlando, FL, USA, 19–24 July 1998; pp. 115–122.
13. Cook, R.L.; Torrance, K.E. A reflectance model for computer graphics. *Comput. Graph.* **1981**, *15*, 307–316.
14. Dabov, K.; Foi, A.; Katkovnik, V.; Egiazarian, K. Colour image denoising via sparse 3D collaborative filtering with grouping constraint in luminance-chrominance space. In Proceedings of the IEEE International Conference on Image Processing, San Antonio, TX, USA, 16–19 September 2007; pp. 313–316.
15. DCRaw. Available online: http:// www.cybercom.net/~dcoffin/dcraw/.
16. Debevec, P. Rendering synthetic objects into real scenes: bridging traditional and image-based graphics with global illumination and high dynamic range photography. In Proceedings of the SIGGRAPH '98, Orlando, Florida USA 19–24 July 1998; pp. 189–198.
17. Debevec, P. Image-based lighting. *IEEE Comput. Graph. Appl.* **2002**, *22*, 26–34.
18. Dorsey, J.; Rushmeier, H.; Sillion, F. *Digital Modeling of Material Appearance*; Morgan Kaufmann: San Francisco, CA, USA, 2007.

19. Dunyach, M.; Vanderhaeghe, D.; Barthe, L.; Botsch, M. Adaptive Remeshing for Real-Time Mesh Deformation. In Proceedings of Eurographics 2013, Girona, Spain, 6–10 May 2013; pp. 29–32.

20. Fernando, R.; Kilgard, M.J. *The Cg Tutorial: The Definitive Guide to Programmable Real-Time Graphics*; Addison Wesley: New York, NY, USA, 2003.

21. Filip, J.; Haindl, M. Bidirectional Texture Function Modeling: A State of the Art Survey, *IEEE Trans. PAMI* **2009**, *31*, 1921–1940.

22. Franceschi, E.; Letardi, P.; Luciano, G. Colour measurements on patinas and coating system for outdoor bronze monuments. *J. Cult. Herit.* **2006**, *7*, 166–170.

23. Gaiani, M. (Ed.). *I portici di Bologna. Architettura, Modelli 3D e Ricerche Tecnologiche*; BUP: Bologna, Italy, 2015.

24. Gaiani, M.; Apollonio, F.I.; Clini, P. Innovative approach to the digital documentation and rendering of the total appearance of fine drawings and its validation on Leonardo's Vitruvian Man. *J. Cult. Herit.* **2015**, *16*, 805–812.

25. Gaiani, M.; Apollonio, F.I.; Ballabeni, A.; Remondino, F. A technique to ensure color fidelity in automatic photogrammetry. *Colour Color. Multidiscip. Contrib.* **2016**, *12*, 53–66.

26. Gaiani, M.; Remondino, F.; Apollonio, F.I.; Ballabeni, A. An Advanced Pre-Processing Pipeline to Improve Automated Photogrammetric Reconstructions of Architectural Scenes. *Remote Sens.* **2016**, *8*, 178.

27. Ginsburg, D.; Gosselin, D. Dynamic Per-Pixel Lighting Techniques. *GP Gems* **2001**, *2*, 452–462.

28. Green, P.; MacDonald, L.W. (Eds.). *Colour Engineering: Achieving Device Independent Colour*; Wiley: New York, NY, USA, 2002.

29. Hong, G.; Luo, M.R.; Rhodes, P.A. A Study of Digital Camera Colorimetric Characterization Based on Polynomial Modeling. *Color Res. Appl.* **2001**, *26*, 76–84.

30. Horn, B.K.P. Obtaining shape from shading information. In *Shape from Shading*; MIT Press: Cambridge, MA, USA, 1989, pp. 121–171.

31. IEC. *Multimedia Systems and Equipment—Colour Measurement and Management—Part 2–1: Colour Management—Default RGB Colour Space—sRGB*. IEC: Geneva, Switzerland, 1999.

32. Imatest. Available online: http://www.imatest.com/products/imatest-master/.

33. Kautz, J. Approximate Bidirectional Texture Functions. In *GPU Gems 2*; Addison-Wesley: New York, NY, USA, 2005; pp. 177–190.

34. Lempitsky, V.; Ivanov, D. Seamless Mosaicing of Image-Based Texture Maps. In Proceedings of the IEEE CVPR, Minneapolis, MN, USA, 17–22 June 2007; pp. 1–6.

35. Letardi, P. Laboratory and field tests on patinas and protective coating systems for outdoor bronze monuments. In Proceedings of the Metal 04, Canberra, Australia, 4–8 October 2004; pp. 379–387.

36. Lu, J.; Georghiades, A.S.; Glaser, A.; Wu, H.; Wei, L.Y.; Guo, B.; Dorsey, J.; Rushmeier, H. Context-aware textures. *ACM Trans. Graph.* **2007**, *26*, Article 3.

37. Mazzeo, R.; Joseph, E. Micro-destructive analytical investigation for conservation and restoration. In *Monumenti in Bronzo all'aperto. Esperienze di Conservazione a Confront*; Letardi, P., Trentin, I., Cutugno, G., Eds.; Nardini: Firenze, Italy, 2004; pp. 53–58.

38. McCamy, C.S.; Marcus, H.; Davidson, J.G. A Color Rendition Chart. *J. Appl. Photograph. Eng.* **1976**, *11*, 95–99.

39. Merillou, S.; Ghazanfarpour, D. A survey of aging and weathering phenomena in computer graphics. *Comput. Graph.* **2008**, *32*, 159–174.

40. Mihaylov, V. Rendering Metals and Worn or Weathered Metallic Objects, M.Sc. Thesis, Technical University of Denmark, Lyngby, Denmark, 2013.

41. Morel, J-M.; Yu, G. ASIFT: A new framework for fully affine invariant comparison. *SIAM J. Imaging Sci.* **2009**, *2*, 438–469.

42. Nehab, D.; Rusinkiewicz, S.; Davis, J.; Ramamoorthi, R. Efficiently combining positions and normals for precise 3D geometry. *ACM Trans. Graph.* **2005**, *24*, 536–543.

43. Nicodemus, F. Directional reflectance and emissivity of an opaque surface. *Appl. Opt.* **1965**, *4*, 767–775.

44. Pascale, D. *RGB Coordinates of the Macbeth ColorChecker*; The BabelColor Company: Montreal, QC, Canada, 2006.

45. Python. Available online: https://www.python.org/.

46. Pratt, W.K. *Digital Image Processing: PIKS Scientific Inside*; Wiley-Interscience: Hoboken, NJ, USA, 2007.

47. RAGS. Available online: http://www.rags-int-inc.com/PhotoTechStuff/.

48. Remondino, F.; Spera, M.G.; Nocerino, E.; Menna, F.; Nex, F. State of the art in high density image matching. *Photogramm. Rec.* **2014**, *29*, 144–166.

49. Rushmeier, H. Computer Graphics Techniques for Capturing and Rendering the Appearance of Aging Materials. In *Service Life Prediction Polymeric Materials*; Martin, J.W., Ryntz, R.A., Chin, J., Dickie, R.A., Eds.; Springer: New York, NY, USA, 2009; pp. 283–292.

50. Schlick, C. An Inexpensive BRDF Model for Physically-Based Rendering. *Comput. Graph. Forum* **1994**, *13*, 233–246.

51. Sharma, G.; Wu, W.; Dalal, E.N. The CIEDE2000 Color-Difference Formula: Implementation Notes, Supplementary Test Data, & Mathematical Observations. *Color Res. Appl.* **2005**, *30*, 21–30.

52. Stamatopoulos, C.; Fraser, C.S.; Cronk, S. Accuracy aspects of utilizing raw imagery in photogrammetric measurement. *ISPRS Ann. Photogramm. Remote Sens. Spat. Inf. Sci.* **2012**, *39*, 387–392.

53. Torrance, K.E.; Sparrow, E. Theory for off-specular reflection from roughened surfaces. *J. Opt. Soc. Am.* **1967**, *57*, 1105–1114.

54. Tuttle, R.J. *The Neptune fountain in Bologna. Bronze, Marble and Water in the Making of a Papal City*; Brepols-Harvey Miller: New York, NY, USA, 2015.

55. Walter, B.; Marschner, S.R.; Li, H.; Torrance, K.E. Microfacet models for refraction through rough surfaces. In Proceedings of EGSR'07, Grenoble, France, 25–27 June 2007; pp. 195–206.

56. Ward, G.; Eydelberg-Vileshin, E. Picture perfect RGB rendering using spectral prefiltering and sharp color primaries. In Proceedings of the 13th EGWR, Pisa, Italy, 26–28 June 2002; pp. 117–124.

57. Westin, S.H.; Arvo, J.; Torrance, K.E. Predicting reflectance functions from complex surfaces. In Proceedings of the SIGGRAPH '92, Chicago, IL, USA, 26–31 July 1992; pp. 255–264.

58. Weyrich, T.; Lawrence, J.; Lensch, H.P.; Rusinkiewicz, S.; Zickler, T. Principles of Appearance Acquisition and Representation. *Found. Trends Comput. Graph. Vision* **2008**, *4*, 75–191.

59. Wu, C. Towards Linear-Time Incremental Structure from Motion. In Proceedings of the 3DV 2013, Seattle, WA, USA, 29 June–1 July 2013; pp. 127–134.

Continuous-Time SLAM—Improving Google's Cartographer 3D Mapping

Andreas Nüchter [a,b], **Michael Bleier** [b], **Johannes Schauer** [a] and
Peter Janotta [c]

[a] Informatics VII – Robotics and Telematics, Julius-Maximilians University Würzburg,
   Germany; (andreas.nuechter, johannes.schauer)@uni-wuerzburg.de
[b] Zentrum für Telematik e.V., Würzburg, Germany;
   (andreas.nuechter, michael.bleier)@telematik-zentrum.de
[c] Measurement in Motion GmbH, Theilheim, Germany; peter.janotta@mim3d.de

Abstract: This paper shows how to use the result of Google's simultaneous localization and mapping (SLAM) solution, called Cartographer, to bootstrap a continuous-time SLAM algorithm that was developed by the authors and presented in previous publications. The presented approach optimizes the consistency of the global point cloud, and thus improves on Google's results. Algorithms and data from Google are used as input for the continuous-time SLAM software. In preceding work, the continuous-time SLAM was successfully applied to a similar backpack system which delivers consistent 3D point clouds even in the absence of an IMU. Continuous-time SLAM means that the trajectory of a mobile mapping system is treated in a semi-rigid fashion, i.e., the trajectory is deformed to yield a consistent 3D point cloud of the measured environment.

Keywords: 3D laser scanner; mapping; SLAM; trajectory optimization; personal laser scanner

1. Introduction

On October 5, 2016, Google released the source code of its real-time 2D and 3D simultaneous localization and mapping (SLAM) library Cartographer[1]. The utilized algorithms for solving SLAM in 2D have been described in a recent paper by the authors of the software (Hess et al., 2016). It can deliver impressive results—especially considering that it runs in real-time on commodity hardware. A publication describing the 3D mapping solution is still missing. The released software however, solves the problem. In addition, Google published a very demanding, high-resolution data set to the public for testing their algorithms. Also, custom data sets are easy to process, as Google's software comes with an

---

[1] https://opensource.googleblog.com/2016/10/introducing-cartographer.html

integration into the robot operating system (ROS) (Quigley et al., 2009). ROS is the de-facto standard middleware in the robotic community. It allows to connect heterogeneous software packages via a standardized inter-process communication (IPC) system and is available on recent GNU/Linux distributions.

Google's sample data set was recorded in the museum "Deutsches Museum" in München, Germany. It is the world's largest museum of science and technology, and has about 28,000 exhibited objects from 50 fields of science and technology. The data set was recorded with a backpack system, which features an inertial measurement unit (IMU) and two Velodyne PUCK (VLP-16) sensors. The processed trajectory was 108 meters long and contained 300,000 single 3D scans from the PUCK sensors.

Due to a high demand on flexible mobile mapping systems, mapping solutions on pushcarts, on trolleys, on mobile robots, and backpacks have recently been developed. Human-carried systems offer the advantage of overcoming doorsteps and the operator can open closed doors etc. To this end, several vendors build human-carried systems which are also often called personal laser scanners.

This article shows how to use the result of Google's SLAM solution to bootstrap our continuous-time SLAM algorithm. Our approach optimizes the consistency of the global point cloud, and thus improves on Google's results. In this article, the algorithms and data from Google are used as input for our continuous-time SLAM solution, which was recently published in (Elseberg et al. 2013). In preceding research, the presented algorithmic solution was successfully applied to a similar backpack system set up by the authors which delivers consistent 3D point clouds even in the absence of an IMU (Nüchter et al., 2015).

In the following, this article will discuss the related work with a focus on unconventional mobile mapping systems and the work on calibration, referencing and SLAM. Then, the focus is shifted towards the registration of 3D scans in more detail and the ICP algorithm described and the globally consistent scan matching is derived. This is finally extended to a continuous-time SLAM solution, which takes Google's Cartographer 3D Mapping as input.

2. Related Works

2.1. Laser Scanner on Robots and Backpacks

Mapping environments has received a lot of attention in the robotics community, especially after the appearance of cost effective 2D laser range finders. Seminal work with 2D profiles in robotics was performed by (Lu and Milios, 1994). After deriving 2D variants of the by now well-known ICP algorithm, they derived a PosegraphSLAM solution (Lu and Milios, 1997) that considers all 2D scans in a global fashion. Afterwards, many other approaches to SLAM were presented, including extended Kalman filters, particle filters, expectation maximization and GraphSLAM. These SLAM algorithms aimed at enabling mobile robots to map the environments where they have to carry out user-specific tasks. Thrun et al. (2000) presented a system, where a horizontally mounted scanner performed FastSLAM—a particle filter approach to SLAM—while an upward looking scanner was used to acquire 3D data, exploiting the robot motion to construct environments in 3D. Lu and Milios' approach was extended to 3D point clouds and possesses six degrees of freedom (DoF) in (Borrmann et al., 2008).

In 2004 an early version of a backpack system was presented. Saarinen et al., 2004, used the term Personal Localization And Mapping (PLAM). They used a horizontally mounted SICK LMS200 scanner in front of the human-carried system and put additional sensors and the computing equipment into a backpack. Chen et al. (2010), presented a backpack system that featured a number of lightweight 2D profilers (Hokuyo scanner) mounted in different viewing directions. In previous work, the authors applied the algorithms to a backpack system without an IMU (Nüchter et al., 2015). The system consists of a horizontally mounted SICK LMS100 scanner and a spinning Riegl VZ400. Similar to the work of Thrun et al. (2000) a horizontal scanner is used to estimate an initial trajectory that is afterwards updated to regard the six DoF motion. The term Personal Laser Scanning System was coined by Liang et al. (2014). They use a single FARO scanner and rely on the global navigation satellite system (GNSS) system. Similarly, the commercially available ROBIN system features a RIEGL VUX scanner and GNSS. In contrast, the Leica Pegasus is a commercially available backpack wearable mobile mapping solution, which is composed of two Velodyne PUCK scanners, cameras and a GNSS. The PUCKs scan 300.000 points per second and have a maximal range of 100 meters. Sixteen profilers are combined to yield a vertical field of view of ±15 degree.

The Google Cartographer backpack was initially presented in September 2014. Back then, the backpack system was based on two Hokuyo profilers and an internal measurement unit (IMU). The current version features two Velodyne PUCK scanners. Figure 1 shows the system from Google and our backpack solution.

2.2. Calibration, Referencing, and SLAM

To acquire high-quality range measurement data with a mobile laser scanner, the position and orientation of every individual sensor have to be known. Traditionally, scanners, GPS and IMU are calibrated against other positioning devices whose pose in relation to the system is already known. The term Boresight calibration is used for the technique of finding the rotational parameters of the range sensor with respect to the already calibrated IMU/GPS unit. In the airborne laser scanning community, automatic calibration approaches are known (Skaloud and Schaer, 2007), and similarly vehicle-based kinematic laser scanning has been considered (Rieger et al., 2010). In the robotics community, there exist approaches for calibrating several range scanners semi-automatically, i.e., with manually labelled data (Underwood et al., 2009) or using automatically computed quality metrics (Sheehan et al., 2011), (Elseberg et al., 2013). Often, vendors do not make their calibration methods public and unfortunately, the authors of this paper have no information on the calibration of the Google Cartographer backpack. In general, calibration inaccuracies can, to some extent, be compensated with a SLAM algorithm.

In addition to from sensor misalignment, timing related issues are second sources of errors. All subsystems on a mobile platform need to be synchronized to a common time frame. This is often accomplished with pure hardware via triggering or with mixes of hardware and software such as pulse per second (PPS) or the network time protocol (NTP). However, good online synchronization is not always available for all sensors. Olson, 2010, has developed a solution for the synchronization of clocks that can be applied after the fact. In ROS, sensor data is time-stamped when it arrives and it is recorded in an open file format (.bag files). Afterwards, one works with the time-stamped data using nearest values or interpolation.

As the term direct referencing or direct Geo-referencing implies, it is the direct measurement of the position and orientation of a mapping sensor, i.e., the laser scanner, such that each range value can be referenced without the need for collecting additional information. This means that the trajectory is then used to "unwind" the laser range measurements to produce the 3D point cloud. This approach has been taken by (Liang et al., 2014).

Some systems employ a horizontally mounted scanner and perform 2D SLAM on the acquired profiles. Thrun et al. (2000), used FastSLAM, whereas Nüchter et al. (2015) used SLAM based on the truncated signed distance function (TSD SLAM), or alternatively, HectorSLAM (Kohlbrecher et al., 2011). These 2D SLAM approaches produce 2D grid maps. Similarly, Google's Cartographer code is for creating 2D grid maps (Hess et al., 2016). Afterwards, the computed trajectory and the IMU measurements are used to "unwind" the laser range measurements to produce the 3D point cloud.

The Google Cartographer library achieves its outstanding performance by grouping scans into probability grids that they call submaps and by using a branch and bound approach for loop closure optimization. While new scans are matched with and subsequently entered into the current submap in the foreground, in the background, the library matches scans to nearby submaps to create loop closure constraints and to continually optimize the constraint graph of submaps and scan poses. The authors differentiate between local scan matching which inserts new scans into the current submap and which will accumulate errors over time and global SLAM which includes loop closing and which removes the errors that have accumulated in each submap that are part of a loop. Both local and global matching are run at the same time.

During local scan matching, the Cartographer library matches each new scan against the current submap using a Ceres-based scan matcher (Agarwal and others). A submap is a regular probability grid where each discrete grid point represents the probability that the given grid point is obstructed or free. These two sets are disjoint. A grid point is obstructed if it contains an observed point. Free points are computed by tracing the laser beam from the estimated scanner location to the measured point through the grid. The optimization function of the scan matcher makes use of the probability grid as part of its minimization function.

During global SLAM, finished submaps (those that no longer change) and the scans they contain are considered for loop closing. As is the case during local scan matching, the problem is passed to Ceres as a nonlinear least squares problem. The algorithm is accurate down to the groups of points defined by the regular probability grid of each submap. By taking the submap grid size as translation step delta and the angle under which a grid point is seen at maximum range as the rotation step delta, a finite set of possible transformations is created. This solution space is searched using a branch and bound approach where nodes are traversed using a greedy depth first search and the upper bound of the inner nodes is defined in terms of computational effort and quality of the bound. To compute the upper bound efficiently, grids are precomputed for tree heights that overlay the involved submaps and store the maximum values of possible scores for each obstructed grid point. This operation is done in $O(n)$ with n being the number of obstructed grid points in each precomputed grid.

Hess et al. (2016), describe the 2D version of the algorithm, which uses the horizontally mounted 2D profiler. The provided data sets also contain data from a setup with Velodyne PUCK scanners (cf. Figure 1). Their algorithm is able to process 3D data and to output poses with six DoF, however, a description of their 3D approach is missing from their paper. Nevertheless, one can understand from their published source code that their 3D implementation is mostly an extension of their 2D approach to three dimensions with a 3D probability grid. Some changes have been made to improve performance. For example, the 3D grid is not fully traversed to find free grid cells but only a configurable distance up to the measured point is checked.

Only a few approaches optimize the whole trajectory in a continuous fashion. Stoyanov and Lilienthal, (2009), presented a non-rigid optimization for a mobile laser scanning system. They optimize point cloud quality by matching the beginning and the end of a single scanner rotation using ICP. The estimate of the 3D pose difference between the two points in time is then used to optimize the robot trajectory in between. In a similar approach, Bosse and Zlot, (2009), use a modified ICP with a custom correspondence search to optimize the pose of six discrete points in time of the trajectory of a robot during a single scanner rotation. The trajectory in between is modified by distributing the errors with a cubic spline. The software of Riegl RiPRECISION MLS automatically performs adjustments of GNSS/INS trajectories to merge overlapping mobile scan data based on planar surface elements. Our own continuous-time SLAM solution improves the entire trajectory of the data set simultaneously based on the raw point cloud. The algorithm is adopted from (Elseberg et al., 2013), where it was used in a different mobile mapping context, i.e., on platforms such as the LYNX mobile mapper or the Riegl VMX-250. As no motion model is required, it can be applied to any continuous trajectory.

Figure 1. Above left: Google's Cartographer system featuring two Hokuyo laser scanners (image: Google blog). Above right and below left: Google's Cartographer system with two Velodyne PUCKs and the Cartographer team (image courtesy of the Cartographer team). Below right: Second author operating Würzburg's backpack scanner.

3. Registration of 3D Laser Scans

3.1. Feature-Based Registration

Many state-of-the-art registration methods rely on initial pose (position and orientation) estimates acquired by global positioning systems (GPS) or local positioning using artificial landmarks or markers as reference (Wang et al. 2008). Pose information is hard to acquire and in many scenarios is prone to errors or not available at all. Thus, registration without initial pose estimates and place recognition are highly active fields of research.

In addition to range values, laser scanners record the intensity of the reflected light. These intensities provide additional information for the registration process. Böhm and Becker, (2007), suggest using SIFT (scale-invariant feature transform) features for automatic registration and present an example of a successful registration on a 3D scan with a small field of view. Wang and Brenner, (2008), extended this work by using additional geometry features to reduce the number of matching outliers in panoramic outdoor laser scans. Kang et al. (2009), propose a similar technique for indoor and outdoor environments. Weinmann et al. (2011), use a method that is based on both reflectance images and range information. After extraction of characteristic 2D points based on SIFT features, theses points are projected into 3D space by using interpolated range information. For a new scan, combining the 3D points with 2D observations on a virtual plane yields 3D-to-2D correspondences from which the coarse transformation parameters can be estimated via a RANSAC (random sample consensus) based registration scheme including a single step outlier removal for checking consistency (Weinmann, et al., 2011). They extend their method (Weinmann and Jutzi, 2011) to calculate the order of the scans in unorganized terrestrial laser scan data by checking the similarity of the respective reflectance images via the total number of SIFT correspondences between them. Bendels et al., (2004), exploit intensity images often recorded with the range data and propose a fully automatic registration technique using 2D-image features. The fine registration of two range images is performed by first aligning the feature points themselves, followed by a so-called constrained-domain alignment step. In the latter, rather than feature points, they consider feature surface elements. Instead of using a single 3D-point as feature, they use the set of all points corresponding to the image area determined by the position and scale of the feature.

Other approaches rely only on the 3D structure. Brenner et al., (2008), use 3D planar patches and the normal distribution transform (NDT) on several 2D scan slices for a coarse registration. Similarly, Pathak et al., (2010), evaluated the use of planar patches and found that it is mostly usable. A solution using the NDT in 3D is given (Magnusson et al., 2009). While this approach computes global features of the scan, several researchers use features that describe small regions of the scan for place recognition and registration (Huber 2002) (Steder et al., 2010) (Barnea and Filin, 2008). Flint et al., (2007), use a key point detector called ThrIFT, to detect repeated 3D structures in range data of building facades.

In addition to coarse, feature-based registration, many authors use the well-known iterative closest point algorithm (ICP) for fine registration.

3.2. Registration without Using Features— The ICP Algorithm

The following method is the *de facto* standard for registration of two 3D point clouds, given a good initial pose estimate. ICP requires no computation of features. Instead, it matches raw point clouds by selecting point correspondences on the basis of smallest distances and by minimizing the resulting Euclidean error. This iterative algorithm converges to a local minimum. Good starting estimates significantly improve the matching results, i.e., they ensure that ICP converges to a correct minimum. The complete algorithm was invented at the same time in 1991 by Besl and McKay, (1992), by Chen and Medioni, (1991) and by Zhang, (1992). The method is called the Iterative Closest Points (ICP) algorithm.

Given two independently acquired sets of 3D points, M (model set) and D (data set) which correspond to a single shape, one wants to find the transformation (R, t) consisting of a rotation matrix R and a translation vector t which minimizes the following cost function

$$E(R, t) = \sum_{i=1} \|m_i - (Rd_i - t)\|^2 \tag{1}$$

All corresponding points can be represented in a tuple (m_i, d_i) where $m_i \in M \subset \hat{M}$ and $d_i \in D \subset \hat{D}$. Two calculations need to be made: First, the corresponding points, and second, the transformation (R, t) that minimizes $E(R, t)$ on the basis of the corresponding points. The ICP algorithm uses the closest points as corresponding points. A sufficiently accurate starting guess enables the ICP algorithm to converge to the correct minimum.

Current research in the context of ICP algorithms mainly focuses on fast variants of ICP algorithms (Rusinkiewicz and Levoy, 2001). If the inputs are 3D meshes, then a point-to-plane metric can be used instead of Equation (1). Minimizing the use of a point-to-plane metric outperforms the standard point-to-point one, but requires the computation of normal metrics and meshes in a pre-processing step.

4. Globally Consistent n-Scan Matching

Chen and Medioni, (1992), aimed at globally consistent range image alignment when introducing an incremental matching method, i.e., all new scans are registered against the so-called metascan, which is the union of the previously acquired and registered scans. This method does not spread out the error and is order-dependent.

Bergevin et al., (1996), Stoddart and Hilton, (1996), Benjemaa and Schmitt, (1997, 1998), and K. Pulli, (1999) present iterative approaches. Based on networks representing overlapping parts of images, they use the ICP algorithm for computing transformations that are applied after all correspondences between all views have been found. However, the focus of research is mainly 3D modelling of small objects using a stationary 3D scanner and a turn table. Therefore, the used networks consist mainly of one loop (Pulli, 1999). These solutions are locally consistent algorithms that retain the analogy of the spring system (Cunnington and Stoddart, 1999), whereas true globally consistent algorithms minimize the error function in one step.

A probabilistic approach was proposed by Williams and Bennamoun, (1999), where each scan point is assigned a Gaussian distribution in order to model the statistical errors made by laser scanners. This causes high computation time due to the large amount of data in practice. Krishnan, et al., (2000), presented a global registration algorithm that minimizes the global error function by optimization on the manifold of 3D rotation matrices.

To register scans in a globally consistent fashion, a network of poses is formed, i.e., a graph. Every edge represents a link $j \rightarrow k$ of matchable poses. The error function is extended to include all links and to minimize for all rotations and translations at the same time.

$$E = \sum_{j \rightarrow k} \sum_{i=1} \left\| (R_j m_i - t_j) - (R_k d_i - t_k) \right\|^2 \tag{2}$$

For some applications, it is necessary to have a notion of the uncertainty of the poses calculated by the registration algorithm. The following is an extension of the probabilistic approach first proposed by Lu and Milios, (1997) to six DoF. This extension is not straightforward, since the matrix decomposition, i.e., Equation (20), cannot be derived from first principles. For a more detailed description of these extensions, refer to Borrmann et al., (2008). In addition to the poses X_j, the pose estimates $\bar{X}_j$ and the pose errors ΔX_j are required.

The positional error of two poses X_j and X_k is described by:

$$E_{k,j} = \sum_{i=1}^{m} \left\| X_j \oplus d_i - X_k \oplus m_i \right\| = \sum_{i=1}^{m} \left\| Z_i(X_j, X_k) \right\|^2 \tag{3}$$

here, $\oplus$ is the compounding operation that transforms a point into the global coordinate system. For small pose differences, $E_{k,j}$ can be linearized by use of a Taylor expansion:

61

$$Z_i(X_j, X_k) \approx \bar{X}_j \oplus d_i - \bar{X}_k \oplus m_i - (\nabla_j Z_i(\bar{X}_j, \bar{X}_k)\Delta X_j - \nabla_k Z_i(X_j, \bar{X}_k)\Delta X_k) \quad (4)$$

where ∇k denotes the derivative with respect to X_j and X_k. Utilizing the matrix decompositions $M_i H_j$ and $D_j H_k$ of the respective derivatives that separate the poses from the associated points gives:

$$\begin{aligned} Z_i(X_j, X_k) &= Z_i(\bar{X}, \bar{X}_k) - (M_i H_j \Delta X_j - D_i H_k \Delta X_k) \\ &= Z_i(\bar{X}, X_k) - (M_i X_j' - D_i X_k') \end{aligned} \quad (5)$$

Appropriate decompositions are given for both the Euler angles and quaternion representation in the following paragraphs. Because M_i as well as D_i are independent of the pose, the positional error $E_{j,k}$ is minimized with respect to the new pose difference E', i.e.,

$$\begin{aligned} E_{j,k} &= (H_j \Delta X_j - H_k \Delta X_k) \\ &= (X_j' - X_k') \end{aligned} \quad (6)$$

is linear in the quantities X_j that will be estimated so that the minimum of $E_{j,k}$ and the corresponding covariance are given by

$$\bar{E}_{j,k} = (M^T M^{-1}) M^T Z \quad (7)$$

$$C_{j,k} = s\char`\^2 (M^T M) \quad (8)$$

where s^2 is the unbiased estimate of the covariance of the identically, independently distributed errors of Z:

$$s^2 = \frac{(Z - M\bar{E})^T (Z - M\bar{E})}{(2m - 3)}. \quad (9)$$

Here, Z is the concatenated vector consisting of all $Z_i(\bar{X}_j, \bar{X}_k)$ and M the concatenation of all M_is.

Up to now, all considerations have been on a local scale. With the linearized error metric $E_{j,k}'$ and the Gaussian distribution $\bar{E}_{j,k}$, $C_{j,k}$ a Mahalanobis distance that describes the global error of all the poses is constructed:

$$\begin{aligned} W &= \sum_{j \to k} (\bar{E}_{j,k} - E_{j,k}')^{-1} C_{j,k}^{-1} (\bar{E}_{j,k} - E_{j,k}') \\ &= \sum_{j \to k} (\bar{E}_{j,k} - (X_j' - X_k'))^{-1} \end{aligned} \quad (10)$$

In matrix notation, W becomes:

$$W = (\bar{E} - HX)^T C^{-1} (\bar{E} - HX). \tag{11}$$

Here, H is the signed incidence matrix of the pose graph, $\bar{E}$ is the concatenated vector consisting of all $\bar{E}'_{j,k}$ and C is a block-diagonal matrix comprised of $C^{-1}_{j,k}$ as submatrices. Minimizing this function yields new optimal pose estimates. The minimization of W is accomplished via the following linear equation system:

$$(H^T C^{-1} H) X = H^T C^{-1} \bar{E} \tag{12}$$

$$BX = A \tag{13}$$

The matrix B consists of the submatrices

$$B_{j,k} = \begin{cases} \sum_{k=0}^{n} C^{-1}_{j,k} & (j = k) \\ C^{-1}_{j,k} & (j = k) \end{cases} \tag{14}$$

The entries of A are given by:

$$A_j = \sum_{\substack{k=0 \\ k \neq j}}^{n} C^{-1}_{j,k} \bar{E}_{j,k}. \tag{15}$$

In addition to X, the associated covariance of C_X is computed as follows:

$$C_X = B^{-1} \tag{16}$$

Note that the results have to be transformed in order to obtain the optimal pose estimates.

$$X_j = \bar{X}_j - H_j^{-1} X'_j \tag{17}$$

$$C_j = (H_j^{-1} C_j^X (H_j^{-1})^T. \tag{18}$$

The representation of pose X in Euler angles, as well as its estimate and error is as follows:

$$= \begin{pmatrix} t_x \\ t_y \\ t_z \\ \theta_x \\ \theta_x \\ \theta_z \end{pmatrix}, \bar{X} = \begin{pmatrix} \bar{t}_x \\ \bar{t}_y \\ \bar{t}_z \\ \bar{\theta}_x \\ \bar{\theta}_x \\ \bar{\theta}_z \end{pmatrix}, \Delta X = \begin{pmatrix} \Delta t_x \\ \Delta t_y \\ \Delta t_z \\ \Delta \theta_x \\ \Delta \theta_x \\ \Delta \theta_z \end{pmatrix} \tag{19}$$

The matrix decomposition $M_i H = \nabla Z_i \overline{X}$ is given by

$$H = \begin{pmatrix} 1 & 0 & 0 & 0 & \bar{t}_z \cos(\bar{\theta}_x) + \bar{t}_y \sin(\bar{\theta}_x) & \bar{t}_y \cos(\bar{\theta}_y)\cos(\bar{\theta}_x) - \bar{t}_z \cos(\bar{\theta}_x)\sin(\bar{\theta}_x) \\ 0 & 1 & 0 & -\bar{t}_z & -\bar{t}_x \sin(\bar{\theta}_x) & \bar{t}_x \cos(\bar{\theta}_x)\cos(\bar{\theta}_x) - \bar{t}_z \sin(\bar{\theta}_x) \\ 0 & 0 & 1 & \bar{t}_y & -\bar{t}_x \cos(\bar{\theta}_x) & \bar{t}_x \cos(\bar{\theta}_y)\sin(\bar{\theta}_x) + \bar{t}_y \sin(\bar{\theta}_y) \\ 0 & 0 & 0 & 1 & 0 & \sin(\bar{\theta}_y) \\ 0 & 0 & 0 & 0 & \sin(\bar{\theta}_x) & \cos(\bar{\theta}_x)\cos(\bar{\theta}_y) \\ 0 & 0 & 0 & 0 & \cos(\bar{\theta}_y) & -\cos(\bar{\theta}_y)\sin(\bar{\theta}_x) \end{pmatrix} \qquad (20)$$

$$M_i = \begin{pmatrix} 1 & 0 & 0 & 0 & -d_{y,i} & -d_{z,i} \\ 0 & 1 & 0 & d_{z,i} & d_{x,i} & 0 \\ 0 & 0 & 1 & -d_{y,i} & 0 & d_{x,i} \end{pmatrix} \qquad (21)$$

As required, M_i contains all point information while H expresses the pose information. Thus, this matrix decomposition constitutes a pose linearization similar to those proposed in the preceding sections. Note that, while the matrix decomposition is arbitrary with respect to the column and row ordering of H, this particular description was chosen due to its similarity to the 3D pose solution given in (Lu and Milios, 1997).

5. Continuous-Time SLAM

Unlike other state-of-the-art algorithms, (Stoyanov and Lilienthal, 2009 and Bosse et al., 2012), our continuous-time SLAM algorithm is not restricted to purely local improvements. Our method makes no rigidity assumptions, except for the computation of the point correspondences. For instance, the method requires no explicit motion model of a vehicle, thus it works well on backpack systems. The continuous-time SLAM for trajectory optimization works in full six DoF. The algorithm requires no high-level feature computation, i.e., it requires only the points themselves.

In the case of mobile mapping, one does not have separate terrestrial 3D scans. In the current state-of-the-art in the robotics community developed by Bosse et al. (2012), for improving overall map quality of mobile mappers, the time is coarsely discretized and the scene is described by features, i.e., local planar patches. This results in a partition of the trajectory into sub-scans that are treated rigidly. Then, rigid registration algorithms such as the ICP and other solutions to the SLAM problem are employed. Obviously, trajectory errors within a sub-scan cannot be improved in this fashion. Applying rigid pose estimation to this non-rigid problem directly is also problematic since rigid transformations can only approximate the underlying ground truth. When a finer discretization is used, single 2D scan slices or single points result that do not constrain a six DoF pose sufficiently for rigid algorithms.

More mathematical details of our algorithm are in the available open-source code and are given in Elseberg et al. (2013). Essentially, the algorithm first splits the trajectory into sections, and matches these sections using the automatic high-precision registration of terrestrial 3D scans, i.e., globally consistent scan matching that is the 6D SLAM core. Here, the graph is estimated using a heuristic that measures the overlap of sections using the number of closest point pairs. After applying globally consistent scan matching on the sections, the actual continuous-time or semi-rigid matching as described in (Elseberg et al., 2013) is applied, using the results of the rigid optimization as starting values to compute the numerical minimum of the underlying least square problem. To speed up the calculations, the algorithm exploits the sparse Cholesky decomposition by Davis (2006).

Given a trajectory estimate, the point cloud is "unwound" into the global coordinate system and uses a nearest neighbour search to establish correspondences at the level of single scans (those that can be single 2D scan profiles). Then, after computing the estimates of pose differences and their respective covariance, the algorithm optimizes the trajectory. In a pre-dependent step, trajectory elements in every k step are considered and l trajectory elements around these steps are fused temporarily into a meta-scan.

A key issue in continuous-time SLAM is the search for closest point pairs. An octree and a multi-core implementation using OpenMP is used to solve this task efficiently. A time-threshold for the point pairs is used, i.e., the algorithm matches only to points if they were recorded at least t_d time steps away.

Finally, all scan slices are joined in a single point cloud to enable efficient viewing of the scene. The first frame, i.e., the first 3D scan slice from the PUCKs scanner, defines the arbitrary reference coordinate system.

6. Bootstrapping Continuous-Time SLAM with Google's SLAM Solution

To improve the Cartographer 3D mapping, the graph is estimated using a heuristics that measures the overlap of sections using the number of closest point pairs. After applying globally consistent scan matching on the sections for several iterations, the actual continuous-time SLAM is started.

The data set provided by Google is challenging in several ways: Due to the enormous amount of data, clever data structures are needed to store and access the point cloud. The trajectory is slit every 300 PUCK-scans and ±150 PUCK-scans are joined into a meta-scan. These meta-scans are processed with an octree where a voxel size of 10 cm is used to reduce the point cloud by selecting one point per voxel. We prefer a data structure that stores the raw point cloud over a highly approximate voxel representation. While the latter one is perfectly justifiable for some use cases, it is incompatible with tasks that require exact point measurements such as scan matching. Our implementation of an octree prioritizes memory efficiency. The implementation uses pointers in contrast to serialized pointer-free encodings in order to efficiently access the large amounts of data. The octree is free of redundancies and is nevertheless capable of fast access operations. Our implementation allows for access operations in $O(\log n)$. The usage of 6 bytes for pointers is already sufficient to address a total of 256 terabyte. Two-bit fields signal if there is a child or leaf node, thus our implementation needs a few bit operations and 8 bytes are sufficient for an octree node.

The pre-processing step of the continuous-time SLAM runs for 20 iterations, where the edges in the graph are added, when more than 400 point pairs between these meta-scans are present. The maximal allowed point-to-point distance is set to 50 cm. Figure 2 and 6 present results, where the consistency of the point cloud has been improved. Figure 3 details the changes in the trajectory's position and orientation. It is an open traverse, thus the changes are mainly at the end of the trajectory. Processing was done on a server featuring four Intel Xeon CPUs E7-4870 with 2.4 GHz (40 cores, 80 threads). The overall computing time for the optimization of the Google data set was 10–12 days (few interruptions).

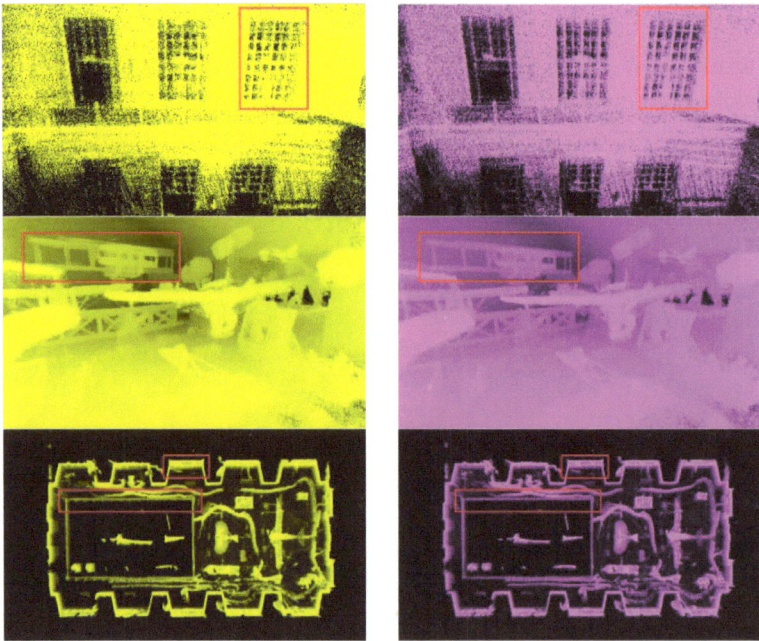

Figure 2. Results of continuous-time simultaneous localization and mapping (SLAM) on Google's Cartographer sample data set from Deutsches Museum in München. Left: input. Right: output of our solution. Shown are three 3D views (perspective) of the scene. Major changes in the point cloud are highlighted in red.

Figure 3. Results of continuous-time SLAM on Google's Cartographer sample data set from Deutsches Museum in München. Left: input. Right: output of our solution. Shown are sections of the point cloud. Major changes in the point cloud are highlighted in red.

In a second experiment, the trajectory from Google has been discretised on a much coarser level in order to process the complete trajectory. The smallest element is now two complete PUCK sweeps. The changes in the trajectory are given in Figure 4. At this level, the trajectory is adjusted at a coarser scale, i.e., the changes of the inner accuracy are smaller, cf. Figure 5.

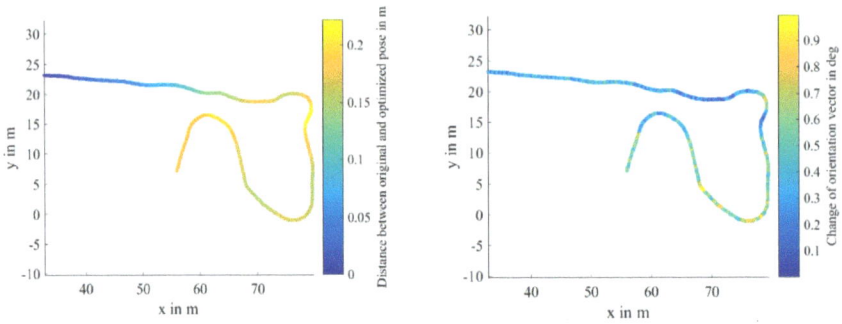

Figure 4. Visualization of the changes in the trajectory computed by our method to bootstrapped trajectory. Left: distance. Right: orientation.

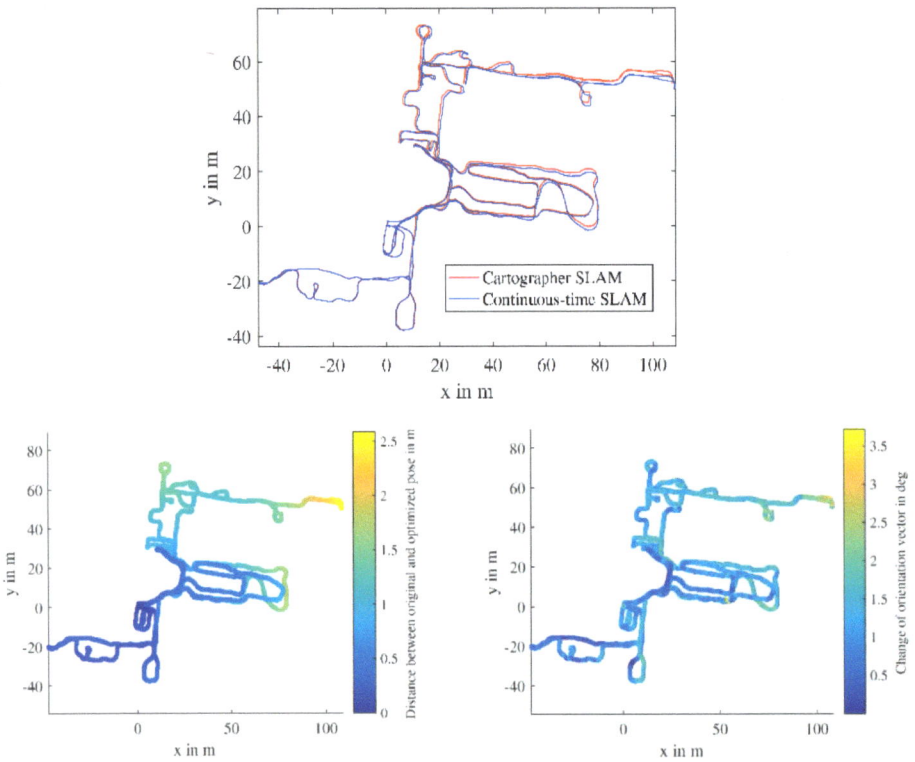

Figure 5. Visualization of the changes in the trajectory over the complete trajectory of the data set. The part presented in Figure 3 is located in the center of the plots.

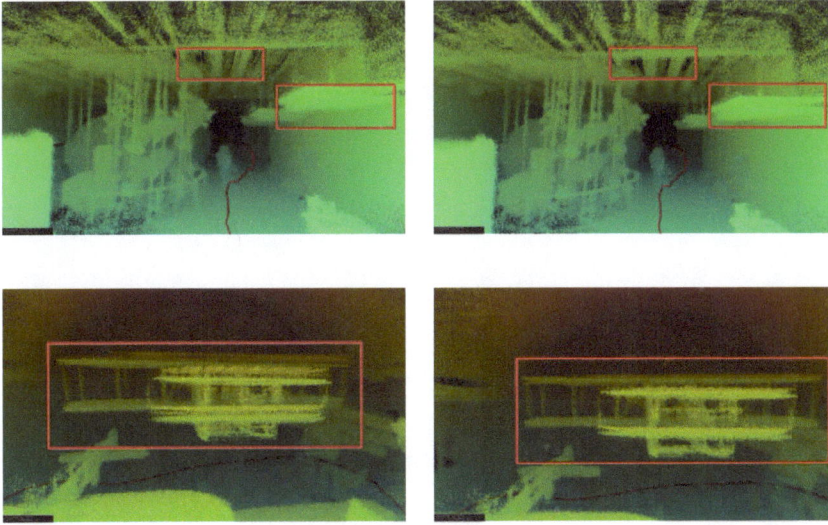

Figure 6. Results of continuous-time SLAM applied to the complete data set. Throughout the data set, the changes are minor, however, in open spaces, the visual impression is still improved.

7. Summary, Conclusions and Future Work

This work revisits a continuous-time SLAM algorithm and its application on Google's Cartographer sample data. The algorithm starts by splitting the trajectory into sections, and matches these sections using the automatic high-precise globally consistent registration of terrestrial 3D scans.

This article has shown how to use the result of Google's novel SLAM solution to bootstrap an existing continuous-time SLAM algorithm. Our approach optimizes the consistency of the global point cloud, i.e. the inner accuracy, and thus improves on Google's results in a local and global fashion. A visual inspection shows the improvements. Personal localization and mapping or personal laser scanning systems will be an emerging research topic in the near future, since the hardware is becoming affordable. The research is also applicable to SLAM systems that work with structured-light scanners such as Microsoft's Kinect, time-of-flight cameras such as the kinectv2, or flash LiDARs (light detection and ranging). An intrinsic challenge remains: How to handle the enormous amount of 3D point cloud data. The time complexity and the memory needs exceed current computing capabilities.

A further conclusion of the article is that the ICP-algorithm with its many variants and extensions is the basic method that is common in the robotics SLAM community and photogrammetry and computer vision community. Bundle adjustment, also called the GraphSLAM method, dominates the mapping methods.

Needless to say, a lot of work remains to be done. First of all, we plan to evaluate the 2D mapping method as we have indicated above. Secondly, as calibration is as crucial as SLAM, we will apply our calibration framework (Elseberg et al., 2013) to the data files provided by Google. Furthermore, we will transfer our continuous-time SLAM to different application areas, e.g., underwater and aerospace mapping applications.

References

1. Agarwal, S.; Mierle, K.; Others. Ceres Solver. Available online: http://ceres-solver.org.
2. Barnea, S.; Filin, S. Keypoint based Autonomous Registration of Terrestrial Laser Point-Clouds. *ISPRS J. Photogramm. Remote Sens.* **2008**, *63*, 19–35.
3. Bendels, G.H.; Degener, P.; Wahl R., Körtgen M.; Klein R. Image-based registration of 3d-range data using feature surface elements. In *The 5th International Symposium on Virtual Reality, Archaeology and Cultural Heritage (VAST 2004)*; Chrysanthou, Y., Cain, K., Silberman, N., Niccolucci, F., Eds.; Eurographics: Lower Saxony, Germany, 2004; pp. 115–124.
4. Benjemaa, R.; Schmitt, F. Fast Global Registration of 3D Sampled Surfaces Using a Multi-Z-Buffer Technique. In Proceedings IEEE International Conference on Recent Advances in 3D Digital Imaging and Modeling (3DIM '97), Ottawa, Canada, 12–15 May 1997.
5. Benjemaa, R.; Schmitt, F. A Solution for the Registration Of Multiple 3D Point Sets Using Unit Quaternions. In Proceedings of the 5th European Conference on Computer Vision Computer Vision—ECCV '98, Freiburg, Germany, 2–6 June 1998; Volume 2, pp. 34–50.
6. Bergevin, R.; Soucy, M.; Gagnon, H.; Laurendeau, D. Towards a general multi-view registration technique. *IEEE Trans. Pattern Anal. Mach. Intell. (PAMI)* **1996**, *18*, 540–547.
7. Besl, P.; McKay, D. A Method for Registration of 3–D Shapes. *IEEE Trans. Pattern Anal. Mach. Intell. (PAMI)* **1992**, *14*, 239–256.
8. Böhm, J.; Becker, S. Automatic marker-free registration of terrestrial laser scans using reflectance features. In Proceedings of 8th Conference on Optical 3D Measurement Techniques, Zurich, Switzerland, 9–13 July 2007; pp. 338–344.
9. Borrmann, D.; Elseberg, J.; Lingemann, K.; Nüchter, A.; Hertzberg, J. Globally consistent 3D mapping with scan matching. *J. Robot. Auton. Syst. (JRAS)* **2008**, *56*, 130–142.
10. Bosse, M.; Zlot, R.; Flick, P. Zebedee: Design of a spring-mounted 3-D range sensor with application to mobile mapping. *IEEE Trans. Robot. (TRO)* **2012**, *28*, 1104–1119.
11. Bosse, M.; Zlot, R. Continuous 3D Scan-Matching with a Spinning 2D Laser. In Proceedings of the IEEE International Conference on Robotics and Automation (ICRA '09), Kobe, Japan, 12–17 May 2009, pp. 4312–4319.
12. Brenner C.; Dold, C.; Ripperda N. Coarse orientation of terrestrial laser scans in urban environments. *ISPRS J. Photogramm. Remote Sens.* **2008**, *63*, 4–18.

13. Chen, G.; Kua, J.; Shum, S.; Naikal, N.; Carlberg M.; Zakhor, A. Indoor Localization Algorithms for a Human-Operated Backpack System. In Proceedings of the International Conference on 3D Data Processing, Visualization, and Transmission (3DPVT '10), Paris, France, 17–20 May 2010.

14. Chen, Y.; Medioni, G. Object Modelling by Registration of Multiple Range Images. In Proceedings of the IEEE Conference on Robotics and Automation (ICRA '91), Sacramento, CA, USA, 9–11 April 1991; pp. 2724–2729.

15. Chen, Y.; Medioni, G. Object Modelling by Registration of Multiple Range Images. *Image Vision Comput.* **1992**, *10*, 145–155.

16. Cunnington, S.; Stoddart, G. N-View Point Set Registration: A Comparison. In Proceedings of the 10th British Machine Vision Conference (BMVC '99), Nottingham, UK, 13–16 September 1999.

17. Davis, T.A. *Direct Methods for Sparse Linear Systems*; SIAM: Philadelphia, PA, USA, 2006.

18. Elseberg, J.; Borrmann, D.; Nüchter, A. Algorithmic solutions for computing accurate maximum likelihood 3D point clouds from mobile laser scanning platforms. *Remote Sens.* **2013**, *5*, 5871–5906.

19. Flint, A.; Dick, A.; van den Hengel, A.J. Thrift: Local 3d structure recognition. In Proceedings of the 9th Biennial Conference of the Australian Pattern Recognition Society on Digital Image Computing Techniques and Applications (DICTA '07), Glenelg, Australia, 3–5 December 2007; pp. 182–188.

20. Hess, W.; Kohler, D.; Rapp H.; Andor D. Real-time loop closure in 2d lidar slam. In Proceedings of the IEEE International Conference on Robotics and Automation (ICRA '16), Stockholm, Sweden, 16–21 May 2016.

21. Huber, D. Automatic Three-dimensional Modeling from Reality. PhD Thesis, Carnegie Mellon University, Pittsburgh, PA, USA, 2002.

22. Kang, Z.; Li J.; Zhang, L.; Zhao, Q.; Zlatanova, S. Automatic Registration of Terrestrial Laser Scanning Point Clouds using Panoramic Reflectance Images. *Sensors* **2009**, *9*, 2621–2646.

23. Kohlbrecher, S.; Meyer, J.; von Stryk, J.; Klingauf, U. A flexible and scalable slam system with full 3d motion estimation. In Proceedings of the IEEE International Symposium on Safety, Security and Rescue Robotics (SSRR' 11), Kyoto, Japan, 1–5 November 2011.

24. Krishnan, S.; Lee, P.Y.; Moore, J.B.; Venkatasubramanian S. Global Registration of Multiple 3D Point Sets via Optimization on a Manifold. In Proceedings of the Eurographics Symposium on Geometry Processing 2005, Vienna, Austria, 4–6 July 2005.

25. Liang, X.; Kukko, A.; Kaartinen, H.; Hyyppä, J.; Xiaowei, Y.; Jaakkola, A.; Wang, Y. Possibilities of a personal laser scanning system for forest mapping and ecosystem services. *Sensors* **2014**, *14*, 1228–1248.

26. Lu, F.; Milios, E. Robot Pose Estimation in Unknown Environments by Matching 2D Range Scans. In Proceedings of the IEEE Computer Society Conference on Computer Vision and Pattern Recognition (CVPR '94), Seattle, WA, USA, 21–23 June 1994; pp. 935–938.

27. Lu F.; Milios, E. Globally Consistent Range Scan Alignment for Environment Mapping. *Auton. Robot.* **1997**, *4*, 333–349.

28. Magnusson, M.; Andreasson, H.; Nüchter, A.; Lilienthal, A.J. Automatic appearance-based loop detection from 3d laser data using the normal distributions transform. *J. Field Robot. (JFR)* **2009**, *26*, 892–914.

29. Nüchter, A.; Borrmann, D.; Koch, P.; Kühn, M.; May, S. A man-portable, imu-free mobile mapping system. *ISPRS Ann. Photogramm. Remote Sens. Spatial Inf. Sci.* **2015**, *II-3/W5*, 17–23.

30. Olson, E. A Passive Solution to the Sensor Synchronization Problem. In Proceedings of the IEEE/RSJ International Conference on Intelligent Robots and Systems (IROS '10), St. Louis, MO, USA, 18–22 October 2010; pp. 1059–1064.

31. Pathak, K.; Borrmann, D.; Elseberg, J.; Vaskevicius, N.; Birk, A.; Nüchter, A. Evaluation of the robustness of planar-patches based 3d-registration using marker-based ground-truth in an outdoor urban scenario. In Proceedings of the IEEE/RSJ International Conference on Intelligent Robots and Systems (IROS '10), Taipei, Taiwan, 11–15 October, 2010; pp. 5725–5730.

32. Pulli, K. Multiview Registration for Large Data Sets. In Proceedings of the 2nd International Conference on 3D Digital Imaging and Modeling (3DIM '99), Ottawa, Canada, 4–8 October 1999; pp. 160–168.

33. Quigley, M.; Gerkey, B.; Conley, K.; Faust J.; Foote, T.; Leibs, J.; Berger, E.; Wheeler, R.; Ng, A. Ros: an open-source robot operating system. In Proceedings of the IEEE International Conference on Robotics and Automation (ICRA '09), Kobe, Japan, 12–17 May 2009.

34. Rieger, P.; Studnicka, N.; Pfennigbauer, M. Boresight Alignment Method for Mobile Laser Scanning Systems. *J. Appl. Geodesy (JAG)* **2010**, *4*, 13–21.

35. Rusinkiewicz, S.; Levoy, M. Efficient variants of the ICP algorithm. In Proceedings of the Third International Conference on 3D Digital Imaging and Modellling (3DIM '01), Quebec City, Canada, 19–21 May 2001; pp. 145–152.

36. Saarinen, J.; Mazl, R.; Kulich, M.; Suomela, J.; Preucil, L.; Halme, A. Methods For Personal Localisation And Mapping. In Proceedings of the 5th IFAC symposium on Intelligent Autonomous Vehicles (IAV '04), Lisbon, Portugal, 5–7 July 2004.

37. Sheehan, M.; Harrison, A.; Newman, P. Self-calibration for a 3D Laser. *Int. J. Robot. Res. (IJRR)* **2011**, *31*, 675–687.

38. Skaloud, J.; Schaer, P. Towards Automated LiDAR Boresight Self-calibration. In Proceedings of the 5th International Symposium on Mobile Mapping Technology (MMT '07), Padua, Italy, 25–29 May 2007.

39. Steder, B., Grisetti, G., and Burgard, W., 2010. Robust place recognition for 3D range data based on point features. In Proceedings of the IEEE International Conference on Robotics and Automation (ICRA '10), Anchorage, AK, USA, 3–7 May 2010; pp. 1400–1405.

40. Stoddart, A.; Hilton, A. Registration of multiple point sets. In Proceedings of the 13th IAPR International Conference on Pattern Recognition, Vienna, Austria, 25–29 August 1996; pp. 40–44.

41. Stoyanov, T.; Lilienthal, A.J. Maximum Likelihood Point Cloud Acquisition from a Mobile Platform. In Proceedings of the IEEE International Conference on Advanced Robotics (ICAR '09), Munich, Germany, 22–26 June 2009; pp. 1–6.
42. Thrun, S.; Fox, D.; Burgard, W. A Real-time Algorithm for Mobile Robot Mapping with Application to Multi Robot and 3D Mapping. In Proceedings of the IEEE International Conference on Robotics and Automation (ICRA '00), San Francisco, CA, USA, 24–28 April 2000.
43. Underwood, J.P.; Hill, A.; Peynot, T.; Scheding, S.J. Error Modeling and Calibration of Exteroceptive Sensors for Accurate Mapping Applications. *J. Field Robot. (JFR)* **2009**, *27*, 2–20.
44. Wang, X.; Toth, C.; Grejner-Brzezinska, D.; Sun, H. Integration of Terrestrial Laser Scanner for Ground Navigation in GPS-challenged Environments. In Proceedings of the XXIst ISPRS Congress: Commission V, WG 3, Berlin, Germany, 1–2 December 2008; pp. 513–518.
45. Wang, Z.; Brenner, C. Point Based Registration of Terrestrial Laser Data using Intensity and Ge- ometry Features. In Proceedings of the ISPRS Congress ('08), Beijing, China, 3–11 July 2008; pp. 583–590.
46. Weinmann, M.; Jutzi, B. Fully automatic image-based registration of unorganized TLS data. *Int. Arch. Photogramm. Remote Sens. Spat. Inf. Sci.* **2011**, *XXXVIII-5/W12*, 55-60.
47. Weinmann, M.; Hinz, S.; Jutzi, B. Fast and automatic image-based registration of TLS data. *ISPRS J. Photogramm. Remote Sens.* **2011**, *66*, 62–70.
48. Williams, J.A.; Bennamoun, M. Multiple View 3D Registration using Statistical Error Models. In Proceedings of the Vision Modeling and Visualization, Erlangen, Germany, November 1999.
49. Zhang, Z. *Iterative Point Matching for Registration of Free–Form Curves*; Technical Report RR-1658; INRI—Sophia Antipolis: Valbonne, France, 1992.

Nüchter, A.; Bleier, M.; Schauer, J.; Janotta, P. Continuous-Time SLAM—Improving Google's Cartographer 3D Mapping. In *Latest Developments in Reality-Based 3D Surveying and Modelling*; Remondino, F., Georgopoulos, A., González-Aguilera, D., Agrafiotis, P., Eds.; MDPI: Basel, Switzerland, 2018; pp. 53–73.

A General Approach for the Reconstruction of Complex Buildings from 3D Pointclouds Using Bayesian Networks and Cellular Automata

Maria Chizhova [a], **Dmitrii Korovin** [b], **Andrey Gurianov** [c],
Ansgar Brunn [d], **Thomas Luhmann** [e] **and Uwe Stilla** [f]

[a] Technical University of Munich/University of Applied Sciences Würzburg-Schweinfurt /Jade University Oldenburg; Germany; mariatschishowa@yahoo.de
[b] Ivanovo State University of Power Engineering, Ivanovo, Russia; dmitriyikorovin@list.ru
[c] Ivanovo State University, Ivanovo, Russia; a.v.gur.2008@mail.ru
[d] University of Applied Sciences Würzburg-Schweinfurt, Würzburg, Germany; ansgar.brunn@fhws.de
[e] Jade University of Applied Sciences, Oldenburg, Germany; luhmann@jade-hs.de
[f] Technical University of Munich, Munich, Germany; uwe.stilla@tum.de

Abstract: Point cloud interpretation and reconstruction of 3D buildings from point clouds has already been addressed for a few decades. There are many articles which consider different methods and workflows of the automatic detection and reconstruction of geometrical objects from point clouds. Each method is suitable for a specific geometry type of object or sensor. General approaches are rare. In our work, we present an algorithm which develops the optimal process sequence of the automatic search, detection and reconstruction of buildings and building components from a point cloud. It can be used for the detection of sets of geometric objects to be reconstructed, regardless of the level of damage. In a real example, we reconstruct a complete Russian Orthodox church starting from a set of detected structural components and reconstruct missing components with high probability.

Keywords: reconstruction; cellular automata; Bayesian networks

---

1. Introduction

1.1. Motivation

The development of science, technology and equipment as well as the human ability to invent new constructions have allowed the creation of a large variety of different unique objects in the real world. Over time, a lot of information about former cultural objects has been lost, and some objects have been severely damaged or ruined. Digital reconstruction makes building principles easier to understand, especially for partly destroyed or no longer existing objects. The reconstruction of lost historical and architectural objects is an up-to-date topic in many researches. Using 3D laser scanning, we can create 3D construction models of real scenes with high precision and completeness. Complex geometry reconstruction from point clouds in the context of the big data problem is not a trivial matter and needs methods and algorithms for the optimization of data processing. Due to the large amount of data, the choice of efficient methods is an important task.

As a standard rule, in the case of building reconstruction with geometrical features, object reconstruction begins with feature detection. The existing methods are mostly appropriate for simple objects such as recognition of planes or spheres. For complex objects, combining different geometric entities, the choice of one particular detection method is not always appropriate because each method offers its own advantages for specific types of geometric entities.

Figure 1. Reconstruction of a destroyed church.

In this article, we develop the mathematical model of a new method, which allows an optimized extraction of geometrical information from laser scanning point clouds and its efficient interpretation for further reconstruction. This method is suitable for object reconstruction from incomplete data, which can be the result of missing object parts (e.g., destroyed object) or a failed measurement process (e.g., due to reflectivity properties of an object).

This work is carried out in the context of recent research in the virtual reconstruction of destroyed orthodox churches, which are characterized by their complex architecture.

1.2. Previous Works

Considering previous work, we focus on two aspects:

1. point cloud interpretation and
2. reconstruction from precise point clouds.

In our case, point cloud interpretation means extraction of analytical, geometrical and semantical information from point clouds. Here, geometry extraction is relevant to our work. The most common techniques of geometry extraction from point clouds are as follows:

- RANSAC and its variations:
  Schnabel et al. (2007) use the RANSAC algorithm in their work for fitting geometric primitives (such as planes, spheres, etc.) from unorganized point clouds. Al-Durgham et al. (2013) search for straight lines in the point clouds for LS-point clouds registration using RANSAC. Rusu et al. (2009) use a RM-SAC algorithm—Randomized M-Estimator Sample Consensus—that can be considered as an extension of RANSAC using the modified minimizing target function. The algorithm finds a best-fit geometrical surface form for the further model-based object reconstruction.
- Hough transformation (HT):
  Vosselman et al. (2004), Overby et al. (2004), Rabbani and Heuvel (2005) extract geometric primitives (planes, cylinders, spheres and cones) using 3D Hough transformation, which can be understood as an extension of the classic 2D Hough transformation. The Hough transformation is quite suitable for the detection of simple geometric forms, but it is not efficient for the detection of more complex geometries due to the increasing time involved and calculation complexity. Maltezos and Ioannidis (2016) use Randomized Hough Transformation (RHT), a probabilistic variation of classic HT based on affinity calculation between the new object (e.g., curve) and the object in the accumulator for the detection of roof planes.
- Least-squares fitting:
  The fitting of parametric curves and surfaces using least squares methods has been considered in detail by Ahn (2004) and elaborated upon by Liu and Wang (2008). Wang et al. (2004) address the problem of correct curve fitting in unorganized noisy point clouds. The algorithm is based on squared distance minimization between points of the LS point cloud and the given B-spline-curve. Fleischmann et al. (2005) present surface fitting using robust Moving Least Squares for further surface reconstruction.
- Methods of differential geometry:

a detailed description of geometrical primitives from point clouds was given in Becker (2005).

Even though these methods are quite robust, some of them are dependent on the number of processing iterations, need prior sampling and are not always correct for complex objects, which can be a combination of different geometries. In the case of complex architectures, which cannot be approximated with only single geometric primitives, the detection method of one geometric object can differ from others. The chosen method determines the further reconstruction of the whole object.

There are different reconstruction approaches based on interpreted information from point clouds. Some reconstruction methods use a strict prototype model. In many articles about reconstruction of cultural heritage objects, a library of typological architectural elements taking into account construction canons is used. In Quattrini et al. (2015), a destroyed architectural object has been completely reconstructed from the TLS point cloud as well as single classified archaeological samples according to practical and theoretical canons of roman architecture. Dore and Murphy (2013) generated digital historical models using Historic Building Information Modelling (HBIM) containing parametric library objects and procedural modelling techniques. Later on, Dore et al. (2015) developed a set of rules and algorithms for the automatic combination of parametric library objects and generation of HBIMs from survey data (historic surveys and a recent laser scan survey as segmented point cloud and cut sections). A conceptual framework is based on the definition of shape grammar, which allows for the automatic generation of 2D and 3D geometries from a basic vocabulary of shapes. The reconstruction is supported by architectural rules and proportions. There are some software applications that allow for semi-automatic modelling of architectural forms according to a library of structural elements, in which the model parameters of a structure element are estimated from user defined key points (Kivilcim and Duran, 2016).

Huang et al. (2011) developed roof decomposition rules for the reconstruction of LoD2 buildings. Based on a predefined library of primitives, generative modelling has been conducted to construct the target roof that fits the data. Extracted primitives from a point cloud were composed and merged. Nguatem et al. (2013) extracted a ridgeline from the highest points of the point cloud—bounded with a ground plan—for roof model fitting using the likelihood principle.

The reconstruction of single architectural models is considered in Canciani et al. (2013). The method is based on the extrusion path modelling of architectural elements from the point cloud section, which have been compared with a knowledge-based model.

In many articles, a building reconstruction is based on the integration of extracted geometrical information about single object parts according to different composition rules.

Extracted straight lines and planes (e.g., using RANSAC, Hough transform) often assist in planar object reconstruction and decomposition of its elements (e.g., roofs—Nizar et al. (2006), Arefi et al. (2010), buildings—Rusu et al. (2009)). Verma et al. (2006) detected planes and rectangular outlines for roof composition using a Roof Topology Graph. Kada and Wichmann (2013) generated complex building shapes using a Boolean intersection of half-spaces, which define convex building components. Xiong et al. (2015) represented roofs with topological graphs and applied the Minimum Cycles Method for roof decomposition using extracted geometrical primitives from airborne LiDAR data.

In recent years, probabilistic approaches have been widely used in recognition processes. Probabilistic graphical models (PGM), developed by Koller and Friedman (2009), integrate schematic graphical object representation (e.g., as a graph) with different stochastic statistic models, such as the Hidden Markov Model (HMM), Conditional Random Field (CRF) and Bayesian nets, allowing for probabilistic decision-making. Ruiz-Sarmiento et al. (2015) integrated PGMs as Conditional Random Fields (CRF) with Semantical Knowledge (SK) for the representation of object relations in the context of an input scene. Xiong and Huber (2010) extracted and classified planar regions for further object recognition using CRF. Anand et al. (2013) used an isomorphic to MRF (Markov Random Fields) model for object recognition and classification in certain scenes (e.g., office, house). Förstner (2013) showed the efficiency of object parameter estimation and classification, which are optimized through the implementation of Bayesian nets and MRF in the context of the flexible construction of graphical models.

In all cases, a big data problem is relevant for point cloud processing. The role of high mechanism complexity and the creation of self-reproducing intelligent automates for the acceleration of technological processes has already been discussed for many years (e.g., v. Neumann, 1966). An additional example from such automates is cellular automaton (CA)—a discrete model for the description of dynamic systems, which can be based on various rules (using, for example, Bayes theory and Bayesian networks (Wolfram (1983), Neapolitan (2004)). Since CA provides accelerated processing, this approach is favoured in our actual work.

2. The New Proposed Method

2.1. Outline of Algorithm

The developed method is based on a probabilistic approach and discrete mathematical methods, namely Bayesian networks and cellular automata theory (see Ch. 2.5). Let us consider an algorithm, which makes an automatic choice of an acceptable method for the detection of different geometrical entities in a complex object and its further automatic reconstruction from incomplete data. Thus, we consider an a priori known object configuration with its pre-defined semantic information, which is organized like a graph. A combination of Bayesian nets with cellular automata allows one to define an optimal

- geometric candidate for object detection;
- means of recognition;
- decision-making process for reconstruction from incomplete data.

The reconstruction algorithm can be described with the following steps.

1. Pre-processing

    - creation of a database from complete churches;
    - manual church description;
    - church representation with a graph, including topological, semantic and geometrical information about some church components;
    - fusion of all church graphs into one joint graph;
    - creation of a Bayesian network based on a joint graph using frequent information about components in the database.

2. Object measurement (e.g., laser scanning)

    - object measurement (laser scanning, photogrammetric approach);
    - point cloud registration;
    - point cloud filtering;
    - cleaning.

3. Segmentation

    - segmentation of rotational objects (domes, tambour);
    - nave segmentation;
    - segmentation of simple geometries and non-rotational objects (walls, roofs etc.)

4. Topology and semantic definition of segmented church components:

    - segments connection into a topological graph;
    - semantic definition of "sure" components (nave, domes on the main roof);
    - semantic definition of "unsure" components through matching with a joint graph.

5. Geometry detection and reconstruction:

    - geometry extraction of components with "sure" semantic;
    - geometry extraction of components with "unsure" semantic and its semantic correction (using cellular automata);
    - reconstruction of destroyed or missing parts using cellular automata.

2.2. Research Object

Our research objects are stone orthodox churches. There are a considerable number of such churches in the territory of Russia, Belarus and Ukraine. Several churches have been destroyed and are not used as religious institutions any more. There is a huge interest in the restoration of these objects.

It is necessary to observe the number of canons, determined by religion during the churches' construction. These canons resulted in the evolution of the culture in that territory (and that time), in which we observe this or that church. Nevertheless, it is possible to claim that each church has a certain topological structure of some elements determined by canons. Each of these elements is geometrically characterized by its properties.

Table 1. Example of a classification library with visual models (not all models are represented).

Component	Model 1	Model 2	Model 3
crucifix (*Kr*)			
dome (*K, HK*)			
cylinder (*Tr*)			
roof (*D*)			
altar (*A, HA, OA*)			
nave (*X*)			

2.3. Object Classification and Representation

A database of a certain amount of churches with a classification of structural elements was built for the derivation of probabilistic relations between elements and acts as a base for further construction of a Bayesian network. In our case, the abbreviations of the church parts are (see example in Table 1): *Kr* - crucifix, *K* - dome (*HK* - main dome), *Tr* - cylinder, *St* - prop, *D* - roof, *X* - nave, *OA$_r$* - sacrifice altar right, *D$_r$* - sacrifice altar roof (right), *HA* - main altar; *D* - main altar roof, *OA$_l$* - sacrifice altar left, *D$_l$* - sacrifice altar roof (left). We model our objects—orthodox churches—with colour-oriented topological graphs *G = (V, E)*. The vertices *V* are the structural components of the church, like a cupola or a crucifix with arbitrary geometries. The edges *E* are the neighbouring relations between the elements of V, which are attributed by conditional probabilities to the edges. The objects differ in their complexity (cf. Figure 2).

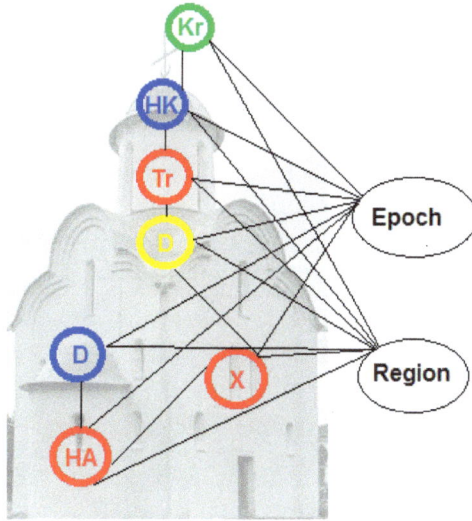

Figure 2. Church representation with a coloured graph.

We take into account the following encoding for the representation of some components:

- The letters in the vertices are the specific structural components with a specific geometry and localization.
- The edges, coded with a line, show the topological and probabilistic connection or the probabilistic influence of the components on each other (for example, if one component were detected, then we can claim, with a defined probability, that another component must be detected or reconstructed and vice versa).
- For visual comfort, we represent the different geometry types with different colours, i.e., each class of detected geometry type has its own colour.

2.4. Mathematical Background

Let us consider a mathematical method solving a decision problem correctly. This means, in the context of our research, that we have checked the information about some parts of our object, we can apply the arbitrary method of recognition and search an object in a certain place. Consider a method based on the use of Bayesian networks as one method to guide this search process.

We use the basic formula (Bayesian Theorem)

$$P(\theta \mid I) = \frac{P(\theta)P(\theta \mid I)}{P(I)} \tag{1}$$

where

- *I* is the information about those parts of an object, which were detected;
- *P(I)* is the probability of *I*, defined as the frequency of objects, identified as *I* in all sets of churches;
- *θ* is the identifiers of those geometrical objects, which we want to detect;
- *P(I|θ)* is the posterior probability: the distribution element of object identifiers under the assumption of already detected objects;
- *P(θ)* is the prior probability: it is the formalization of our intuition about the possibility of the detection of an object. In our case, the value of this probability is defined statistically after the quantitative data analysis, after the supervision of such parts in orthodox churches.

All conditional probabilities are learnt from training datasets.

The expression *P(I|θ)* is called likelihood; it is a probability of a known data supervision by the fixation of certain identifiers. Thus, our task is to find the maximum a posteriori hypothesis *θ*, by which *P(θ|I)* is maximized.

If we set *P(I|θ)* of likelihood functions according to Equation (1), the decision problem is solved correctly. However, the establishment process of these functions is not a trivial procedure. These distributions can be rather difficult.

If we add some additional restrictions, the problem of *P(I|θ)* - function calculation can be simplified. By the conditional independence *a₁, a₂, ... aₙ* of the object identifiers, which define *I*, we have:

$$P(a_1, a_2, \ldots a_n \mid \theta = \theta k) = P(a_1 \mid \theta k)P(a_2 \mid \theta k)\ldots P(a_n \mid \theta k) \tag{2}$$

If *I* contains information in our case that $X = X_1$ (an event *a₁*), $HK - St = HK - St_1$ (an event *a₂*), and $\theta = \theta_k$ means that $D = D_k$, then $P(I \mid \theta = \theta_k)$ is the multiplication of the two probabilities of $P(a_1 \mid \theta = \theta_k)$ and $P(a_2 \mid \theta = \theta_k)$. Each of these probabilities will be defined from the set of those churches, at which X_1, $HK - St_1$ respectively among those churches, at which $D = D_k$.

In our example, we can find the subgraphs shown in Figure 3.

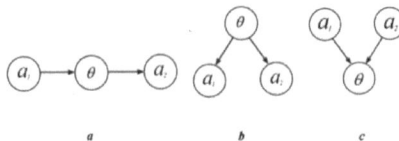

Figure 3. Subgraphs of the probability connections.

The pointers on the graph edges are suspended in most of the cases of this article. The graph of Figure 3a corresponds to the expression in Equation (2):

$$P(a_1, \theta, a_2) = P(a_1)\, P(\theta\,|\,a_1)\, P(a_2\,|\,\theta)$$

From this follows

$$P(I\,|\,\theta) = P(a_1, a_2\,|\,\theta) = \frac{P(a_1,\theta,a_2)}{P(\theta)} = \frac{P(a_1)P(\theta\,|\,a_1)P(a_2\,|\,\theta)}{P(\theta)} = P(a_1\,|\,\theta)P(a_2\,|\,\theta) \qquad (2a)$$

Because

$$\frac{P(a_1)P(\theta\,|\,a_1)}{P(\theta)} = P(a_1\,|\,\theta)$$

according to the Bayes Theorem (cf. Equation (1)).

The graph of Figure 3b defines the following equality:

$$P(\theta, a1, a2) = P(\theta)\, P(a1\,|\,\theta)\, P(a2\,|\,\theta) \qquad (3)$$

resulting in

$$P(I\,|\,\theta) = P(a_1, a_2\,|\,\theta) = \frac{P(\theta,a_1,a_2)}{P(\theta)} = \frac{P(\theta)P(a_1\,|\,\theta)P(a_2\,|\,\theta)}{P(\theta)} = P(a_1\,|\,\theta)P(a_2\,|\,\theta) \qquad (3a)$$

This simplifies the likelihood functions. When $a1$ and $a2$ influence θ at the same time (Figure 3c), it is possible to express this by

$$P(a1, a2, \theta) = P(a1)\, P(a2)\, P(\theta\,|\,a1, a2) \qquad (4)$$

In this case, we receive a likelihood function which depends on the intended value, which, obviously, does not yield the correct decision. The likelihood function will be calculated directly.

A similar approach will allow us to optimize the procedure of this method choice for object extrapolation.

2.5. Processing Principles

Let us consider the graph which describes an input church as a special type of cellular automaton. As described above, cellular automaton is a discrete model for the description of dynamic systems consisting of spatial discrete cells, for which the state of time $t + 1$ depends on the states of cells in the neighbourhood and its own state of time t.

A cellular automata has the following characteristics:

- Cellular lattice/space R (Bayesian network with a certain number of churches)
- Final neighbourhood B (number of vertices in the Bayesian network)
- States amount k (probable geometries of church components in the network)

- Local transition function σ

The church graph is a lattice of cellular automaton, in which vertices are the automata cells. The cellular automaton can be defined as a set of final automatas with concrete state in discrete time t:

$$\sigma \in \Sigma = \{0, 1, 2 \ldots k - 1, k\} \tag{5}$$

A detection of component geometries changes the automaton states of each cell in the neighbourhood according to the transition rule:

$$\sigma i, j(t+1) = \emptyset(\sigma k(t) \mid \sigma k(t) \in N) \tag{6}$$

N is the set of automatas that constitute a neighbourhood. At time t, we have three kinds of cells:

1. cells with an unambiguously defined state (e.g., detected geometry);
2. "pending" cells which incident (but do not belong) to defined cells;
3. "empty" cells which do not incident and do not belong to defined cells.

In our case, we set an initial state randomly. It is reasonable to set the most probable state as the initial state: it means that the search is statistically driven, starting with the most likely geometry for each component. The transition function will be defined; if the vertex B does not belong to N, there are edges from the vertex B incidental to the vertices from N ("pending cells"):
The state of the vertex is

$$B = \begin{cases} B_1, P_1 = P(B_1 \mid N) \\ B_2, P_2 = P(B_2 \mid N) \\ \ldots \\ B_k, P_k = P(B_k \mid N) \end{cases} \tag{7}$$

If the vertex B belongs to N, the state of a vertex remains the same with the probability of 1 (almost sure event). If the vertex B has no edge with any vertex from N, the state of a vertex remains the same with the probability of 1.

Applying the Maximum-Likelihood method, we choose the state in the pending vertices with respect to the maximum of the corresponding probabilities $B_k = \{B_i, max(P_i) = P_k\}$ and check for the correctness. It defines appropriate processes in all pending vertices, in which we detect the geometry of the church structural element by the method chosen according to a state. The following results are possible:

1. The geometry in the pending vertex has been detected. In this case, we attach a vertex to the N-set.

2. The geometry in the pending vertex has not been detected with the method chosen according to a state (we define this state as B_1 without loss of generality).

Let us transform the transition function: If the vertex B does not belong to N, there are edges from the vertex B incidental to vertices from N ("pending vertices") and in the previous step this vertex was not a pending vertex.

The state of the vertex is:

$$B = \begin{cases} B_1, P_1 = P(B_1 \mid N) \\ B_2, P_2 = P(B_2 \mid N) \\ \dots \\ B_k, P_k = P(B_k \mid N) \end{cases} \tag{7a}$$

If the vertex B does not belong to N, there are edges from the vertex B incidental to the vertices from N ("pending vertices") and in the previous step this vertex was a pending vertex.

The state of the vertex is:

$$B = \begin{cases} B_2, P_2 = P(B_2 \mid N, not \quad B_1) \\ \dots \\ B_k, P_k = P(B_k \mid N, not \quad B_2) \end{cases} \tag{7b}$$

If the vertex B belongs to N, the state of a vertex remains the same with the probability of 1. If the vertex B has no edge with any vertex from N, the state of a vertex remains the same with the probability of 1.

This iterative process yields a distinct solution since all cells will finally change their state to "unambiguously defined". In our case, it is convenient to visualize the state of a vertex with a colour. To each state of B_k, a specific colour is assigned. Thus, the presented probabilistic automaton presents the process of vertex colouring in the graph.

3. Tests

In this section, we will present the test results of our probabilistic automaton, which is based on the real data. For the test, we have chosen the Russian Orthodox Church in Wiesbaden (Germany) that completely conforms to the orthodox construction canons and Russian church typology. The Russian Orthodox Church in Wiesbaden was built in 1855. The church has not been destroyed and is well maintained. However, the golden domes and roof have incorrectly reflected laser beams and therefore there is a gap in the church point cloud (see Figure 4). It was a suitable case to verify our algorithm.

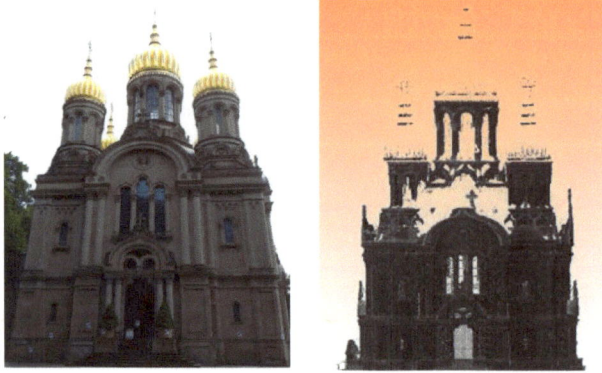

Figure 4. Russian church in Wiesbaden (left) and its point cloud with gaps (right).

To realize the algorithm, we applied a database of 50 orthodox churches that contains information about the semantics and geometry of church components, which have been manual entered. A joint graph, which includes church component frequencies (probability of its occurrence among 50 churches in the database), is calculated based on information in the database (see Figure 5). Further, the data in a joint graph serves to detect geometries of destroyed or lacking components.

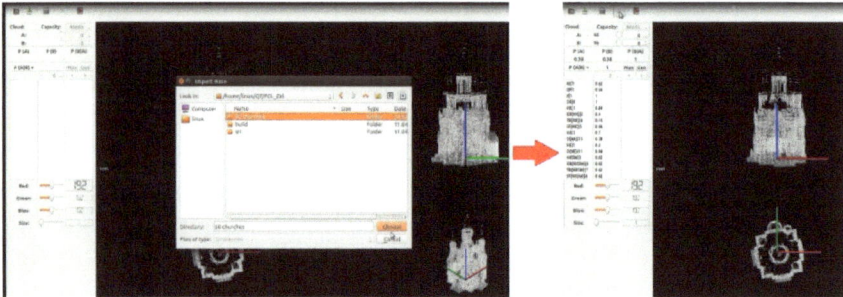

Figure 5. Generation of a joint graph with component probabilities.

As previously stated, only the reconstruction algorithm is realized in this article, which does, however, require some pre-processing steps, such as segmentation, semantic definition etc., that are not automatically realized at this moment. Therefore, these steps will be made manually. According to the classification library (see above), some church components can be visually detected in the point cloud of the Wiesbaden church (see Table 2, Figure 6):

Table 2. Visual detection of church component geometries corresponding to the classification library.

PC Sample	Pattern	ID	Vertex Type
		Tr 3	Tr
		*X*1	X
		HA 2	HA
		*D*1	D

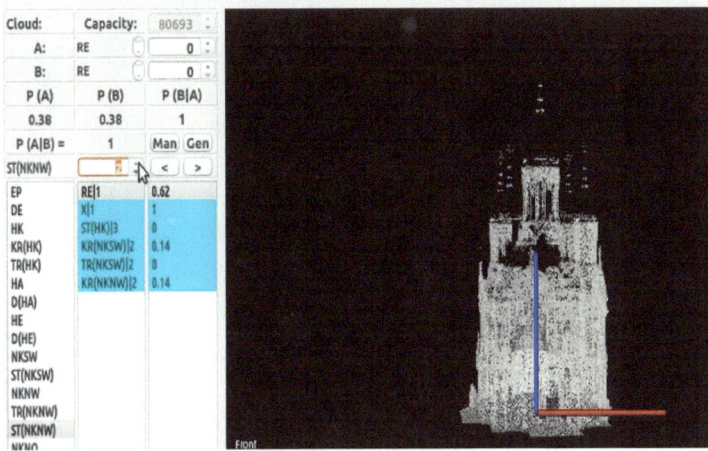

Figure 6. Manual generation of a test graph.

Certain components, e.g., domes and crucifixes, were not scanned correctly and make a gap in the church point cloud. Let us try to reconstruct the geometries of these components using our algorithm. The information about geometries and the semantic of "detected" church components, that we have visually defined, will be entered manually according to our graph construction rules. Based on such "detected" components, the automation completes the test graph of the input church with the most probable missed components, which can be presented through the schema of graph colouring (Figure 7).

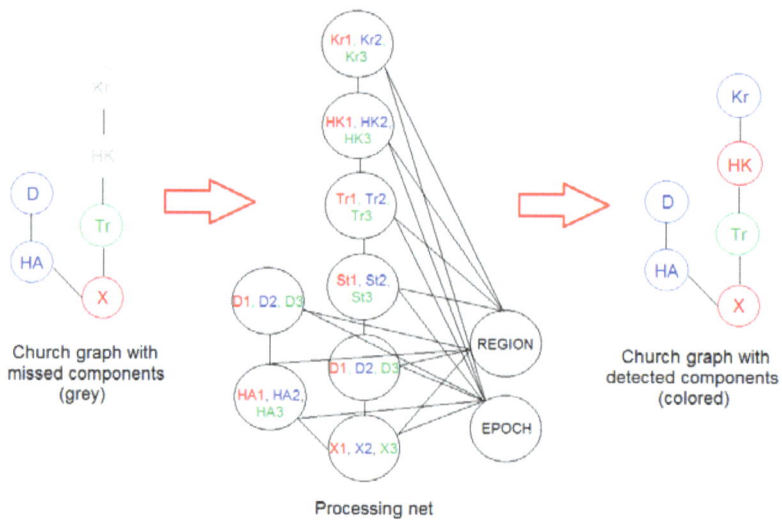

Figure 7. Schematic representation of the detection process.

To verify the algorithm, we used a one-step automaton, which generated the most probable geometries of missing components. In this article, we have not imitated the conflicts in the case of an unsuccessful geometry detection from a real point cloud, because our goal is to compare the results proposed from an automaton with the original church. Based on "detected" components, the algorithm proposed the most probable component geometries, which conform to the component geometries of the original church (Figures 8 and 9).

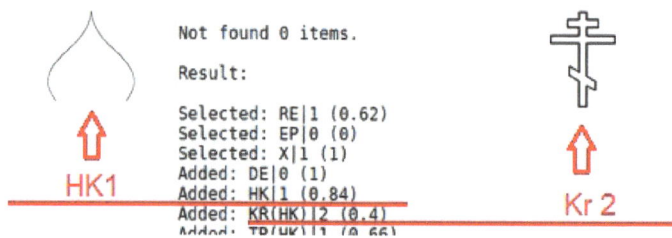

Figure 8. Correct detected geometries of missing parts.

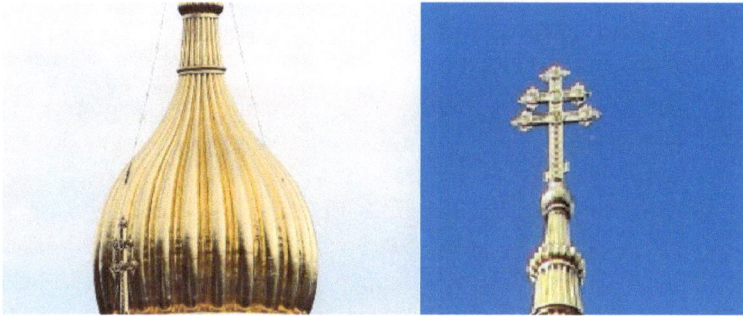

Figure 9. Original church parts.

4. Conclusions

We developed a mathematical theory and showed the feasibility of an approach to simulated data of stone Russian Orthodox churches. The presented method allows an automatic reconstruction of the whole complex object with missing components on the basis of the iterative detected geometries of its components. The method's principles are universal and applicable for other types of objects (this would require additional libraries of structural components and component relations in the graph). The results of this algorithm depend on the quality of pre-processing steps, especially correct geometry detection and semantic definition of the church parts.

Further work will be concentrated in two directions: On the one hand, empirical work on the correct segmentation and semantic definition of the church parts. On the other hand, the cellular automaton will be realized for larger and more complex building models with consideration of conflicts such as the presence of new objects, which were not described in our classification library, or the absence of some components that were not planned for in some church constructions.

Acknowledgements: We thank the Free State of Bavaria, who made this research possible by a grant for the visit of Prof. Korovin at the FHWS.

References

1. Ahn, S. *Least Squares Orthogonal Distance Fitting of Curves and Surfaces in Space*; Springer: Berlin, Germany, 2004.
2. Al-Durgham, K.; Habib, A.; Kwak, E. Ransac approach for automated registration of terrestrial laser scans using linear features. *ISPRS Ann. Photogramm. Remote Sens. Spat. Inf. Sci.* **2013**, *II*, 13–18.

3.   Anand, A.; Koppula, H.; Joachims, T.; Saxena, A. Contextually guided semantic labeling and search for three-dimensional point clouds. *Int. J. Robot. Res.* **2013**, *32*, 19–34.

4.   Arefi, H.; Hahn, M.; Reinartz, P. Ridge based decomposition of complex buildings for 3d model generation from high resolution digital surface models. *Int. Arch. Photogramm. Remote Sens. Spat. Inf. Sci.* **2010**, *34*, 15–22.

5.   Becker, R. Differentialgeometrische Extraktion von 3D-Objektprimitiven aus terrestrischen Laser- scannerdaten. Ph.D. Thesis, Veröffentlichungen des Geodätischen Instituts der Rheinisch-Westfälischen Technischen Hochschule, Aachen, Germany, 2005. Available online: https://www.deutsche-digitale-bibliothek.de/binary/WI6IH566CY75KXJY3QX JBTG- MXWKJI5TS/full/1.pdf.

6.   Canciani, M.; Falcolini, C.; Saccone, M.; Spadafora, G. From point clouds to architectural models: algorithms for shape reconstruction. *Int. Arch. Photogramm. Remote Sens. Spat. Inf. Sci.* **2013**, *XL*, 27–34.

7.   Dore, C.; Murphy, M. Semi-automatic modeling of building facades with shape grammars using historic building information modeling. *Int. Arch. Photogramm. Remote Sens. Spat. Inf. Sci.* **2013**, *XL*, 57–64.

8.   Dore, C.; Murphy, M.; McCarthy, S.; Brechin, F.; Casidy, C.; Dirix, E. Structural simulations and conservation analysis of historic building information model (HBIM). *Int. Arch. Photogramm. Remote Sens. Spat. Inf. Sci.* **2015**, *XL*, 351–357.

9.   Fleischmann, S.; Cohen-Or, D.; Silva, C. Robust moving least-squares fitting with sharp features. *Proc. ACM SIGGRAPH* **2005**, *24*, 544–552.

10.   Förstner, W. Graphical models in geodesy and photogrammetry. *PFG Photogramm. Fernerkund. Geoinform.* **2013**, *4*, 255–267.

11.   Huang, H.; Brenner, C.; Sester, M. 3d building roof reconstruction from point clouds via generative models. In Proceedings of the 19th ACM SIGSPATIAL International Conference on Advances in Geographic Information Systems, Chicago, IL, USA, 1–4 November 2011; pp. 16–24.

12.   Kada, M.; Wichmann, A. Feature-driven 3d building modeling using planar. *Int. Arch. Photogramm. Remote Sens. Spat. Inf. Sci.* **2013**, *II-3/W3*, 37–42.

13.   Kivilcim, C.; Duran, Z. A semi-automated point cloud processing methodology for 3d cultural heritage documentation. *Int. Arch. Photogramm. Remote Sens. Spat. Inf. Sci.* **2016**, *XLI*, 293–296.

14.   Koller, D.; Friedman, N. *Probabilistic Graphical Models: Principles and Techniques*; MIT Press: Cambridge, MA, USA, 2009.

15.   Liu, Y.; Wang, W. Advances in Geometric Modeling and Processing. In *Springer Berlin Heidelberg, chapter A Revisit to Least Squares Orthogonal Distance Fitting of Parametric Curves and Surfaces*; Springer: Berlin/Heidelberg, Germany, 2008; pp. 384–397.

16.   Maltezos, E.; Ioannidis, C. Automatic extraction of building roof planes from airborne LiDAR data applying an extended 3d randomized Hough transform. *Int. Arch. Photogramm. Remote Sens. Spat. Inf. Sci.* **2016**, *III*, 209–216.

17.   Neapolitan, R.E. *Learning Bayesian Networks*; Pearson Prentice Hall: Upper Saddle River, NJ, USA, 2004.

18. Nguatem, W.; Drauschke, M.; Mayer, H. Roof reconstruction from point clouds using importance sampling. *Int. Arch. Photogramm. Remote Sens. Spat. Inf. Sci.* **2013**, *II*, 73–78.

19. Nizar, A.A.; Filin, S.; Doytsher, Y. Reconstruction of buildings from airborne laserscanning. In Proceedings of the ASPRS Annual Conference, Reno, NV, USA, 1–5 May 2006; pp. 106–115.

20. Overby, J.; Bodum, L.; Kjems, E.; Ilsoe, P. M. Automatic 3d building reconstruction from airborne laserscanning and cadastral data using Hough transform. *Int. Arch. Photogramm. Remote Sens. Spat. Inf. Sci.* **2004**, *XXXV*, 296–301.

21. Quattrini, R.; Malinverni, E.S.; Clini, P.; Nespeca, R.; Orlietti, E. From TLS to HBIM: high quality semantically-aware 3d modelling of complex architecture. *Int. Arch. Photogramm. Remote Sens. Spat. Inf. Sci.* **2015**, *XL*, 367–374.

22. Rabbani, T.; Heuvel, F. Efficient Hough transform for automatic detection of cylinders in point clouds. *Int. Arch. Photogramm. Remote Sens. Spat. Inf. Sci.* **2005**, *XXXVI*, 60–65.

23. Ruiz-Sarmiento, J.; Galindo, C.; Gonzalez-Jimenez, J. Scene object recognition for mobile robots through semantic knowledge and probabilistic graphical models. *Exp. Syst. Appl.* **2015**, *42*, 8805–8816.

24. Rusu, R.; Blodow, N.; Marton, Z.C.; Beetz, M. Close-range scene segmentation and reconstruction of 3d point cloud maps for mobile manipulation in domestic environments. In Proccedings of the International Conference on Intelligent Robots and Systems, St. Louis, St. Louis, MO, USA, 10–15 October 2009; pp. 1–6.

25. Schnabel, R.; Wahl, R.; Klein, R. Efficient RANSAC for point-cloud shape detection. *Comput. Graph. Forum* **2007**, *26*, 214–226.

26. Neumann, J. *Theory of Self-Reproducing Automata*; University of Illinois Press: Champaign, IL, USA, 1966.

27. Verma, V.; Kumar, R.; Hsu, S. 3d building detection and modeling from aerial LiDAR data. In Proceedings of the 2006 IEEE Computer Society Conference on Computer Vision and Pattern Recognition, New York, NY, USA, 17–23 June 2006; Volume 2, pp. 2213–2220.

28. Vosselman, G.; Gorte, B.; Sithole, G.; Rabbani, T. Recognising structure in laser scanner point clouds. *Int. Arch. Photogramm. Remote Sens. Spat. Inf. Sci.* **2004**, *XXXVI*, 33–38.

29. Wang, W.; Pottmann, H.; Liu, Y. *Fitting B-Spline Curves to Point Clouds by squared Distance Minimization*; Technical Report, HKU CS Tech Report TR-2004-11; 2004. Available online: http://www.cs.hku.hk/research/techreps/document/TR-2004-11.pdf.

30. Wolfram, S. Statistical mechanics of cellular automata. *Rev. Modern Phys.* **1983**, *55*, 601–644.

31. Xiong, B.; Jancosek, M.; Elberink, S.O.; Vosselman, G. Flexible building primitives for 3d building modeling. *ISPRS J. Photogramm Remote Sens.* **2015**, *101*, 275–290.

32. Xiong, X.; Huber, D. Using context to create semantic 3d models of indoor environments. In Proceedings of the British Machine Vision Conference (BMVC), Wales, UK, 30 August–2 September 2010; pp. 1–11.

Chizhova, M.; Korovin, D.; Gurianov, A.; *et al.* A general Approach for the Reconstruction of Complex Buildings from 3D Pointclouds Using Bayesian Networks and Cellular Automata. In *Latest Developments in Reality-Based 3D Surveying and Modelling*; Remondino, F., Georgopoulos, A., González-Aguilera, D., Agrafiotis, P., Eds.; MDPI: Basel, Switzerland, 2018; pp. 74–92.

Chapter 2

Sensor Fusion and Data Integration

Three-Dimensional (3D) Modelling and Optimization for Multipurpose Analysis and Representation of Ancient Statues

Elisabetta Donadio [a], **Luigi Sambuelli** [b], **Antonia Spanò** [a] and **Daniela Picchi** [c]

[a] DAD Department, Politecnico di Torino, Italy; (elisabetta.donadio antonia.spano)@polito.it

[b] DIATI Department, Politecnico di Torino, Italy; luigi.sambuelli@polito.it

[c] Civic Archaeological Museum, Via dell'Archiginnasio 2, 40124 Bologna (Italy); daniela.picchi@comune.bologna.it

Abstract: The technological advances that have developed in the field of three-dimensional (3D) survey and modelling allow us to digitally and accurately preserve many significant heritage assets that are at risk. With regard to museum assets, extensive digitalization projects aim at achieving multilingual digital libraries accessible to everyone. A first trend is geared to the use of 3D models for further specialized studies, acquiring and processing virtual detailed copies as close as possible to the shape and contents of the real one. On the other hand, many museums look today for more interactive and immersive exhibitions, which involve the visitors' emotions, and this has contributed to the increase in the use of virtual reality and 3D models in museums installations. In this paper, we present two case studies that belong to these scenarios. Multisensor surveys have been applied to some archeological statues preserved in two museums for multipurpose analyses and representation: a UTI test, which required high detailed data about the geometry of the object, and a communicative application, which needed instead a high level of model optimization, poor geometry, but very good representation that was achieved through remeshing tools and normal maps.

Keywords: 3D models; normal map; remeshing; 3D Ultrasonic Tomographic Imaging (3D UTI); cultural heritage

---

1. Introduction

Many significant heritage assets around the world face different and continuous challenges due to the need of continuously evaluate the state of conservation for future generations, or, even more for sites than for museum assets, due to the severe natural and hand-made hazards.

The technological advances that have developed in the field of three-dimensional (3D) survey and modelling allow us to digitally preserve the historical memory of such built heritage thanks to feasible, accurate, and portable tools and methodologies.

The development and refinement of 3D survey techniques, as well as their integration, involve the small objects, artistic and archaeological heritage, as well as many other types of cultural assets, sites, and landscapes. (Lerma et al., 2011, Remondino et al. 2014). With regard to museum assets, recent extensive digitalization projects aimed at achieving multilingual information archives that make digital libraries accessible to everyone. This is the main mission of Europeana project, but also of other institutions as the Getty Foundation or the Smithsonian institution, which are devoted to the preservation of Cultural Heritage (CH), promoting research, culture, and educational activities, or directing efforts towards the achievement of standards for CH and the realization of digital inventories.[1]

A first trend in techniques development aims to respond to the requests of the institutions, such as the museums that deal with conservation, arranging the massive detailed digitization of their preserved objects. About this, Hindmarch offers a very thorough research on the digitization of CH small assets, with a point of view that is centred on institutions requirements, such as museums (Hindmarch 2015), and an example currently much observed is the cultlab3D system (Santos et al., 2014). This trend is geared to the widespread use of 3D models, ranging from their possible use for further specialized studies to the acquisition and processing of virtual detailed copies that are as close as possible to the shape and contents of the real one. In this case, the models must be particularly dense and rich in information, with a very high accuracy and high geometrical and radiometric resolution. The first case study presented in paper falls into this first set of models and is applied to an Egyptian statue of the Archaeological Museum of Bologna, whose processed 3D model was subjected to Ultrasonic Tomographic Imaging (UTI) investigations (Di Pietra et al, 2017).

On the other hand, many museums look today for more interactive exhibitions that involve and increase the visitors' emotions enhancing the museum interactive experience (Kersten et al. 2017, Spanò et al., 2016). In recent years, in fact, museums' role changed from a mere "container" of cultural objects to a "narrative space" able to explain, describe, and revive the historical material

---

[1] http://www.europeana.eu/portal/
http://www.getty.edu/about/whoweare/mission.htl
http://3d.si.edu/about

(Martina, 2014). The combination of these factors has contributed to the increase in the use of virtual reality and 3D models in museums installations that are offering to the visitors more dynamic and immersive experiences, and, in some cases, giving them the possibility to interact and select the information according to their interests. In this second scenario, 3D models must be suitably processed to make them easy to be handled by portable devices or even in web systems. In this case, the information density can decrease depending on the communication purpose, and, with the aim of maintaining high adhesion to the original object, the optimization of surfaces, involving remeshing techniques and textures maps, seems to be a very promising perspective.

The second case study of this paper is particularly connected to this study area. Two statues, known as "busti loricati of Susa" and representing two roman emperors, have been surveyed with a photogrammetric method with the aim to process two 3D models, representing the statues as they are now and as they were before the nineteenth century restoration, according to archive sources. Such models provided the base for a video installation for a museum, which simulates a holographic projection and explains the different armour parts highlighting them in sequence.

The structure of the article is as follows: in the next paragraph the employed methods for the 3D survey process are presented, together with the tested optimization techniques and the UTI method. Section 3 presents both case studies followed by conclusions and observations.

2. Methods Outline

2.1. 3D Survey Methods

In the field of ancient Cultural Heritage, 3D accurate models of archaeological objects and sculptures have a significant role for their documentation, maintenance, and restoration. The digitization of such heritage ensures the store of the information about the shape and appearance of an object against its possible lost and damage over time by natural or accidental causes. The collected data allow also the dissemination of digital media collections for a large audience via virtual museums and enable the creation of replicas via 3D prints.

In the last decade, the development in 3D survey techniques was rapid and continuous in producing new, more effective methods in terms of automation, acquisition, and processing speed, and the quality and precision of the output data. By now, it is well recognized that the integration of multisensor methods provide more complete and detailed data than standalone acquisition (Ramos et al., 2015; Trinks et al., 2017)

The employed sensors are generally distinguished in active and passive ones, basing on the emission of an electromagnetic signal.

Active sensors, as used in laser scanner (TOF) and range cameras (based on the triangulation principle), measure angles, and distances emitting signals (laser beams, infrared lights, etc.) and recording the reflected answers (Adami et al., 2015). The results of a laser scanner acquisition are point clouds that are constituted by millions of points, which can be coloured thanks to an integrated RGB camera, and from which it is possible to process very detailed textured 3D model and other two-dimensional (2D) more traditional representation. (Donadio et al., 2015).

Passive sensors, which are used in photogrammetric applications, use the ambient light to make measurements, recording the electromagnetic energy, i.e. visible light, emitted by the objects. In recent years, the development made on digital cameras and in calibration technology, in connection with very competitive costs, and especially thanks to the integration of *image matching* and *Structure from Motion* (SfM) algorithms, derived from the Computer Vision field, meant that photogrammetry is more and more used to recover objects with high accuracy (Samaan, et al. 2013). The final result is a dense 3D point cloud, which can be integrated with LiDAR data and processed to obtain a 3D textured model. In this process, it is also possible to generate true orthophotos in an increasingly automatic way, which are very useful metric products where radiometric information is combined with real measures allowing for a complete representation of the analysed object (Chiabrando et al., 2015). Such products are usable also as texture for mapping materials, deteriorations, or other important damaging effects. (Koska, et al., 2013; Rijsdijk, 2014).

In the very close range field, still among CH requirements, system such as the Time of Flight (ToF) cameras and structured light 3D scanners allow us to acquire with accuracies not lower than one or few millimetres. In the field of very high accuracy metrology systems ($\pm 25 \div 30\mu m$), new developed scanning systems, such as arm scanners, combine arms together with high definition hand-held scanners. Such systems, even though they are still expensive, are very useful in rapid prototyping and reverse engineering fields.

To identify the most appropriate technology for 3D digitalization, several aspects must be considered:

- characteristics of the object (shape, dimension, colour, reflectivity and homogeneity of the material, etc.),
- the acquisition place (internally or externally, with natural or artificial light, with the possibility to move the object or not),
- the aim of the survey (documentation, analysis, dissemination, virtual reality applications, real time applications), and
- time and budget constraints.

In the following test cases, the statues have been surveyed by means of the photogrammetric technique, which allowed to pursue the level of detail requested by the UTI analysis and communication purposes. In addition, the Egyptian statue was also acquired by a hybrid hand-held Freestyle scanner, in order to test its potentialities and carry out some considerations concerning the quality of the acquired data in comparison to the photogrammetric ones.

2.2. 3D Model Optimization

As mentioned before, the modelling of object surfaces and points clouds produces detailed 3D models, consisting of millions of polygons and high resolution images, with very large data volumes, which are difficult to be handled and visualized by common computers, portable devices with low performance (e.g., Smartphones and tablets), or on the Internet. (Kersten et al., 2016). Despite technological developments, the management of large polygonal datasets is, in fact, influenced by several technical problems, i.e. very long processing and editing time or real-time visualization constrained by the graphics card performance (Manferdini and Remondino, 2012).

Nowadays, the scientific community is trying to find new ways to optimize and share such 3D digital contents, in order to guarantee fast access to the data and their effective communication (e.g., in dissemination and valorization application of Virtual and Augmented Reality), providing low resolution 3D models that are easy navigable and viewable (Guarnieri et al., 2010; Cipriani et al., 2017).

Some critical considerations about the contradiction between the great increase in digitization projects of cultural assets in the last decade and the problems that today are an obstacle to their massive sharing on the web can be found in Scopigno et al. (2017). They are attributable to intellectual property issues and of improper use fear, the confidentiality of ongoing projects, the recurring problem of scarcity of financial resources, and of high expertise required.

In all of the cases, it is necessary to establish the needs that the 3D model have to fulfill, which determine the final level of geometrical and radiometric accuracy, size, and visualization, but with the constant aim of not losing quality and information. Different purposes led to different processing choices and different optimization levels, depending on the final aim to pursue, whether it is a specialized analysis (Guidi et al., 2017), which require to maintain the geometric detail in the final mesh or a communication application, which needs a lower data volume (Kai-Browne et al., 2016).

To achieve this, various methods that were borrowed from the field of gaming technology can be employed for drastically reducing the file size while maintaining a very high degree of detail. Game engine and entertainment software, in fact, provide several useful tools to process and generate Low Poly models, which show however a good detail on the surfaces in a very low byte files.

The whole process of computing, given a 3D mesh, a new mesh whose elements satisfy some quality requirements, while approximating the input acceptably, is called "re-meshing" (Alliez et al., 2008).

Among the available tools of the remeshing we mention the decimation and retopology algorithms, the SubDivision surfaces, and Polynomial texture maps (PTM). The decimation module automatically decreases the number of faces in the model, deleting the vertices depending on the specific decimation ratio as specified by the user. The retopology technique, which requires a skilled and experienced operator, consists in manually tracing the High Poly mesh with a square mesh of lower density, called "quad-dominant mesh". The final quadrangular mesh can support the conversion into Catmull-Clark SubDivision surface, which is a variable-detail model that may be interactively increased in different output scales by adjusting the level of subdivision (Merlo et al., 2013)

PTM is an image-based relighting technique for visualizing the appearance of surface under variable lighting condition. Pan et al. (2016) diffuse color maps, normal maps, and displacement maps are some examples. In order to compute such 2D representation, it is essential to firstly process the UV map of each 3D model by unwrapping it in a 2D coordinate system (u, v), which ensures the spatial link between the x, y, z coordinates of the mesh polygons and the image. After the UV mapping different textures can be computed, projecting surface information from the high resolution mesh onto the Low Poly mesh, converting geometrical characteristics into pixel information in a so-called baking process.

A normal map, for example, contains the surface normal vector of the high-resolution mesh, which is stored as a RGB value in the pixel image. Once calculated and applied to the Low-poly mesh, the normal map simulates the behavior of light reflections from the starting model, adjusting the shading of the low-detail model, which looks again like the original Hi-poly model.

The displacement map is, instead, a grey scale image, in which each shade of grey stands for a deviation of the background mesh from the optimized one. This is because, unlike the normal map, the displacement map, once applied to the model, displaces the points of the Low-poly mesh basing on the deviations calculated in the baking process (Merlo et al., 2013)

In this paper, the optimization process has been realized within the software: 3DReshaper, MeshLab, and Blender.3D Reshaper was used for the mesh generation from the points clouds. MeshLab was chosen for the effectiveness of its tools for mesh editing, cleaning and remeshing whereas Blender was chosen due to its capabilities in re-meshing, UV unwrapping, and baking.

2.3. UTI Method

Micro geophysics techniques can contribute to facilitating the restoration of artworks or historical building elements, also evaluating, with respect to the management of a museum, the possibility and the precautions that must be taken when moving artefacts (Piro et al., 2015).

The 3D Ultrasonic Tomographic Imaging (UTI) method consists in the estimation of the variation of the apparent velocity of an ultrasonic (US) pulse send within the volume of interest (VOI) (e.g., Sambuelli et al., 2015). According to the theory of elasticity, in fact, the velocity of propagation of a mechanical pulse in a given material decrease, for example, with the increasing of the number of fracture in unit of volume. It is estimated that dividing the Euclidean distance d between a transmitter probe, TX, and a receiver probe, RX, by the shortest time t (time break - TB) that the US impulse need to travel from TX to RX. This velocity is called apparent because the US pulse does not really travel through the low velocity volume (Wieland, 1987), but rather, according to the Fermat principle, it travels around it, taking a longer time. Dividing the straight path by a longer time we get a lower velocity, and this is the sign that between TX and RX there is an inhomogeneity, which can be due to the presence of weaker volume of rock, mortar, or a more densely fractured medium.

3. Experiments

3.1. Case Study 1: Integrated Investigations on a Damaged Egyptian Statue

The work aimed at highlighting the ability of methods that are devoted to the 3D geometry acquisition of small objects when applied to diagnosis performed by geophysical investigation (Figure 1). The data acquisition consisted in a photogrammetric survey and in two type of laser scanner: a laser Faro Focus 3D and the hybrid hand-held Freestyle scanner, in order to carry out some considerations concerning the quality of the acquired data in relation also to the Ultrasonic Tomographic analysis needs (Di Pietra et al., 2017). Since the statue was in a laboratory, with a bad light condition, we decided to add some artificial properly oriented lights.

3.1.1. The Egyptian Naophorous Statue of Amenmes and Reshpu

The survey is applied to the Egyptian *naophorous* statue of Amenmes and Reshpu, which dates to the reign of Ramses II (1279-1213 BC) or later, and is now preserved in the Civic Archaeological Museum in Bologna (Inv. no. MCABo EG 1821) (Kminek-Szedlo, 1895). The statue was dedicated to the gods Osiris, Isis, and Horus by two high officials of the Amun temple in Thebes, Amenmes and Reshpu.

The statue presents a worrying deterioration of its limestone, consisting in stone material chips and cracks, especially on the inscribed base. In order to interrupt the increasingly rapid and progressive deterioration, and to clean and consolidate the limestone, the Civic Archaeological Museum has sought the cooperation of prestigious Italian institutions and a restorer expert in stone conservation [2] (Picchi, 2016).

Figure 1. (**Left**) Worrying state of deterioration of the statue with visible fractures, (**right**) the ultrasonic measurements to reconstruct the fracturing state of the statue needs to know several vector size between points on the damaged surface.

3.1.2. The Integrated 3D Survey Technique

The survey phase consisted in a photogrammetric and laser acquisition of the statue, with the aim to generate a 3D model, on which it can make a variety of detailed 3D measures. In particular, it was necessary to evaluate on the processed 3D model the locations of the source and receiver points of the Ultrasonic Tomographic test, their coordinates in a fixed reference system, and their mutual distance. Before any acquisition, eight checkerboard targets have been placed on the base of the statue, to be used for registering all of the data in a unique reference system, and 71 white numbered stickers have been arranged on the statue to form a 3D network of max $150 \times 150 \times 150$ mm³/voxler, as a 3D mesh for ultrasonic tomography (Figure 1).

The photogrammetric survey consisted in the acquisition of 74 images, captured at three different heights all around the object with a Nikon D800E (7360×4912 CMOS sensor, Zeiss optic system and 50 mm focal length) at a distance of 1 m. The acquired images have been processed by means of the Structure from Motion (SfM) technique and image matching algorithms, thus generating a dense point cloud that is constituted by some six mil. points. The high quality of the bundle adjustment process was secured by the re-projection error limited to 1.2 pixel. After these steps, a reference system has been assigned to the

---

[2] Engel (*Environmental-Engineering Geophysics Laboratory*) and Geomatics Laboratory of Politecnico in Turin, Opificio delle Pietre Dure (OPD) in Florence, IUAV University in Venice and the restorer Cristina Del Gallo.

point cloud, scaling also the model according to measures between targets placed on the crankcase. The targets positions have been obtained by the terrestrial fixed scanning (Faro Focus 3D scanner, featured by measurement accuracy of ±2mm up to 10m), which is specially used to make the photogrammetric model that is independent from the next one generated by the freestyle hybrid scanner. Subsequently, a textured 3D model has been processed, ensuring an average accuracy on the final model of less than 2 mm (accuracy evaluated measuring some control points on the mesh and on the original point cloud) (Figure 2).

The second step of the survey test applied to the statue employed a handheld 3D laser scanning (the FARO Scanner Freestyle3D: http://www.faro.com/products/3d-documentation/faro-scanner-freestyle-3d/overview).

Such an instrument offers fast data acquisition and real-time visualisation; it uses a structured light technology consisting in two infrared cameras that create a "stereo pair" of images that are looking at the structured light pattern, which allows for the determination of the moving centre. A laser sensor ensures the measurement of the surveyed surface with a range of acquisition of 0.5-3 m, (3D point accuracy 1.5 mm) and an RGB camera acquires the radiometric data. Its relatively low cost makes such technique suitable and reliable for small objects modelling and addressable for many different uses. The acquisition phase consisted in five scans collected at a distance from the object of 0.5 m, providing a resolution of 0.2 mm and with a predicted noise of 0.7 mm. Certainly, the ambient light is a determining factor for the accuracy of the clouds, and, in fact, the registration process that is based on the automatic target recognition ended with an average residual of 2–3 mm, which is insufficient for the intended purposes. The final point cloud, obtained after a progressive alignment according to the shape of clouds, is constituted by some nine mil points. The amount of cloud points refers to the exact surface extension used for the photogrammetric model so as to give value to the comparison of the clouds.

One of the aim of the work was the evaluation of the accuracy of both the surveys, mostly regarding the possibility for the freestyle laser point cloud to provide a sufficient level of detail for the Ultrasonic Tomographic needs. As reported in Table 1, the laser Freestyle cloud is constituted by a highest number of points than the photogrammetric one. The roughness of both the points clouds has been computed in Cloud Compare, which estimates the distance between each point and the best fitting surface computed on its nearest neighbours. As shown in Figure 3 and Table 1, the photogrammetric cloud has minimal noise with a mean value of 0.269 mm, while the freestyle cloud presents a higher noise, with a mean value of 0.552 mm (to make such estimation a radius of a sphere centred on each point of 0.005 m has been considered) (Figure 3). Once aligned, several section profiles have been also extracted from both the cloud. Measuring samples of residuals among the photogrammetric cloud, assumed as the reference cloud, and the freestyle one, we have been able to estimate that the second one has discards of 1–2 mm due to alignment operation and its roughness. As it is possible to notice in Figure 4, we also computed the level of points density: the freestyle cloud has a higher number of points and a higher density, but the points distribution is less uniform (to make such estimation a radius of a sphere centred on each point of 0.01 m has been considered). This is due surely to the acquisition process related to the operator motion.

Figure 2. Photogrammetric point cloud, mesh and textured mesh.

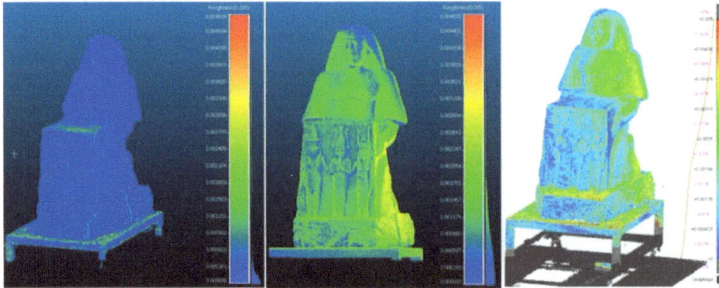

Figure 3. Evaluation of the level of roughness of the clouds; (**left**) insignificant level of roughness computed on the photogrammetric cloud (less than 0.5 mm almost throughout the cloud); (**middle**) laser cloud presenting a higher percentage of points affected by noise (several at 2.5–3 mm); comparison between clouds (**right**) about 60% of the freestyle point cloud is at a distance of less than 2 m from the photogrammetric cloud.

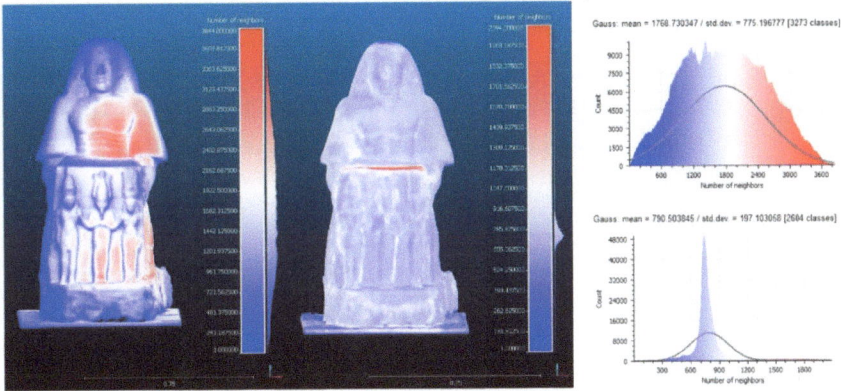

Figure 4. Points density: freestyle cloud (**left**); photogrammetric cloud (**middle**); histogram distribution of the points density - freestyle cloud (**right above**); histogram distribution of the points density – photogrammetric cloud (**right below**).

Table 1. Number of points, level of roughness and level of points density computed in both the clouds in Cloud Compare.

	Number of Points	Roughness Mean Value	Density Mean Value (in a Spherical Neighbourhood with Radius of 0.01 m)	Std. Dev.
Photogrammetric data	6.011.037	0.269 mm	790 points	197 points
Laser Freestyle data	9.195.056	0.552 mm	1768 points	775 points

3.1.3. UTI Data Acquisition and Processing

The 3D UTI on Amenmes was restricted to the lower part of the statue, which was the part giving major concern about its mechanical properties. Around this volume, 71 points have been materialized with sticky circular stamps (diameter 15 mm), on which the central point and a progressive number were marked (Figure 1).

The point positions were planned so that, on average, the straight rays connecting each couple of points would have crossed the VOI with a dense 3D net and would have not left volumes greater than 15 × 15 × 15 cm not scanned. To acquire, store and process the geophysical data the following instruments and software's were used. A PUNDIT pulse generator with exponential US probes with a nominal frequency of 50 kHz; a 7 dB fixed gain amplifier, a variable gain (1, 3, 10 dB) amplifier; a Le Croy Wave Jet 20 MHz digital oscilloscope; Matlab proprietary scripts for time reading, straight rays tracing and velocity interpolation; and, the software GeoTomCG by GeoTom LLC for 3D UTI and the software Voxler by Golden Software for 3D rendering of the velocity volumes. Across the VOI, 226 measurements were made according to the scheme in Figure 6, right. Before the measurements, the time delay introduced by the exponential probes was measured and it was 20.7 µs. This time was subtracted from each TB. In Figure 5 (left), an example of the graph on which the TB's were read is shown.

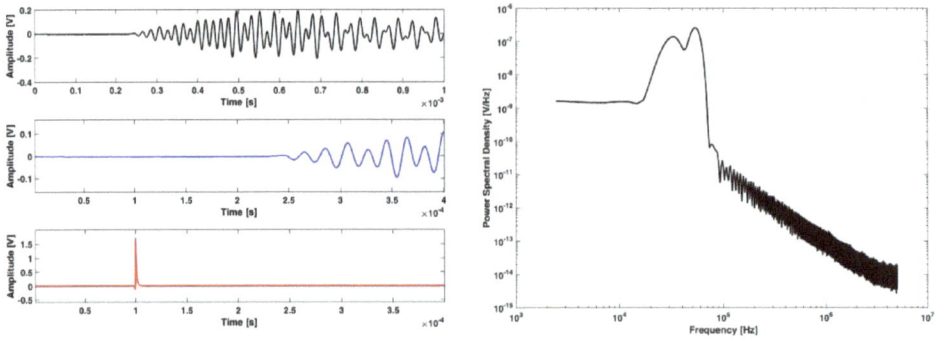

Figure 5. Top raw ultrasonic (US) signal (**left above**); zoomed US signal to read the TB (**left center**); triggering signal from which the time-zero (the time the US pulse entered the rock) was read (**left below**); PSD of a transmitted US pulse (**right**).

The spatial resolution, i.e., the smaller detectable anomalous volume, of a 3D UTI is roughly related to the dominant wavelength λ of the US pulse. An estimation of λ can be obtained dividing, for example, the average velocity of US pulse in the medium by the dominant frequency of the US pulse itself. To evaluate an order of magnitude of the achievable spatial resolution in the 3D UTI, the power spectral density (PSD) of some long-path and short-path signals have been evaluated. The PSD graph, as shown in Figure 5 (right), can be considered as representative of most of the transmitted signals. From the graph, the main energy content is roughly within 30 and 60 kHz.

When considering an uncertainty on d equal to 5 mm and on t equal to 15 µs, an uncertainty of 60 m/s can be supposed on apparent velocities. An elementary statistical analysis on the apparent velocities gives: minimum velocity 300 m/s (this value is not realistic and it's likely due to a missing FB); maximum velocity 2800 m/s; mean velocity 1330 ± 60 m/s; velocity standard deviation 630 m/s. From the velocity values and the PSD frequency band, a dominant wavelength of 5 cm could be safely assumed. In Figure 6 (left) a histogram of the apparent velocities is shown. The histogram clearly shows that many apparent velocities are around and even below 1000 m/s; that is a quite low velocity range for a limestone. In figure 6 (middle) a plot of d versus TB is shown.

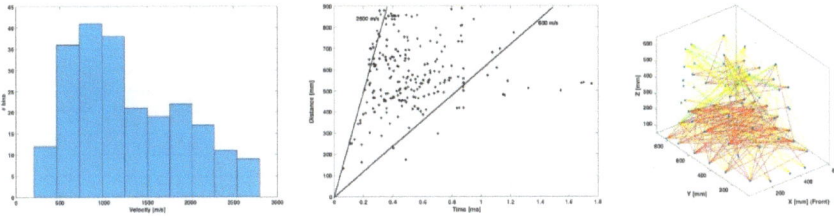

Figure 6. Histogram of the apparent velocities (**left**); distance d versus Time TB plot (**center**); the Euclidean straight paths used in three-dimensional (3D) Ultrasonic Tomographic Imaging (UTI) processing. Ray colour codes refer to apparent velocities: red v < 1300 m/s; yellow 1300 < v < 2500 m/s; green v > 2500 m/s (**right**).

The slope of each straight line on the graph in Figure 6 (middle) is an apparent velocity. The two continuous lines represents two velocities. The points relative to a sound limestone should lie around or above the line at 2500 m/s. This graph shows that within the VOI, a few rays have an apparent velocity that is compatible with a sound limestone.

In Figure 6 (right), a 3D view of the straight ray paths used in the 3D UTI is shown. Each ray path is drawn with a colour related to its apparent velocity. Even in this representation, before the 3D UTI processing, it is easy to guess that the lower part of the statue has worse mechanical properties than the higher part.

3.1.4. 3D UTI and Improved Graphical Representation

UTI analysis has been performed with the software GeoTomCG (by GeoTom, LLC, http://dev.geotom.net), which allows for the tomographic processing of seismic data in a volume with sources and receivers located anywhere on a 3D Cartesian grid. GeoTomCG uses a SIRT (Simultaneous Iterative Reconstruction Tomography) algorithm to produce, from the position (x, y, z) of transmitters and receivers and from the TB's, a velocity distribution in the VOI (Trampert and Leveque, 1990). The volume is firstly discretized in prismatic cells, and then an iterative procedure is used to calculate the velocity of the US pulse within each cell. The VOI of Amenmes was discretized in 5 × 8 × 6 prismatic cells with size: dx = 114 mm, dy = 108 mm, dz = 101 mm. Then, on average, the cell size was about two times the dominant wavelength of the US pulse along each direction. From the final model, only the cells crossed by more than three rays have been considered, and a 3D rendering of the resulting velocity distribution has been plotted.

From the 3D UTI analysis, it is possible to extract a standard file format that typically used in computational fluid dynamics (CFD), the PLOT3D file. It is a standard format that usually includes two different files, the grid file and the solution file. The first contains the coordinate of the solution grid, while the second contains information typical of a CFD solution, in our case, the correspondent velocities. With the aim to improve the visualization of the results of the 3D UTI analysis and to make volumetric measurements on the statue, the PLOT3D file format has been converted in a standard format that is usable in a 3D point cloud processing software. In particular, an easy function, developed in Matlab was used to transform the structure of the grid file in a .xyz file, where the coordinate of the grid could be visualized as a sparse point cloud and the velocity value as a scale bar. The VOI of Amenmes extracted from GeoTomCG software was discretize in 30 × 45 × 35 prismatic cells with size of 20 mm. The correspondent point cloud that was extracted with Matlab was too sparse so it was oversampled with the use of the Point Cloud Library (PCL), an open project (Rusu, Cousins, 2011), integrated as plugin in the open source software Cloud Compare. The tool 'Smoothing MLS' allows to interpolate the point cloud with the moving last square method so that the initial cloud can move from 47.250 to 37.469.250 points. At the end of the process, it has obtained a 3D representation of the velocity distribution in the VOI of Amenmes statue. The graphical representation was like the one already used in the UTI analysis with: red colour v < 1500 m/s, yellow 1501 < v < 2000 m/s, green 2001 < v < 2500 m/s, dark green 2501 < v < 3000 m/s, and blue v > 3001 m/s. The VOI represented with this scale bar, was combined with the statue points cloud and this has made possible the representation of different section level to better understand the region that was affected by anomalies (Figure 7, left).

The iso-surface was created, extrapolating the points of the cloud belonging to the separation surface between the velocity defined before, and generating the corresponding mesh. Once defined, the iso-surfaces and the points cloud between two steps level, it was possible to compute the volume of statue for each average velocity, and so the volume was characterized by the worse mechanical proprieties. The volumes were defined from a meshed closed shell that was generated with the portion of points cloud included between two iso-surfaces and was measured with the tool Measure/Volume of the software 3DReshaper (Figure 7, right – Table 2).

Table 2. The volumes of the statue for each range of velocity.

Average Velocities [m/s]	Interested Volumes [m³]	Percentage Compared to Investigated Volume [%]
v < 1500	0.077 m³	38.31%
1500 < v < 2000	0.053 m³	26.37%
2000 < v < 2500	0.039 m³	19.40%
2500 < v < 3000	0.025 m³	12.44%
V > 3000	0.007 m³	3.48%

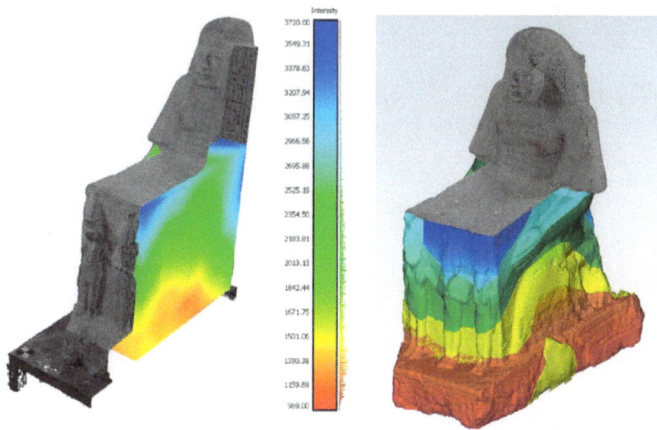

Figure 7. Improved representation of average velocities in the VOI of the Ramessid statue and the closed mesh extracted from the classified point cloud.

In conclusion, the survey applied to the Egyptian sculpture proved that a little less than 40% of the volume of the crankcase presents a state of conservation particularly worrying (Table 2), since the overall measured average velocity is lower than 1500 m/s, due to damages and fractures.

3.2. Case Study 2: Optimization Surface Methods on a Roman Statue for Communicative Purpose

The second case study is aimed at the dissemination of an archeological asset producing multimedia contents from multi-sensor surveyed 3D data.

3.2.1. The Roman Loricate Busts of Susa

The ruins of the two "loricati" busts (so named because of the decorated armour), representing two Julio-Claudians emperor, (Figure 8) were found in 1802 in the city of Susa and in the nineteenth century a restoration action completed them adding the missing heads, low parts of the legs, and the arms.

These busts and their decorated armours are an exceptional witness of the north Italian sculpture during the first imperial age (Cadario, 2005). In the first bust, the decoration invokes the mythical origins of the *gens Iulia* (Figure 8, left), whereas in the second *torso* (Figure 8, right), the representation had to summon the victory against Middle East areas after the Parti population' submission. (Cadario, 2005).

Figure 8. The roman statues placed inside the Civic Archaeological Museum in Turin.

3.2.2. The Photogrammetric Survey

The aim of the 3D survey was to process two detailed textured models of the busts, as they are today and the way they were before the addition of the nineteenth century restoration, to be used for a multimedia installation in the new Civic Museum of Susa.

Both of the statues have been surveyed by means of a photogrammetric application with a Canon EOS-1Ds Mark II with a 50 mm focal length. About 200 images have been captured at three different heights all around the statues at a distance of about 1 m and processed by means of the Structure from Motion (SfM) technique and image matching algorithms. After the image alignment, a dense point cloud constituted by about 12 mil. (bust with the baldric) and 15 mil. (bust without the baldric) of points have been generated, as well as a 3D mesh model of about 2.5 mil. and 3.4 mil. triangles, textured with the images acquired. After these steps, the data have been scaled according to measures taken between targets placed in the scene. The surface resolution was very high, respectively, of 0.253mm/pix and 0.221mm/pix (Figure 9).

3.2.3. The Bust 1: 3D Models Optimization

In the optimization process, we tested several decimation and remeshing tools in order to make some evaluation on different levels of decimation, by comparing the geometrical accuracy that was preserved and the byte of the output files. This test was also suggested by the final use of the 3D models which was a multimedia installation for the Museum.

A first step of the optimization was done within the software Photoscan where the 3D model has been generated.

Even though the parameters for the mesh processing were set very high and with the interpolation disabled, the mesh presented a bad topology, especially in the arm and the neck part (Figure 9).

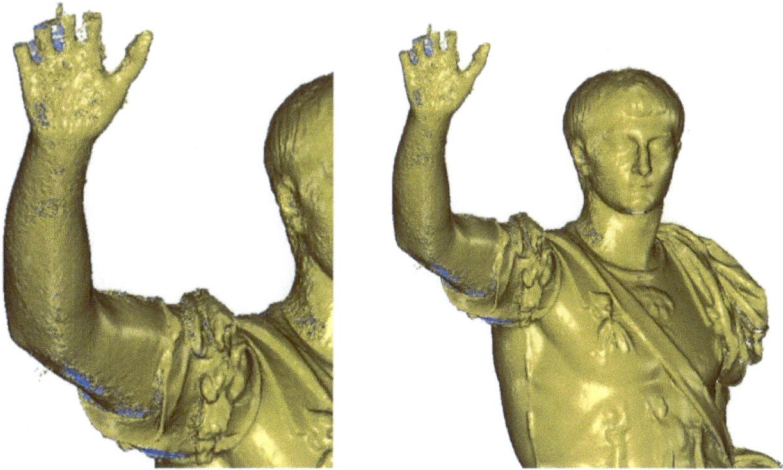

Figure 9. Mesh processed in Photoscan; it is possible to notice the topological error that affect the surface.

For these reasons, we decimated the mesh with the Photoscan tools from 2.457.747 triangles to 1.800.000 triangles (a lower number made the mesh too smoothed), and we also ran the command "Fix Topology", which is active if there are any topological problems and can be clicked to solve the problems (Figure 10).

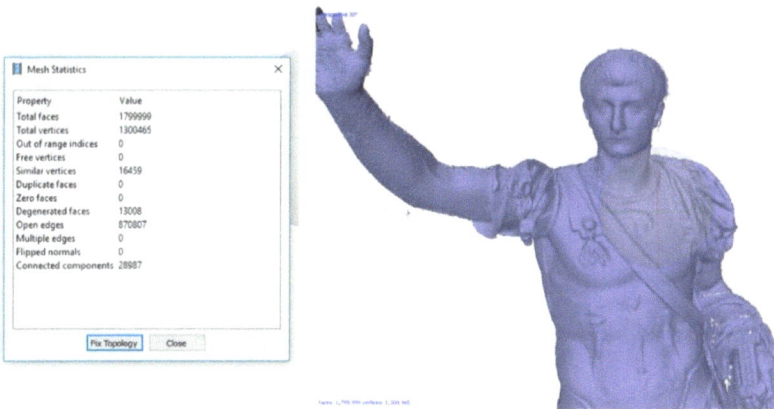

Figure 10. Mesh after the decimation filter to 1.800.000 faces.

Since the mesh so decimated did not satisfy the geometrical requests, due to the continued presence of noise, we decided to make a new one exporting the dense point cloud within the software 3D Reshaper.

The mesh has been processed in 3D Reshaper setting a triangle size of 0.003 m and refining it taking new points from the cloud with a deviation error of 0.002 m. In Figure 11 and in Table 3, it is possible to notice that the final mesh has a greater quality when compared to the Photoscan ones, a very good geometrical detail, fewer topological errors and also a lower byte file.

Table 3. Comparison between the mesh computed within the photogrammetric software and the one realized from the photogrammetric cloud in 3D Reshaper.

	N. Points	N. Triangles	N. of Independent Pieces	N. of Holes or Free Contours	KB File .obj.
Mesh Photoscan	1.720.221	2.457.747	30.405	71.075	110.309
Mesh 3DReshaper	374.774	741.327	2	264	37.390

Figure 11. Original high detail mesh model processed in 3D Reshaper.

In order to process, as mentioned before, an extreme optimization of the model, testing the potentialities of the remeshing tools, we imported the model in MeshLab in order to test several levels of decimation.

After the application of the filters of cleaning and repairing, which allow us to remove duplicate faces and vertices, non manifold edges and vertices, we ran the command *Simplification: Quadric Edge Collapse Decimation* to reduce the number of vertices setting the minimum number of faces of the final remeshed mesh.

In order to identify different steps of decimation, which can respond to different purposes, we tested and compared three type of decimation and optimization (Figures 12 and 13, Table 4):

- In a first level, the mesh is decimated indiscriminately and automatically with the *Quadric Edge Collapse Decimation* tool. In this process, it is not possible to reduce that much the mesh if you do not want to completely lose the geometric and visible detail. For this reason, we set the decimation from 741.327 to 100.000 faces. The file size changed from 37.390 KB to 2.483 KB, but visually the mesh still has a good level of detail, although it is more smoothed (Figure 12, left).
- In a second level, we tried to maintain the original high level of detail in the high detailed portions (the decorated armour and the *pterigi* part), hardly decimating just the flat part, such as the legs and the arm. In this way, the file size is reduced a little less but the detail on the decorated portion is still very high. (Figure 12, middle–right).
- In the third level of optimization, we tried to reduce at most the mesh making use of the Polynomial texture maps (PTM) to just visually simulate the detail. The mesh has been decimated to 7000 faces and then imported in Blender for the UV mapping and baking process. In order to minimize the work time, we used the UV map that was automatically generated by the unwrapping tool and then we calculate the UV map, also importing the original High Poly mesh. The final result is a 3D model that looks like the original one but with a very low size in terms of byte. This representation is very suitable for entertainment application, such as the one aimed for the museum, but it is important to highlight that it has completely lost its geometric detail, making it useless for other kind of specialized geometrical analysis (Figure 13).

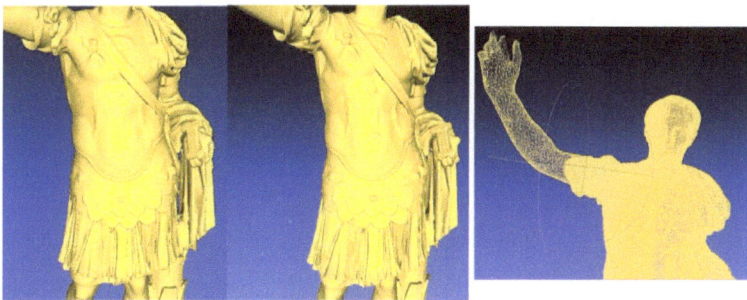

Figure 12. Level 1: decimated surface mesh; Level 2: surface model and wireframe view.

113

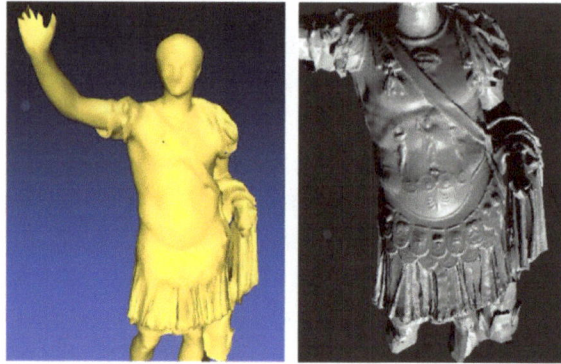

Figure 13. Level 3: Decimated mesh before and after the projection of the Normal Map calculated in Blender.

Table 4. Comparison between different optimization level.

	N. Vertices	N. of Faces	KB file .obj	% KB Reduction	% Faces Reduction
Original mesh	374.774	741.327	37.390	-	-
Level 1	52.622	100.000	8.599	77%	86,5%
Level 2	147.754	288.165	25.200	32,6%	61%
Level 3	3.874	7.000	575	98%	99%

4. Conclusions

As shown in this article, Cultural Heritage assets can have great benefits from 3D data and reality-based models for multipurpose aims. It is possible to outline different considerations pertaining the cases study and the methods.

The survey applied to the Egyptian sculpture prove that a little less than 40% of the volume of the crankcase presents a state of conservation that is particularly worrying (Table 2), since the overall measured average velocity is lower than 1500 m/s, due to damages and fractures. That means that the integration of the two methods is capable of investigating the magnitude of the internal damage of stone objects, which is a necessity that is encountered in most cases of cultural heritage and small objects too.

The operative outcomes of the second application are summarized in Table 4: the conclusion of the optimization process leads to having a model reduced by 99% with respect to the original one, which is therefore more suitable for communication projects that require good perception, even though it may be not featured by high level of geometric resolution and accuracy.

In the field of methods evaluation, the versatility of the photogrammetric survey, after the development of the SfM technique, is more than established, although it must be considered that to obtain high accuracies, such as those achieved in the presented test cases, high-quality cameras and optics must be available and a high level of expertise in the modelling phase is necessary. When compared to the most popular and tested hand-held Kinect scanner, the Faro Freestyle present advantages that may make it attractive in cultural heritage documentation: it is portable, it provides full colour point clouds, it is easy to use and is versatile, reaching the extension of small rooms survey, and it is cheaper than many other systems.

Concerning the UTI test, in addition to the proof of the possibility for the processed 3D models to provide useful detailed data, the improvement of the graphic representation of the results is another key result. On this point, it is necessary to highlight that we met many difficulties in the interoperability of digital data formats and many operations and transformations of formats were made with laborious manual processes.

With the optimization methods that are involved in the second test case, we proved how reality-based 3D modelling methodologies might satisfy the requests of dissemination, sharing, and access of Cultural Heritage information. Innovative digital applications provide great potentialities for institutions and museums in helping them to preserve heritage, ensuring the ability to reach multimedia contents with educational purposes, and then to promote themselves. Digital contents accessible time and location independently, immersive experiences, and virtual visits represent new methodologies to increase the communication and transmission of culture, today and into the future generations. Pursuing this aim, we proved how procedures borrowed from reverse modelling and mapping allow us to generate extremely precise visualization, with a very low computing cost.

Acknowledgments: We would like to thank warmly the two institutions that involved us in the preservation or dissemination of the cultural values of the statues covered by this study: the Archaeological Museum of Bologna and the Archaeological Museum of Torino through the official Federico Barello.
We also recall that D. Franco (ENGEL) participated to UTI surveys and Paolo Maschio (Geomatics Lab) made the photogrammetric shooting at the data acquisition stage in Bologna. Vincenzo Di Pietra participated with A. Spanò to the graphical improvement of the 3D UTI analysis and Giulia Sammartano participated in the data acquisition of the statues in Torino.

References

1.   Adami, A.; Balletti, C.; Fassi, F.; Fregonese, L.; Guerra, F.; Taffurelli, L.; Vernier, P. The bust of Francesco II Gonzaga: From digital documentation to 3d printing. *ISPRS Ann. Photogramm. Remote Sens. Spat. Inf. Sci.* **2015**, *2*, doi:10.5194/isprsannals-II-5-W3-9-2015.

2.   Alliez, P.; Ucelli, G.; Gotsman, C.; Attene, M. Recent advances in remeshing of surfaces. In *Shape Analysis and Structuring*; Springer: Berlin/Heidelberg, Germany, 2008; pp. 53–82, doi:10.1007/978-3-540-33265-7_2.

3.   Cadario, M. Ipotesi sulla circolazione dell'immagine loricata in età imperiale: I torsi giulio-claudi di Susa. In La Scultura Romana Dell'italia Settentrionale. Quarant'anni Dopo la Mostra di Bologna. Atti del Convegno Internazionale di Studi. Pavia, 22–23 September 2005; Slavazzi, F., Maggi, S., Eds.; All'Insegna del Giglio: Firenze, Italy, 2005.

4.   Campana, S. 3D Modelling in Archaeology and Cultural Heritage. Theory and Best Practices. In *3D Recording and Modelling in Archaeology and Cultural Heritage. Theory and Best Practices: Archaelogical Needs*; Remondino, F., Campana, S., Eds.; BAR International Series 2598; Archaeopress: Oxford, UK, 2014; pp. 7–12.

5.   Cipriani, L.; Fantini, F. Digitalization Culture VS Archaeological Visualization: Integration of Pipelines and Open Issues. *Int. Arch. Photogramm. Remote Sens. Spat. Inf. Sci.* **2017**, 42-2(W3), 195–202, doi:10.5194/isprs-archives-XLII-2-W3-195-2017.

6.   Chiabrando, F.; Donadio, E.; Fernandez-Palacios, B.; Remondino, F.; Spanò, A. Modelli 3D multisensore per l'acropoli segusina. In *L'arco di Susa e i Monumenti Della Propaganda Imperiale in età Augustea*; Del Vecchio, P., Eds.; Susa, Italy, 2015; pp. 217–232.

7.   Di Pietra, V.; Donadio, E.; Picchi, D.; Sambuelli, L.; Spanò, A. Multi-source 3d models supporting ultrasonic test to investigate an egyptian sculpture of the archaeological museum in Bologna. *Int. Arch. Photogramm. Remote Sens. Spatial Inf. Sci.* **2017**, XLII-2/W3, 259–266, doi:10.5194/isprs-archives-XLII-2-W3-259-2017.

8.   Donadio, E.; Chiabrando, F.; Sammartano, G.; Spanò, A. Reality Based Modeling Training. Photomodelling and LiDAR Techniques for the St. Uberto Church in Venaria Reale. In Proceedings of the Disegno & Città, Torino, Italy, 17–19 settembre 2015.

9.   Gonizzi Barsanti, S.; Guidi, G. A Geometric Processing Workflow for Transforming Reality-Based 3d Models in Volumetric Meshes Suitable for Fea. *Int. Arch. Photogramm. Remote Sens. Spatial Inf. Sci.* **2017**, XLII-W3, 331–338, doi:10.5194/isprs-archives-XLII-2-W3-331-2017.

10.   Guarnieri, A.; Pirotti, F.; Vettore, A. Cultural heritage interactive 3D models on the web: An approach using open source and free software, *J. Cult. Heritage* **2010**, 11, 350–353, doi:10.1016/j.culher.2009.11.011.

11.   Hindmarch, J. Investigating the Use of 3D Digitisation for Public Facing Applications in Cultural Heritage Institutions. Ph.D. Thesis, University College London, London, UK, 2016.

12.   Kai-Browne, A.; Kohlmeyer, K.; Gonnella, J.; Bremer, T.; Brandhorst, S.; Balda, F.; Plesch, S.; Lehmann, D. 3D Acquisition, Processing and Visualization of Archaeological Artifacts. In Proceedings of the Euro-Mediterranean Conference, Nicosia, Cyprus, 31 October–5 November 2016; Springer International Publishing: Berlin, Germany, 2016; pp. 397–408, doi:10.1007/978-3-319-48496-9_32.

13.   Kersten, T.P.; Lindstaedt, M. Potential of Automatic 3D object reconstruction from multiple Images for applications in Architecture, Cultural Heritage and Archaeology. *Int. J. Heritage Digit. Era Multi. Sci. Publ.* **2012**, 1, 399–420, doi:10.1260/2047-4970.1.3.399.

14. Kersten, T.P.; Hinrichsen, N.; Lindstaedt, M.; Weber, C.; Schreyer, K.; Tschirschwitz, F. Architectural Historical 4D Documentation of the Old-Segeberg Town House by Photogrammetry, Terrestrial Laser Scanning and Historical Analysis. In Proceedings of the Euro-Mediterranean Conference, Limassol, Cyprus, 3–8 November 2014, Springer International Publishing: Berlin, Germany, 2014; pp. 35–47, doi:10.1007/978-3-319-13695-0_4.

15. Kersten, T.P.; Omelanowsky, D.; Lindstaedt, M. Investigations of Low-Cost Systems for 3D Reconstruction of Small Objects. In Proceedings of the Euro-Mediterranean Conference, Nicosia, Cyprus, 31 October–5 November 2016, Springer International Publishing: Berlin, Germany, 2016; pp. 521–532, doi:10.1007/978-3-319-48496-9_41.

16. Kersten, T.P.; Tschirschwitz, F.; Deggim, S. Developement of a virtual museum including a 4D presentation of building history in virtual reality. *Int. Arch. Photogramm. Remote Sens. Spatial Inf. Sci.* **2017**, 42-2(W3), 361–367, doi:10.5194/isprs-archives-XLII-2-W3-361-2017.

17. Museo Civico; Kminek-Szedlo, G. *Museo Civico di Bologna: Catalogo di Antichità Egizie*; Stamp. Reale Della Ditta G. B. Paravia e C. Edit.: Turin, Italy, 1895.

18. Kyriakaki, G.; Doulamis, A.; Doulamis, N.; Ioannides, M.; Makantasis, K.; Protopapadakis, E.; Hadjiprocopis, A.; Wenzel, K.; Fritsch, D.; Klein, M.; et al. 4D Reconstruction of Tangible Cultural Heritage Objects from Web-Retrieved mages. *Int. J. Heritage Digital Era* **2014**, 3, 431–452, doi:10.1260/2047-4970.3.2.431.

19. Lerma, J.L.; Navarro, S.; Cabrelles, M.; Villaverde, V. Terrestrial laser scanning and close range photogrammetry for 3D archaeological documentation: The Upper Palaeolithic Cave of Parpalló as a case study. *J. Arch. Sci.* **2010**, 37, 499–507, doi:10.1016/j.jas.2009.10.011.

20. Manferdini, A.M.; Remondino, F. A review of reality-based 3D model generation, segmentation and web-based visualization methods. *Int. J. Heritage Digital Era* **2012**, 1, 103–123, doi:10.1260/2047-4970.1.1.103.

21. Martina, A. Virtual Heritage: New Technologies for Edutainment. Ph.D. Thesis, Politecnico di Torino, Torino, Italy, 2014.

22. Merlo, A.; Fantini, F.; Lavoratti, G.; Aliperta, A.; Hernández, J.L. Texturing e ottimizzazione dei modelli digitali reality based: La chiesa della Compañía de Jesús. *DisegnareCon* **2013**, 6, 1–14.

23. Pan, R.; Tang, Z.; Xu, S.; Da, W. Normals and texture fusion for enhancing orthogonal projections of 3D models. *J. Cult. Heritage* **2017**, 23, 33–39, doi:10.1016/j.culher.2016.07.009.

24. Picchi, D. Statua naofora di Amenmes e Reshpu. In *Restituzioni 2016. Tesori d'arte Restaurati*; Bertelli, G., Bonsanti, G., Eds.; Exhibition cataloque, Milan 1 April–17 July 2016; Marsilio: Venezia, Italy, 2016; pp. 30–37.

25. Piro, S.; Negri, S.; Quarta, T.A.M.; Pipan, M.; Forte, E.; Ciminale, M.; Cardarelli, E.; Capizzi, P.; Sambuelli, L. Geophysics and cultural heritage: A living field of research for Italian geophysicists. *First Break* **2015**, 33, 43–54.

26. Remondino, F.; Spera, M.G.; Nocerino, E.; Menna, F.; Nex, F. State of the art in high density image matching. *Photogramm. Rec.* **2014**, 29, 144–166, doi:10.1111/phor.12063.

27. Rusu, R.B.; Cousins, S. 3D is here: Point Cloud Library (PCL). In Proceedings of the IEEE International Conference on Robotics and Automation (ICRA), Shanghai, China, 9–13 May 2011; pp. 1–4, doi:10.1109/ICRA.2011.5980567.

28. Samaan, M.; Héno, R.; Pierrot-Deseilligny, M. Close-range photogrammetric tools for small 3d archaeological objects. *Int. Arch. Photogramm. Remote Sens. Spat. Inf. Sci.* **2013**, *15-5(W2)*, 549–553.

29. Sambuelli, L.; Böhm, G.; Capizzi, P.; Cardarelli, E.; Cosentino, P. Comparison between GPR measurements and ultrasonic tomography with different inversion algorithms: An application to the base of an ancient Egyptian sculpture. *J. Geophys. Eng.* **2011**, *8*, 106–116, doi:10.1088/1742-2132/8/3/S10.

30. Sambuelli, L.; Böhm, G.; Colombero, C.; Filipello, A. Photogrammetry and 3-D Ultrasonic Tomography to Estimate the Integrity of Two Sculptures of the Egyptian Museum of Turin. In Proceedings of the Near Surface Geoscience 2015—21st European Meeting of Environmental and Engineering Geophysics, Non-destructive Tests and Prospections for Cultural Heritage, Turin, Italy, 6–10 September 2015; doi:10.3997/2214-4609.201413675.

31. Santos, P.; Ritz, M.; Tausch, R.; Schmedt, H.; Monroy, R.; De Stefano, A.; Posniak, O.; Fuhrmann, C.; Fellner, D.W. CultLab3D: On the Verge of 3D Mass Digitization. In Proceedings of the GCH 2014—Eurographics Workshop on Graphics and Cultural Heritage, Darmstadt, Germany, 6–8 October 2014; pp. 65–73, doi:10.2312/gch.20141305.

32. Scopigno, R.; Callieri, M.; Dellepiane, M.; Ponchio, F.; Potenziani, M. Delivering and using 3D models on the web: Are we ready? *Virtual Archaeol. Rev.* **2017**, *8*, 1–9, doi:10.4995/var.2017.6405.

33. Trampert, J.; Leveque, J. Simultaneous iterative reconstruction technique: Physical interpretation based on the generalized least square solution. *J. Geophys. Res.* **1990**, *95*, 12553–12559, doi:10.1029/JB095iB08p12553.

34. Verstockt, S.; Gerke, M.; Kerle, N. Geolocalization of Crowdsourced Images for 3-D Modeling of City Points of Interest. *IEEE Geosci. Remote Sens. Lett.* **2015**, *12*, 1670–1674, doi:10.1109/LGRS.2015.2418816.

35. Wielandt, E. On the validity of the ray approximation for interpreting delay times. In *Seismic Tomography*; Springer: Dordrecht, The Netherlands, 1987; pp. 85–98, doi:10.1007/978-94-009-3899-1_4.

Donadio, E.; Sambuelli, L.; Spanò, A.; Picchi, D. Three-Dimensional (3D) Modelling and Optimization for Multipurpose Analysis and Representation of Ancient Statues. In *Latest Developments in Reality-Based 3D Surveying and Modelling*; Remondino, F.; Georgopoulos, A.; González-Aguilera, D.; Agrafiotis, P.; Eds.; MDPI: Basel, Switzerland, 2018; pp. 95–118.

4D Reconstruction of Cultural Heritage Sites

Pablo Rodríguez-Gonzálvez [a,b], **Angel Luis Muñoz-Nieto** [a], **Susana del Pozo** [a], **Luis Javier Sanchez-Aparicio** [a], **Diego Gonzalez-Aguilera** [a], **Jon Mills** [c], **Karolina Fieber** [c], **Ian Haynes** [d], **Gabriele Guidi** [e], **Laura Micoli** [e] and **Sara Gonizzi Barsanti** [e]

[a] TIDOP Research group, High School of Ávila, University of Salamanca, Avila, Spain; (pablorgsf, almuni, s.p.aguilera, luisj, daguilera)@usal.es

[b] Department of Mining Technology, Topography and Structures, Universidad de León, Ponferrada, Spain; p.rodriguez@unileon.es

[c] School of Civil Engineering and Geosciences, Newcastle University, Newcastle upon Tyne, UK; (jon.mills, karolina.fieber)@ncl.ac.uk

[d] School History, Classics and Archaeology, Newcastle University, Newcastle upon Tyne, UK; ian.haynes@ncl.ac.uk

[e] Departament of Mechanical Engineering, Politecnico di Milano, Milan, Italy; (gabriele.guidi, laura.micoli, sara.gonizzi)@polimi.it

Abstract: Multi-temporal three-dimensional (3D) reconstructions are fundamental for the preservation and maintenance of all forms of tangible Cultural Heritage (CH) and often provide the basis for decisions related to interventions and promotion. Introducing the fourth dimension of time into 3D geometric modelling of real-world data allows the creation of a multi-temporal representation of a site to plan maintenance and promotion. This chapter aims to provide a comprehensive approach for CH time-varying representations, to integrate heterogeneous information derived from a range of sources to help inform understanding of temporal aspects of change across different working scales and environments. Two landscape-scale study cases, Hadrian's Wall (UK) and Milan Roman Circus (Italy), are presented based on a methodological approach for CH time-varying representations proposed by the JPI-CH European Project Cultural Heritage Through Time (CHT²). CHT² aims to provide a new set of tools and working methods to support the study of the evolution of CH sites.

Keywords: 4D modelling; cultural heritage; data fusion; historical data; knowledge representation

---

1. Introduction

Cultural Heritage (CH) encompasses both tangible assets (e.g., monuments, archaeological remains, artefacts, etc.) and intangible ones (e.g., traditions, social practices, rituals, etc.). This article focuses on tangible heritage and the creation of graphic representations that can significantly aid monitoring, management, routine maintenance, study and promotion of a CH site. In such a way, knowledge about heritage objects can be transmitted to future generations.

There is an increasing commitment to preserve and restore CH, thereby fostering its better management, study or promotion. CH is a rich legacy for the current generation who have an undeniable responsibility to preserve it. Tangible heritage becomes extremely important as a cultural, social and economic resource in modern societies. It is therefore necessary to continuously develop techniques in order to achieve a better understanding of its evolution through time and improve maintenance approaches. Research, conservation and restoration of CH assets are complex tasks that are being addressed from a multidisciplinary perspective: archaeologists, architects, art historians, surveyors, tourist promoters and advertising agents, amongst others. Since the footprint of time sometimes imposes terrible consequences on CH, it often becomes necessary to not only recover the memory of original features of historical buildings, urban and landscape environments, but also understand its likely evolution. In this way, heritage legacy can be safeguarded for present and future generations, preventing future damage and aiding understanding of the current remains as an evolution of its original state.

Due to the advancement in technology, research and innovation improvements are increasingly noticeable, not only in data acquisition, but also in the ability to include multiple complementary fields. Digital methods and techniques can link historical documentation data and disseminate them for a better understanding and perception of their evolution through time. With the aim of studying the current state and geometry of CH elements, numerous different geospatial technologies can be used, from airborne to ground level, such as Airborne Laser Scanning (ALS), Mobile LiDAR Systems (MLS) and Terrestrial Laser Scanning (TLS), aerial and terrestrial photogrammetry, Global Navigation Satellite Systems (GNSS), etc. Historical datasets are extremely heterogeneous in terms of chronology, shape, style and structure, appearing as texts, paintings, engravings, old photographs, maps, etc., and in analogue or digital formats. Therefore, it is necessary to establish schemes to order and clarify the current status regarding multi-source data acquisition and fusion for CH management at its different scales. It is also necessary to consider different data sources and their nature (e.g., metric or non-metric), as well as the final aim of four-dimensional (4D) reconstruction and visualization.

1.1. Related Works

4D analysis based on real-world data captured on-site has been carried out mainly at urban levels due to the general availability of historical aerial imagery that allows the analysis of urban transformation and 4D modelling (Patias et al., 2011; Adami, 2015). The automatic processing of historical aerial imagery is not a trivial task, since it involves the recovery of unknown parameters (Redecker, 2008), which could yield geometric errors (Nocerino et al., 2012), and therefore mislead subsequent 4D analysis. Besides reality-based models, in recent years reconstructive models have also assumed an interesting role, for the possibility to visualize architecture that no longer exists. This is achieved through an analytical process based on the integrated knowledge of historical sources and real three-dimensional (3D) data. Crowdsourced imagery is also now offering ways to create 3D models of different epochs from web-retrieved images of lost or altered CH assets (Kyriakaki et al., 2014) (Stathopoulou et al., 2015). For more than two decades, researchers have discussed the use of virtual reconstructions of environments that no longer exist as an instrument for the interactive interpretation of archaeological ruins or heavily stratified archaeological sites (Stanco and Tanasi, 2011), for the presentation of generic cultural sites (Rua and Alvito, 2011) and even for new archaeological discoveries (Frischer et al., 2008).

Several 4D projects have integrated 3D data capture of a contemporary scene with 3D data of the same site reconstructed from paintings representing rigorous perspective views (El-Hakim et al., 2008), or from plans giving the horizontal footprint of a building and drawings for reconstructing elevations (Russo and Guidi, 2011). Photogrammetry and TLS provide similar products in terms of accuracy, the point clouds being complementary (Bastonero et al., 2014). However, there are some inherent problems in TLS and digital image integration, such as accuracy preservation when different geometric resolutions are involved (Ramos and Remondino, 2015). An example of such multi-source data combination to generate a 3D metric reconstruction of an urban environment is given in Balsa-Barreiro and Fristch (2015), where the 4D component was added by historical image wrapping. Moreover, 3D data has been derived from written sources that describe different historical stages of a building (Micoli et al., 2013; Kresten et al., 2014), or using the size of actual excavated decorations and the knowledge of specific rituals for adding geometrical constraints to the reconstruction of a religious building (Guidi et al., 2014).

1.2. Paper Aims

The main focus of the research reported in this paper is to bring together heterogeneous information and expertise to offer a better understanding of 4D (3D + time) digital products of CH assets. This workflow and recommendations are framed into the Cultural Heritage Through Time (CHT²) project (CHT² 2017) (Rodríguez-Gonzálvez, et al., 2017a). CHT² not only aims at achieving the full integration of the temporal dimension and its management and visualization for evaluating CH elements through time, but also at creating a protocol to produce and optimize 3D/4D digital models of CH sites useful for architectural studies and analysis and research purposes (Rodríguez-Gonzálvez, et al., 2017b).

2. Theoretical Background and Methodological Approach

The analysis of CH elements or sites from a 4D perspective depends on both the working scale and the historical data available for each particular case study. Both the definition of working scale (Section 2.1) and the time varying representations (Section 2.2) are therefore reviewed in the following subsections.

2.1. Working Scale Definition

For large CH sites, at landscape scale, the optimal solution is invariably derived from multi-source data sensor integration (Guidi et al., 2009), where the different methods and techniques balance their own drawbacks to reach an efficient solution (Gonizzi Barsanti et al., 2012). An example of the spatio-temporal analysis of a rural landscape scale is provided by Modica et al. (2011) where several geomatics technologies were employed, one of the most efficient being remote sensing (Ratcliffe et al., 2004). For CH elements of reduced dimensions at architectural scale, such as buildings or archaeological remains, the reconstruction of temporal evolution is carried out on the basis of metric/non-metric data, such as historical drawings (Nocerino et al., 2014).

Multi-source data fusion is one of the main challenges in 4D reconstruction and visualization of CH. This implies providing a solution for the combination of mixed data sources (both metric and non-metric) with the aim of creating time-varying representations. For this purpose, the suitability of the different sources of metric data should be systemized according to the CH object size and its complexity. For a better understanding of the approach proposed by this paper to perform 4D CH analysis, it is helpful to group CH assets according to their characteristics. One of the most common classifications of CH studies (Kraak and Ormeling, 2011) is based on categorization according to the size of the element under study or scale range at the following levels: artefact, architectural, urban and rural landscapes (Table 1). This classification may seem unsophisticated, but the addition of other variables would complicate the classification and would result in confusion when describing the approach.

Table 1. Typical scale ranges for each of the cultural heritage (CH) categories commonly established (adapted from Kraak and Ormeling, 2011).

CH Category	Scale Ranges
Artefact	From 1:1 to 1:5
Architectural	From 1:10 to 1:100
Urban landscape	From 1:100 to 1:1 000
Rural landscape	From 1:100 to 1:5 000

Here, the different categories of CH assets are specified in order to assist inventory compilers and users in determining the appropriate procedures to be followed. Leaving aside "artefact" scale (although commonly considered in the field of CH, it has not been included within the range of scales studied in this paper), three main categories of CH assets have been established: "rural landscape", "urban landscape" and "architectural" scales. The term "rural landscape" is applied to those cases where an extensive rural area exists. If the study case has a similar extension area to rural landscape cases, but instead of a rural environment it exists in urban space, the term "urban landscape" is adopted. Finally, "architectural" scale refers to those cases where the object in question focuses on a larger scale than in the two previous cases and where the Z axis is predominant (buildings, monuments, sculpture statues, etc.).

2.2. Time-Varying Representations

CH reconstruction can cover a large variety of situations, depending on the type of temporal analysis that is required (Figure 1).

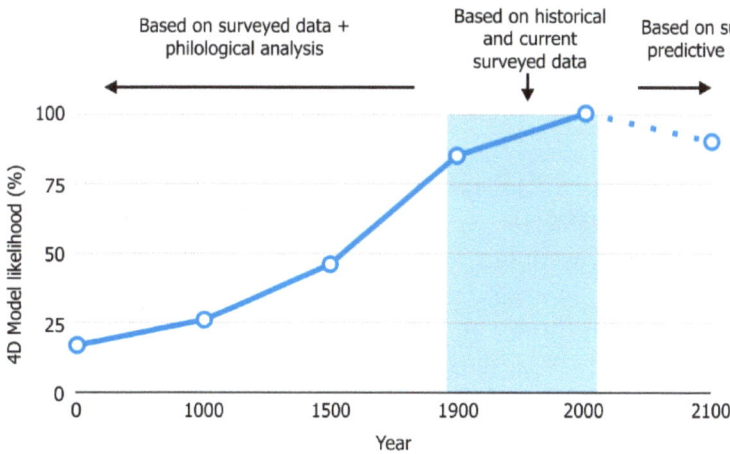

Figure 1. Three cases of time-dependent 3D analysis covered within the framework of the analysis of our legacy through time.

With reference to Figure 1, it is possible to identify three types of 4D models that can be produced for any CH asset:

a. Reconstruction of the diachronic evolution of a structure or environment that does not exist any longer, possibly with the exclusion of a few archaeological remains. This is generally associated with lost heritage, where partial traces arrived in the current age and will require the combination of surveyed data and philological analysis to reconstruct;

123

b. Reconstruction of the diachronic evolution of a structure or environment based on the historical analysis of data acquired on physically accessible assets at the current or a previous time (the blue highlighted area in Figure 1);

c. Prediction of the diachronic evolution of a structure or environment into the future based on the historical analysis of data acquired on physically accessible assets at different times.

Although all cases involve diachronic representation, the first and second cases (a and b) differ in how 3D data can be collected, implying different concepts of 3D data integration:

1. In case a, the creation of time-varying documents or representations is based on the use of mixed sources, both metric and non-metric. For example, 3D reconstruction of a Roman building at different stages according to both the 3D scanning of its remains and historical sources;

2. In cases b and c, the creation of the time-varying representations is based on rigorous metric 2D or 3D data, derived from a wide variety of sensors in physically accessible scenarios at different stages. For example, 3D models acquired at different times of a collapsing building surveyed during its deterioration.

2.3. Multi-Source Data Integration

As commented previously, the methodological procedure required to analyze CH elements or sites from a 4D perspective depends on both the working scale and the historical data available for each particular case study. Figure 2 shows the processing tasks of the workflow approached from a spatial dimension perspective (2D or 3D data format). However, the data collection step, which includes both historical and current data, is described from a different point of view based on the its main source of data. Thus, as detailed in the following point, data collection has been classified into (i) geometrical, (ii) thematic, (iii) historical/cultural, and (iv) non-conventional sources.

A total of four main phases were considered, comprising data collection, processing, fusion, and analysis of results. For an in-depth description of the different phases, please refer to CHT² methodology definition (CHT², 2017).

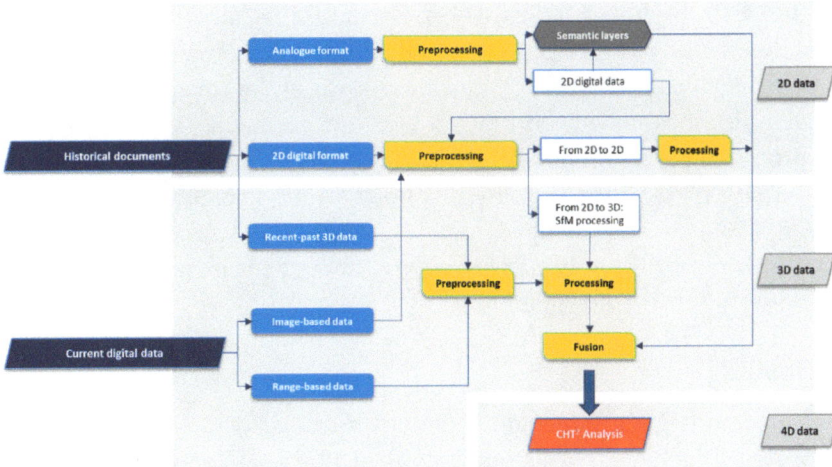

Figure 2. Simplified diagram of the methodology proposed for the integration of CH data in order to perform 4D analysis (CHT² 2017).

Due to the variability of data sources, and in order to provide a more compact summary than is shown in Figure 2, all available information sources can be classified from a quantitative and qualitative point of view and with regard to its geometrical character (Rodríguez-Gonzálvez, et al., 2017a). In the case of the rural landscape, geometrical data is provided mainly by active or passive airborne techniques. On the one hand, aerial photogrammetry can be considered as the main source due to its longest temporal span dating back to the mid-19th Century up until the present day. On the other hand, although relatively new, ALS may provide more robust measures of a rural landscape as it can penetrate vegetation cover, thereby providing more detailed information about terrain. In addition, it is possible to obtain information from satellite imagery, which can provide information at different spectral wavelengths whilst covering vast territories. Moreover, the use of miniature Unmanned Aerial Vehicles (UAVs) since the beginning of the 21st Century provides flexibility (mission planning, height and frequency of flights) and high resolution data. For its part, the common attribute of historical and cultural sources is the subjectivity of the information with no rigorous geometrical representation (graphical information such as map, drawing or paintings) or even any kind of associated geometry (literature, texts, etc.).

The urban landscape scale offers the possibility of using a greater variety of sensors to perform data acquisition to document the current state of CH assets, and with the option of mounting instruments on both terrestrial and aerial platforms. It is possible to obtain information from ALS and TLS, as well as from more innovative systems such as MLS. MLS should be highlighted within the active techniques for urban landscape scales due to its exclusivity of use for the working scales selected in Section 2.1 (Rodríguez-Gonzálvez, et al., 2017b). However, the temporal range of available historical data is limited. Meanwhile, and with greater temporal significance, it is possible to collect a variety of images (single, stereo or multi-image networks) from both aerial and terrestrial platforms for subsequent off-line processing. As in the rural landscape scale, the common attribute here is the subjectivity of the information with no rigorous geometrical representation. Compared with rural landscapes, besides historical documents, drawings and paintings, it is noteworthy that in urban environments it is more probable to find engravings as graphical information.

3. Case Studies

The methodology and recommendations to collect actual and historical data, and its fusion to achieve an historic reconstruction is tested at landscape scales, according to two different CHT[2] case studies and two different perspectives: Hadrian's Wall (Fieber et al., 2017) and Milan Roman Circus (Micoli et al., 2017). In the first case, a 3D approach was employed due to the available historical sources, while in the second study, the nature of available historical information favored a 2D analysis on the basis of geographical information systems.

3.1. Rural Landscape: Case of Hadrian's Wall

Around AD 122, the Roman Emperor Hadrian ordered a wall to be built, dividing Britain in two. It stretches over 117 km (80 Roman miles) from Bowness on the River Solway on the north-west coast of England to Wallsend on the River Tyne in the north-east. Hadrian's frontier system is complemented by a sophisticated system of outposts and coastal watch stations, thereby offering a remarkable glimpse of ancient society. The Wall today (Hadrian's Wall, 2017) is a designated UNESCO World Heritage Site and its surrounding landscape is very different to that of Roman times. Although significant portions are still visible, the remaining fabric and landscape of the Wall are subject to various modern-day stresses, for example tourism, urban development and natural hazards. Due to the length of the Wall, three different sites that characterize natural hazard phenomena were selected as study cases to apply the strategies developed in Section 2:

- *Beckfoot Roman Fort*: The fort (Bibra), civilian settlement and cemetery at Beckfoot (Historic England, 2017a), located to the south-west of the main Wall, is preserved as a cropmark. Analysis of aerial photographs reveals clear details of building outlines (Figure 3a) and partial excavation indicates extensive survival of building and defensive wall foundations. The Beckfoot site has been subject to significant coastal erosion, with archaeology buried under sand dunes being frequently exposed by coastal processes.

- *Birdoswald Roman Fort*: Birdoswald (Historic England, 2017b) is one of the best preserved of the 16 forts built as part of the Hadrian's Wall frontier system. The monument includes the Roman Fort (Figure 3b) and the section of Hadrian's Wall and vallum between the River Irthing to the east and the field boundaries east of Harrow Scar milecastle 50 to the west. The excavated area of the fort is open to visitors but is at critical risk to landslides instigated by fluvial erosion from the River Irthing running along the south of the site.
- *Corbridge Roman Station*: The Roman town of Corbridge (Historic England, 2017c) marks the site of the most northerly urban settlement in the Roman Empire. The streets of the town have been excavated and are open to the public. The landscape is subject to fluvial flood hazard from the River Tyne to the south (Figure 3c).

(a)   (b)   (c)

Figure 3. (a) Beckfoot Fort outlines seen from the air, with a beach and sand dune complex in the foreground, viewed from the west c.1949 (Ref: NY/0848/A). (b) Birdoswald Fort c. 1947 (Ref: NY/6166/A). (c) Corbridge Roman Town seen from the air, viewed from the northeast c. 1945 (NY/9864/A). All images are from University of Newcastle upon Tyne Air Photograph Collection, copyright Cambridge University.

In the first instance, an extensive archive search was conducted in order to identify existing materials, with particular focus on historic aerial photographs, including the Historic England Archive and Newcastle University's archives. Although more data were available, three epochs of aerial photograph datasets were initially selected for each of the sites: one from the 1940s, one from the 1980s or 1990s and one modern dataset acquired in 2016.

The archival photography was subsequently digitized at high resolution (600 dpi for the earliest datasets, 2000 dpi for 1980s/90s) by the Historic England Archive (note that digitisation was not performed using a photogrammetric scanner, and the resultant imagery is therefore likely subject to distortions; see, for example, Thomas et al., 1995). Archival aerial photography was complemented by other data provided by English Heritage and Newcastle University, including historic topographic maps, topographic and geophysical surveys and drawings.

Furthermore, LiDAR point cloud data were obtained for all three sites from the Environment Agency's repositories (Environment Agency, 2017) and an additional LiDAR survey of Birdoswald acquired in 2008, being in possession of Newcastle University (Miller et al., 2012), was added to the time series (Figure 4).

Figure 4. Airborne laser scan of Birdoswald Roman Fort and surroundings (2008).

The 2016 photography was provided by Historic England in digital format for both the Birdoswald and Corbridge sites. Due to the fact that such photography did not exist for Beckfoot at the time of the archive search, an UAV survey was performed in order to complete the time series for that site. The Beckfoot UAV survey was conducted using a Quest300 fixed-wing UAV (QuestUAV Ltd., Amble, UK). This UAV can carry a maximum 5 kg of payload with approximately 15-minute flight time (Peppa et al. 2016). Two gimballed Panasonic DMC-LX5 digital cameras (Panasonic UK Ltd.) were mounted on-board the UAV at the time of survey. One Panasonic camera was a standard camera sensitive to visible light (RGB), whilst the second was modified to be sensitive to near infrared (NIR) wavelengths (three NIR bands). The camera's main specifications are shown in Table 2. The flight height of 91 m and with nominal pixel size of 2 μm enabled a 4 cm ground sampling distance for the acquired images.

Table 2. Camera and survey specifications.

Camera	Sensor Size [mm]	Sensor Size [pixels]	Focal Length [mm]	Average GSD [mm]	Images	Approx. Area [km]
DMC-LX5	8.07 × 5.56	3648 × 2736	5	~40	546	1.0 × 0.5

With a maximum of 15-minute flight time, and operating within Civil Aviation Authority restrictions relating to the use of UAVs, the immediate Beckfoot survey area was divided into two overlapping parts (north and south) and surveyed under two separate sorties. The camera was set up with a fixed shutter speed of 1/800 s to decrease image blurring, a fixed aperture of f/2 and ISO 100. The exposure interval was set to 2 seconds and the side overlap to 80%. A total of 27 circular control point targets of 0.40 m in diameter were evenly distributed over the site (except for the north-east corner where no access to the fields was granted) and were surveyed in GNSS rapid static mode (five-minute observations), which delivered 3D accuracy at mm-level relative to the GNSS base station. The GNSS base station was established on one of the fields and observed in GNSS static mode for approximately four hours, delivering 1 cm planimetric and 2 cm vertical absolute accuracy. Moreover, Ground Reflectance Calibration Targets were also placed on site at the time of survey to enable calibration of RGB and NIR imagery.

Both archival and current-day photography were processed to generate 3D point clouds. Since archival imagery was not scanned on a photogrammetric scanner and the camera calibration information was unavailable, the structure-from-motion (SfM) pipeline was adopted in order to obtain digital surface and terrain models for each site and to facilitate multi-temporal landscape comparisons. Agisoft Photoscan software was used for this task. Point clouds for each individual epoch were geocoded using selected terrain features, considered to have remained stable over the years, or measured Ground Control Points (GCPs) in the case of the Beckfoot 2016 survey data. As can be seen from the internal Photoscan quality assessment of the absolute orientation results, presented in Table 3, the initial orientations of the archival imagery were generally found to be worse than for current-day datasets (particular evident in Table 3 as illustrated for the Beckfoot site since the 2016 epoch comprised the UAV survey). This can be attributed to a number of factors, including the degradation of the image quality over time, the lack of camera calibration data, quality of A/D conversion, sub-optimal network configurations for SfM processing, availability of fewer ground control points, and so on. Moreover, the dense image matching routines used in the SfM pipeline often performed poorly on archival datasets. This can be seen, for example, in the 1946 Corbridge dataset where data voids are clearly visible in the model (Figure 5a). Such data voids are particularly apparent in the middle of open fields where high quality image texture is lacking.

(a)

(b)

Figure 5. (a) 1946 dense point cloud of Corbridge landscape and (b) Visualisation of 2016 Corbridge landscape generated in Geovisionary. Both cases were generated from archive photographs supplied by Historic England/English Heritage.

Table 3. Photoscan quality assessment of absolute orientation for three epochs of data at Beckfoot. .

Year	X Rmse [m]	Y Rmse [m]	Z Rmse [m]	3D Rmse [m]	No. of GCPs Used in Orientation
1948 B/W	1.425	1.512	1.755	2.720	16
1991 B/W	0.731	1.177	2.741	3.071	15
2016 RGB	0.016	0.015	0.020	0.029	27

As a result of the sub-optimal archival dataset orientation results, alignments were refined using various data fusion approaches, including both in-house surface matching algorithms (Miller et al., 2008) and ICP (Besl and McKay, 1992) routines in the OPALS software (Pfeifer et al., 2014), to ensure rigorous registration from epoch to epoch, thereby generating spatially consistent 3D time series for subsequent 4D CH analysis. Whilst the gently undulating landscapes of the Beckfoot and Corbridge sites lack relief and therefore do not readily lend themselves to such data fusion approaches, the application of surface matching was found to improve the registration of multiple epochs in the majority of cases. An example is illustrated in Figure 6, where the improvement in the co-registration of the 1946, 1984 and 2016 epochs for the Corbridge site is visually evident (the section is taken through an area believed to be stable, i.e., unchanged through time). It is noteworthy that the processing of the 1984 Corbridge dataset was particularly problematic, the DEM displaying artefacts characteristic of SfM and dense image matching application such as apparent "doming" (see, e.g., James and Robson, 2014) and data voids in areas of low image texture. As a result, some residual errors are still apparent even after application of the ICP algorithm. Finally, data were collated in ArcGIS (ESRI Inc.) and Geovisionary 3 (Virtalis) software to enable visualisations (e.g., Figure 5b) and 4D CH analysis.

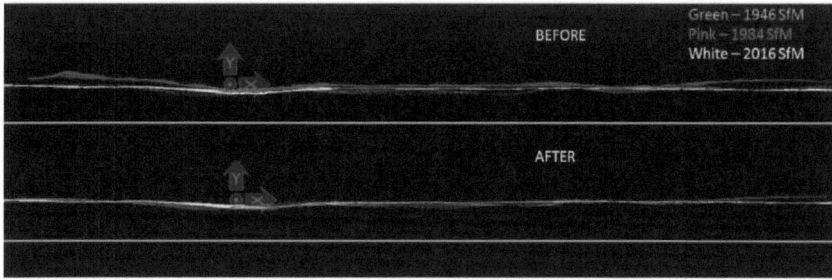

Figure 6. Example cross-section taken through stable ground at the Corbridge study site, before and after application of ICP-based matching.

3.2. Urban Landscape: Case of Milan Roman Circus

The case study covers a south-west area of the city centre of Milan (Italy) that corresponds to the Roman Circus. On this zone, it is possible to see several traces of the different historical periods from ancient times until the densely urbanized structure of the present day. According to archaeological finds, Milan has been inhabited since the 5th Century BC in the area corresponding to the current Via Meravigli, Via Valpetrosa, Piazza del Duomo. In this zone, some protohistoric tracks converge, with traces of the following Roman roads still recognizable in the city plan.

In 2nd to 1st Centuries BC, excavation works and levelling were undertaken, in order to adapt the ground to the Roman urban model. At that time, the first urban planning was conceived, probably maintaining the road network. In 286 AD, with the tetrarchy subdivision, Milan became the capital of the Western Roman Empire, under the Emperor Maximian. During the imperial period, up to 402 AD, the area was modified by the construction of major buildings such as the Imperial Palace, the Circus and the defensive walls (Mirabella Roberti, 1984). The Circus was the open-air venue for chariot and horse races, or rather the place dedicated to the celebration of the Emperor's greatness and for this reason it was generally located near the Imperial Palace (Humphrey, 1986). Milan's Circus was also adjacent to the defensive walls with which it shared the western part. Although the Circus of Milan was one of the most important of the empire, today only few traces are still visible: a tower of the city walls, a tower of the Carceres reused as a bell tower of the Monastero Maggiore after 1500, and some sections of the walls or foundations in the private properties nearby, sometimes hidden in their interiors or in the basements (De Capitani D'arzago, 1939).

Historical sources report the existence of the Circus until Longbard's era. From that period, as with other monuments in Milan, the materials of the Roman structures were re-used in the construction of other buildings. Many questions are

131

still open about the building's elevation and its relation to the surrounding area: the imperial palace and the town's fortification walls.

The area of the Circus includes the Church of San Maurizio; during the 1800s, its main cloister and the buildings connected to the opening of two new roads (Via Luini and Via Anasperto) were destroyed. Moreover, further serious damage occurred during the World War II bombings of August 1943. Nowadays, only small portions of the monument remain visible and a lot of historical documentation was lost in a fire during World War II. Since the 1960s, the area of the complex has been occupied by the Archeological Museum of the Municipality of Milan (Civico Museo Archeologico di Milano) and residential buildings (Capponi, 1998). Small remains of the Circus are still visible in the basements of modern buildings in that area.

The information gathering process involved different sources to be integrated in order to hypothesize a reconstruction of the area:

- *Historical*: involves both non-graphic (mainly bibliographic resources) and graphical ones (drawings and historical painting, maps, and photographs).
- *Current*: 3D survey.

The first kind of source taken into account is the *bibliographic resources* to document the history of the city and the monument. In particular, historical texts were considered, from which it is possible to infer useful information about life, state of the monument and reports of archaeological excavations past and present. Especially important for the study of the monument was the text of the archaeologist De Capitani D'Arzago, who thoroughly studied the Roman Circus of Milan in the late 1930s, confirming the existence, location and essential size of it thanks to the discovery of the parallel walls, some portion of the foundations and a large part of the curve. His planimetric reconstruction of the monument is shown in Figure 7.

Another step was a collection of maps, drawings and images concerning the various topics covered in the research. *Drawings and historical paintings* are fundamental to gather information no longer available today. When possible, data for individual monuments are searched. If these are not available, as in the case of the studied area, historical representations of elements of the same typology and age are searched.

This kind of approach is useful to have typological indications and to validate the reconstructive hypotheses proposed by scholars. Unfortunately, in the case of Milan, only poor graphical representations of the monuments involved were available, with reference to their active period. With regard to more recent times, all the drawings of survey campaigns carried out in the area have been collated.

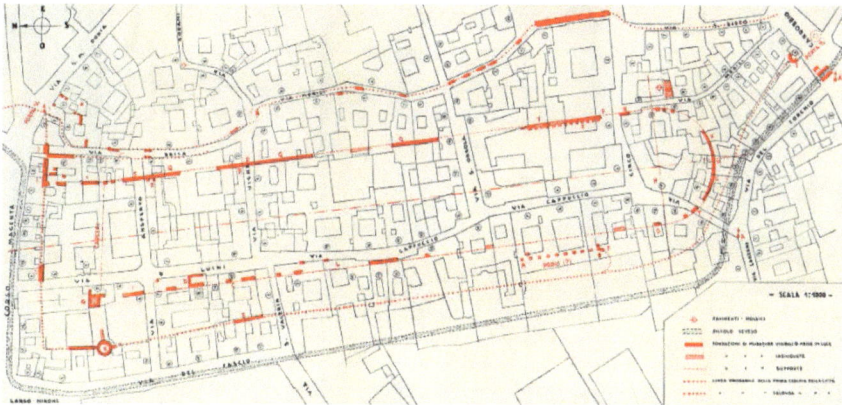

Figure 7. Plan of Milan Circus according to archaeological studies conducted by De Capitani D'Arzago, 1939. Legend: full red pattern: foundations of visible walls or put into light; obliquely red dashed line pattern: identified masonry foundations; red dashed lines pattern: supposed masonry foundations.

In the post-war period, many buildings destroyed by bombing were rebuilt. In this phase, the excavations for the foundations of modern buildings have, in some cases, revealed archaeological findings that were used as the basement for the analyzed building itself. In other cases, for example in correspondence of new roads or other unbuilt areas, such findings were simply covered underground. In the latter cases, the survey drawings made during excavations, as for the example shown in Figure 8.a, were fundamental to derive information about archaeological remains no longer accessible today.

A deep iconographic research was then carried out, collecting also different *maps* from various periods that highlight the urban structure of the area. About 60 city maps representing different historical periods from the Renaissance to the present day have been identified at the Civica Raccolta delle Stampe Achille Bertarelli and analyzed to study the evolution of the urban area.

Another type of data are photographs taken mainly during the post-WW2 excavations. Images of artefacts and structures inside the urban area, taken from different perspectives and sometimes referring to two or more different periods of their life, provide valuable support for the 3D reconstruction process. Specifically, a search of the photographic archive at the Superintendence's office was made with regard to the area of interest. About one thousand images were found and about 100 of them were selected for inclusion in the study. This selection regards artefacts visible during construction projects (e.g., the metro, new skyscrapers) or inspections of the Superintendent. These images provide valuable documentary heritage because many artefacts are no longer visible, embedded in the foundations of modern buildings (Figure 8b).

(a)   (b)

Figure 8. (a) Drawing of an archaeological excavation carried out at Via Circo 14 in 1949 and (b) Elevated view of the remains of the three pillars of the Circus foundations, Via Circo in 1959. Both images are courtesy of Soprintendenza Archeologia della Provincia di Milano.

Another stage of the work relates to the 3D survey of all the remains still visible in order to have a starting point for the reconstruction (Table 4). Please note that the full image resolution was not employed in the 3D photogrammetric reconstruction due to memory limitations. Currently, the survey works of the visible portions of the monument have been performed only inside the archaeological museum in Milan. The results, shown in Figure 9, consist of the 3D digitization of one of the towers of the "Carceres" of the Circus, nowadays used as the bell tower of the church dedicated to San Maurizio, and the so-called "polygonal tower", belonging to the city walls when the Circus was in activity.

Table 4. 3D sensors and survey specifications used for the Roman Circus of Milan.

Building	Building Size [w, d, h] [m]	Measurement Principle	Sensor	Images	GSD [mm]	Average 3D Resolution [mm]
Tower of the *Carceres*	8.6 × 7.8 × 25.8	Phase-shift laser scanning	Faro Focus 3D	26	-	~50
Polygonal tower of the city walls	11.3 × 11.3 × 19.8	Photogrammetry	Canon 5D Mk II	187	2.5 to 16	~35

Due to the wide extent of the site, the presence of remains, and their state of conservation, the opportunity to perform a 3D survey is being assessed with the Superintendent. All the remains identified by the archaeologists will be deemed relevant to the Circus and useful to its digital reconstruction. Depending on the conditions of operation, such 3D digitization will be made with terrestrial photogrammetry or laser triangulation, depending on the available conditions such as lighting, working space, etc., according to the CHT² methodology.

The monument portions detected, suitably georeferenced, will serve to validate the archaeological excavations of the past and will provide the main constraints with which to create the 3D reconstruction. In addition to the validation of historical plans, the 3D portions are fundamental as elements of proportion, in relation to the examples of the same type of monument highlighted by other sources, to define the trend elevation of the building, typically the most critical parameter in the reconstruction of any ancient building that no longer exists.

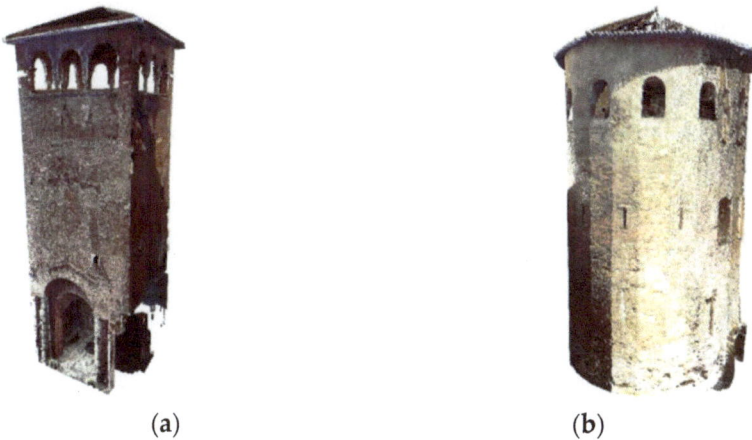

(a)   (b)

Figure 9. 3D point clouds: (a) tower of Carceres, (b) polygonal tower of city walls adjacent to the Circus, still visible in the garden of the Civico Museo Archeologico di Milano.

Integrating such different data as those available for the reconstruction of the Roman Circus of Milan needed to start from a metric framework, adding all the various information in a controlled destination environment. In addition, the same environment should include functions for georeferencing each part. For this reason, QGIS was used as such a destination environment, using the georeferenced survey data of the municipality of Milan as the starting point. The reorientation of the raster image representing the scanned archaeological map was made with a Helmert transformation that performs simple scaling and rotation, similar to the process carried out in the reconstruction of the San Giovanni in Conca Basilica (Guidi and Russo, 2009; Russo and Guidi, 2011). With the same approach, the footprints of the two towers shown in Figure 9 have been georeferenced.

135

This starting material has been imported in the Rhinoceros CAD system where the contours of the Circus and the key elements of the surroundings have been redrawn according to the previous and most recent surveys. The presence of such 3D surveys, in addition to the "architectural grammar", typical of analogous buildings designed and built in the same period, provide crucial information regarding the elevation of the whole structure, the number of seats possibly present inside the hippodrome, the consequential best estimate of access points for letting the crowd enter and exit the structure, the positioning of the Emperor's special seats, from which he could see the most crucial points of the whole Circus for supervising the various operations before and after races (Mirabella Roberti, 1984; Humphrey, 1986).

This 3D reconstruction hypothesis is therefore built up by adding layers of information to the starting raw reconstruction, according to the best match of archaeological evidence. This produces a loop between the phases of modelling and archaeological checking, that guarantee—as already proved in previous projects (Guidi et al., 2014)—a productive interdisciplinary exchange between the various professionals involved, ranging from surveyors, 3D modellers and archaeologists.

Although, at present, such reconstruction is being created for the time period when the Circus was fully functioning as the most important structure for horse races in the city (2nd to 3rd Century AD), another two periods will be considered: the pre-Maximian era (pre-2nd Century AD), and the current time, that will help in understanding the relationship between this important piece of Milan's CH, and its current structure.

4. Conclusions and Future Perspectives

Representing the relationship between time and space provides a powerful mechanism to visualize and communicate CH. It can be useful not only to study and analyze the past, but also to foresee possible risks in the future. The aim is to provide a comprehensive overview for conducting studies of CH assets over time for different working scales and environments. It serves as an initial guide for organizing all tasks, from collecting historical documentation, acquiring current data, processing to visualizing results. It is oriented to perform studies through time, including the monitoring of CH assets that may or may not still exist but whose temporal evolution has left remains on the current landscape.

The developed methodology for integrating multifarious data has been applied to two different case studies at the landscape scale: rural and urban. The rural study, due to the nature of CH assets, necessitates the use not only of geospatial technologies such as aerial photogrammetry, but also geophysical approaches such as ground penetrating radar. Moreover, current data acquisition is aimed not only at providing 4D reconstructions that inform cultural dissemination, but also for the preservation of the present remains in the face of coastal and fluvial evolution of the landscape. To this end, the 4D models produced will help inform appropriate numerical models to predict landscape evolution in the future. In the urban case, the geo-referencing of the hidden assets plays a relevant role in obtaining precise measurements of the shape and length of the structures. By merging data related to the current state of the monument with the vast archival material collected, a rearrangement of the historical representations will be made, for example the normalization of historical plans into a uniform scale. From such an integrated base of information, a 4D reconstruction will be carried out together with archaeologists, in order to better identify the true reconstruction of the ancient building and any changes that affected the area from their origin until the present time.

Meanwhile, due to the wider scope of the topic addressed, the presented methodology is open to augmentation. This is especially the case in those methodology stages that contain greater ambiguity and variability in order to accommodate the requirements of each specific case study and data to be collected. Moreover, application at other working scales, such as the architectural level, is also possible.

Acknowledgments: The authors wish to acknowledge the support of the European Union's Joint Programming Initiative on Cultural Heritage (JPI-CH), for funding the project Cultural Heritage Through Time (CHT²), through national research authorities: UK Arts and Humanities Research Council (AHRC, award number AH/N504440/1), Spanish Ministry of Economy and Competitiveness, and Polish Ministry of Culture and National Heritage. The authors would also like to acknowledge the support of English Heritage and Historic England in the provision of aerial photography and other archival materials for the Hadrian's Wall case study. Furthermore, special acknowledgement is addressed to the Soprintendenza Archeologia della Provincia di Milano and to the Civico Museo Archeologico di Milano for their valuable cooperation in searches.

References

1. Adami, A. 4D city transformations by time series of aerial images. *Int. Arch. Photogramm. Remote Sens. Spat. Inf. Sci.* **2015**, *40-5(W4)*, 339–344.
2. Balsa-Barreiro, J.; Fritsch, D. Generation of 3D/4D photorealistic building models. The testbed area for 4D Cultural Heritage World project: The historical center of Calw (Germany). In Proceedings of the 11th International Symposium on Visual Computing, Las Vegas, NV, USA, 14–16 December 2015; Springer: Berlin, Heidelberg, Germany, 2015; pp. 361–372.
3. Bastonero, P.; Donadio, E.; Chiabrando, F.; Spanò, A. Fusion of 3D models derived from TLS and image-based techniques for CH enhanced documentation. *ISPRS Ann. Photogramm. Remote Sens. Spat. Inf. Sci.* **2014**, *2*, 73–80.

4.   Besl, P.J.; McKay, N.D. A method for registration of 3-d shapes. *IEEE Trans. Pattern Anal. Machine Intell.* **1992**, *14*, 239–256.

5.   Capponi, C. *San Maurizio al Monastero Maggiore in Milano*; Silvana Editoriale: Milan, Italy, 1998.

6.   CHT², 2017. Cultural Heritage through Time. Available online: http://cht2-project.eu.

7.   De Capitani D'arzago, A. *Il Circo Romano*; Casa Editrice Ceschina: Milano, Italy, 1939.

8.   El-Hakim, S.; Lapointe, J.-F.; Whiting, E. Digital Reconstruction and 4D Presentation through Time. In Proceedings of the SIGGRAPH 2008, Los Angeles, CA, USA, 11–15 August 2008.

9.   Environment Agency, 2017. Lidar Data. Available online: https://data.gov.uk/publisher/environment-agency.

10.   Fieber, K.D.; Mills, J.P.; Peppa, M.V.; Haynes, I.; Turner, S.; Turner, A.; Douglas, M.; Bryan, P.G. Cultural heritage through time: A case study at Hadrian's Wall, United Kingdom. *Int. Arch. Photogramm. Remote Sens. Spat. Inf. Sci.* **2017**, *42-2(W3)*, 297–302.

11.   Frischer, B.; Dakouri-Hild, A. *Beyond Illustration. 2D and 3D Digital Technologies as Tools for Discovery in Archaeology*; BAR International Series 1805; Archeopress: Oxford, UK, 2008.

12.   Gonizzi Barsanti, S.; Remondino, F.; Visintini, D. Photogrammetry and Laser Scanning for Archaeological Site 3D Modelling—Some Critical Issues. In Proceedings of the 2nd Workshop on 'The New Technologies for Aquileia', Aquileia, Italy, 25 June 2012.

13.   Guidi, G.; Russo, M. Diachronic representation of ancient buildings: Studies on the "San Giovanni in Conca" Basilica in Milan. *Disegnare Con* **2009**, *2*, 69–80.

14.   Guidi, G.; Remondino, F.; Russo, M.; Menna, F.; Rizzi, A.; Ercoli, S. A multi-resolution methodology for the 3D modeling of large and complex archaeological areas. *Int. J. Arch. Comput.* **2009**, *7*, 39–55.

15.   Guidi, G.; Russo, M.; Angheleddu, D.; 3D Survey and Virtual Reconstruction of Archeological Sites. *Digit. Appl. Archaeol. Cult. Heritage* **2014**, *1*, 55–69.

16.   Historic England, 2017a. Beckfoot Roman Fort. Available online: https://historicengland.org.uk/listing/the-list/list-entry/1007170.

17.   Historic England, 2017b. Birdoswald Roman Fort. Available online: https://historicengland.org.uk/listing/the-list/list-entry/1010994.

18.   Historic England, 2017c. Corbridge (Corstopitum) Roman Station. Available online: https://historicengland.org.uk/listing/the-list/list-entry/1006611.

19.   Humphrey, J.H. *Roman Circuses: Arenas for Chariot Racing*; B.T. Batsford Ltd.: London, UK, 1986; pp. 18–53, 579–638, 687–691.

20.   James, M.R.; Robson, S. Mitigating systematic error in topographic models derived from UAV and ground-based image networks. *Earth Surf. Process. Landf.* **2014**, doi:10.1002/esp.3609.

21.   Kersten, T.P.; Hinrichsen, N.; Lindstaedt, M.; Weber, C.; Schreyer, K.; Tschirschwitz, F. Architectural Historical 4D Documentation of the Old-Segeberg Town House by Photogrammetry, Terrestrial Laser Scanning and Historical Analysis. In Proceedings of the 5th Euro-Mediterranean Conference, Limassol, Cyprus, 3–8 November 2014; pp. 35–47.

22. Kraak, M.J.; Ormeling, F. *Cartography: Visualization of Spatial Data*; Guilford Press: New York, NY, USA, 2011.

23. Kyriakaki, G.; Doulamis, A.; Doulamis, N.; Ioannides, M.; Makantasis, K.; Protopapadakis, E.; Hadjiprocopis, A.; Wenzel, K.; Fritsch, D.; Klein, M.; et al. 4D reconstruction of tangible cultural heritage objects from web-retrieved images. *Int. J. Heritage Digit. Era* **2014**, *3*, 431–451.

24. Micoli, L.; Guidi, G.; Angheleddu, D.; Russo, M. A multidisciplinary approach to 3D survey and reconstruction of historical buildings. In Proceedings of the Digital Heritage International Congress (DigitalHeritage), Marseille, France, 28 October–1 November 2013; pp. 241–248.

25. Micoli, L.; Gonizzi Barsanti, S.; Guidi, G. Interdisciplinary data fusion for diachronic 3D reconstruction of historic sites. *Int. Arch. Photogramm. Remote Sens. Spat. Inf. Sci.* **2017**, *42-2(W3)*, 489–494.

26. Miller, P.; Mills, J.; Edwards, S.; Bryan, P.; Marsh, S.; Mitchell, H.; Hobbs, P. A robust surface matching technique for coastal geohazard assessment and management. *ISPRS J. Photogramm. Remote Sens.* **2008**, *63*, 529–542.

27. Miller, P.; Mills, J.P.; Barr, S.L.; Birkinshaw, S.J.; Hardy, A.J.; Parkin, G.; Hall, S.J. A remote sensing approach for landslide hazard assessment on engineered slopes. *IEEE Trans. Geosci. Remote Sens.* **2012**, *50*, 1048–1056.

28. Roberti, M.M. *Milano Romana*; Rusconi Immagini: Santarcangelo di Romagna, Italy, 1984; pp. 8–40, 63–68, 78–84.

29. Modica, G.; Vizzari, M.; Pollino, M.; Fichera, C.R.; Zoccali, P.; Di Fazio, S. Spatio-temporal analysis of the urban–rural gradient structure: An application in a Mediterranean mountainous landscape (Serra San Bruno, Italy). *Earth Syst. Dyn.* **2012**, *3*, 263–279.

30. Nocerino, E.; Fiorillo, F.; Minto, S.; Menna, F.; Remondino, F. A Non-Conventional Procedure for the 3D Modeling of WWI Forts. *Int. Arch. Photogramm. Remote Sens. Spat. Inf. Sci.* **2014**, *40*, 457–464.

31. Nocerino, E.; Menna, F.; Remondino, F. Multi-temporal analysis of landscapes and urban areas. *Int. Arch. Photogramm. Remote Sens. Spat. Inf. Sci.* **2012**, *39*, 85–90.

32. Patias, P.; Kaimaris, D.; Stylianidis, E. Change Detection in Historical City Centers Using Multi-Source Data: The Case of Historical Center of Nicosia—Cyprus. In Proceedings of the XXIII CIPA Symposium, Prague, Czech Republic, 12–16 September 2011.

33. Peppa, M.V.; Mills, J.P.; Moore, P.; Miller, P.E.; Chambers, J.E. Accuracy assessment of a UAV-based landslide monitoring system. *Int. Arch. Photogramm. Remote Sens. Spat. Inf. Sci.* **2016**, *42*, 895–902.

34. Pfeifer, N.; Mandlburger, G.; Otepka, J.; Karel, W. OPALS—A framework for Airborne Laser Scanning data analysis. *Comput. Environ. Urban Syst.* **2014**, *45*, 125–136.

35. Ramos, M.M.; Remondino, F. Data fusion in cultural heritage-a review. *Int. Arch. Photogramm. Remote Sens. Spat. Inf. Sci.* **2015**, *50-5(W7)*, 359–363.

36. Ratcliffe, I.C.; Henebry, G.M. Using Declassified Intelligence Satellite Photographs with Quickbird Imagery to Study Urban Land Cover Dynamics: A Case Study from Kazakhstan. In Proceedings of the ASPRS Annual Conference, Denver, CO, USA, 23–28 May 2004.
37. Redecker, A.P. Historical aerial photographs and digital photogrammetry for impact analyses on derelict land sites in human settlement areas. *Int. Arch. Photogramm. Remote Sens. Spat. Inf. Sci.* **2008**, *37*, 5–10.
38. Rodríguez-Gonzálvez, P.; Muñoz-Nieto, A.L.; del Pozo, S.; Sanchez-Aparicio, L.J.; Gonzalez-Aguilera, D.; Micoli, L.; Gonizzi Barsanti, S.; Guidi, G.; Mills, J.; Fieber, K.; et al. 4D reconstruction and visualization of cultural heritage: Analyzing our legacy through time. *Int. Arch. Photogramm. Remote Sens. Spat. Inf. Sci.* **2017**, *42-2(W3)*, 609–616.
39. Rodríguez-Gonzálvez, P.; Jiménez Fernández-Palacios, B.; Muñoz-Nieto, A.L.; Arias-Sanchez, P.; Gonzalez-Aguilera, D. Mobile LiDAR System: New Possibilities for the Documentation and Dissemination of Large Cultural Heritage Sites. *Remote Sens.* **2017**, *9*, 189.
40. Rua, H.; Alvito, P. Living the past: 3D models, virtual reality and game engines as tools for supporting archaeology and the reconstruction of cultural heritage—The case-study of the Roman villa of Casal de Freiria. *J. Arch. Sci.* **2011**, *38*, 3296–3308.
41. Russo, M.; Guidi, G. Diachronic 3D reconstruction for lost Cultural Heritage. *Int. Arch. Photogramm. Remote Sens. Spat. Inf. Sci.* **2011**, *38-5(W16)*, 371–376.
42. Stanco, F.; Tanasi, D. Experiencing the past: Computer graphics in archaeology. In *Digital Imaging for Cultural Heritage*; CRC Press: New York, NY, USA, 2011; pp. 1–37.
43. Stathopoulou, E.K.; Georgopoulos, A.; Panagiotopoulos, G.; Kaliampakos, D. Crowdsourcing lost cultural heritage. *ISPRS Ann. Photogramm. Remote Sens. Spat. Inf. Sci.* **2015**, *2*, 295–300.
44. Thomas, P.R.; Mills, J.P.; Newton, I. An investigation into the use of Kodak Photo CD for digital photogrammetry. *Photogramm. Rec.* **1995**, *15*, 301–314.

Rodríguez-Gonzálvez, P.; Muñoz-Nieto, A.L.; del Pozo, S.; *et al.* 4D Reconstruction of Cultural Heritage Sites. In *Latest Developments in Reality-Based 3D Surveying and Modelling*; Remondino, F.; Georgopoulos, A.; González-Aguilera, D.; Agrafiotis, P.; Eds.; MDPI: Basel, Switzerland, 2018; pp. 119–140.

Optimization of Three-Dimensional (3D) Multi-Sensor Models For Damage Assessment in Emergency Context: Rapid Mapping Experiences in the 2016 Italian Earthquake

Giulia Sammartano

Politecnico di Torino, Department of Architecture and Design, Torino, Italy;
giulia.sammartano@polito.it

Abstract: Geomatics techniques offer the chance to manage very cost-effective solutions for three-dimensional (3D) modelling, from both the aerial and terrestrial point of view, with the help of range and image-based sensors. 3D spatial data that is based on integrated documentation techniques, featured by a very high-scale and an accurate metric and radiometric information nowadays are proposed here as metric databases that are applicable for assisting the operative fieldwork in the case of rapid mapping strategies. In sudden emergency contexts for damage and risk assessment, the structural consolidation and the security measures operations meet the problem of the danger and accessibility constraints of areas, for the operators, as well as to the tight deadlines needs in first aid. The use of Unmanned Aerial Vehicles (UAVs) equipped with cameras are more and more involved in aerial survey and reconnaissance missions; at the same time, the ZEB1 portable Light Detection and Ranging (LiDAR) mapping solution implemented in handle tools helped by Simultaneous Localization And Mapping (SLAM) algorithms can help for a quick preliminary survey. Both of these approaches that are presented here in the critical context of a post-seismic event, which is Pescara del Tronto (AP), deeply affected by the 2016-2017 earthquake in Central Italy. The Geomatics research group and the Disaster Recovery team (DIRECT—http://areeweb.polito.it/direct/) is working in collaboration with the Remotely Piloted Aircraft Systems (RPAS) group of the Italian Firefighter.

Keywords: emergency UAV mapping; SLAM; cultural heritage risk; multi-sensor documentation; multi-scale modelling

1. Introduction

Currently, the contribution of the latest Geomatics research in the field of rapid mapping strategies offers the chance to manage very cost-effective solutions for three-dimensional (3D) metric and radiometric documentation, with very high-scale from both the aerial and the terrestrial point of view. The recently implemented approaches in the Geomatics research, together with the commonly explored ones, can be developed for an effective and operative fieldwork, especially in sudden emergency contexts for damage and risk assessment, security measures, and structural consolidation intervention. In fact, in most of the urban centers where the built heritage retrieves the demand of damage documentation and data management and sharing as efficiently as possible, are now encountering the problem of the danger limitation and accessibility constraints of the areas for the operators as well as to the tight deadlines needs in first aid. The solutions to be investigated seem increasingly to be the ones that allow for rapid mapping using both image-based and range-based approaches.

Imagery methods based on sets of collected images following photogrammetric methods that can now rely on the exploitation of image matching and Structure from Motion (SfM) algorithms, which are derived from the computer vision field, can be considered nowadays as a quick means for a low-cost mapping and 3D reconstruction.

The use of Unmanned Aerial Vehicles UAVs equipped with cameras are more and more involved in aerial survey and reconnaissance missions, and they are behaving in a very cost-effective way in the direction of 3D documentation and preliminary damage assessment. More and more UAV equipment with low-cost sensors must become, in the future, suitable in every situation of documentation, but above all, in damages and uncertainty frameworks. Rapidity in acquisition times and low-cost sensors are challenging marks, and they could be taken into consideration maybe with time spending processing.

Even range-based methods based on Light Detection and Ranging (LiDAR) scans are an increasingly used solution but considering the time needed for recording, registering, filtering, and decimation of points clouds, as well as the construction of the 3D mesh, the Terrestrial Laser Scanning (TLS) technique is used when strictly necessary. TLS provides extremely detailed and accurate outputs, but is definitely heavy and less sustainable in relation to photogrammetry. Some mobile scanning solutions are being recently developed for indoor/outdoor environment mapping. These portable systems are based on Simultaneous Localization And Mapping (SLAM) technology and allow for quickly collecting a big amount of 3D metric data, in the form of point cloud, not only for the building scale, but also for surrounding context objects in an environmental documentation scale.

This paper wants to base its proposal on the possibility of analysis, interpretation, and classification of metric and non-metric withdrawable information, starting from 3D aerial and terrestrial high-scale models. These attained starting metric data offer the possibility to carry out a global assessment on them and on their ability to reach a level of detail and consequently meet the most suitable scale. After their metric accuracy control, it should be possible to use these 3D models and their geometries as suitable for further uses as the structural analysis one.

This process is planned in the direction of quick surveys and targeted operation helpful to assess the state of conservation and seismic damage in these types of building belonging to a precarious context. The selection of survey methods depends primarily on these issues, and then it is influenced by the needs of information detection on those models. Metric data and its extraction constitutes the unavoidable base on which to base non-metric information, as visual and qualitative ones. The aim of this research is to evaluate the contribution of terrestrial and aerial quick documentation based on image-based and range-based techniques for quick 3D modelling, gathered specifically here by a SLAM based portable LiDAR and a multirotor UAV equipped by camera, for a first reconnaissance inspection and modelling in terms of level of details, metric, and non-metric information.

The test case is an experience of Cultural Heritage documentation specifically in disaster areas in the center of Italy, carried out by Politecnico di Torino from August 2016 up to now. In these areas, a strong earthquake occurred in August 2016, and current ongoing seismic shocks still take place. The moment magnitude of this event is listed as 6.0 by INGV (Istituto Nazionale di Geofisica e Vulcanologia, http://cnt.rm.ingv.it/), which places the hypocenter depth of the event at 8 km. Two earthquakes occurred after the first event in 24 August: an event on 26 October, with a 5.9 moment magnitude, and another one (the largest event) on 30 October with a 6.5 moment magnitude. In this scenario, quick surveys, damage assessment, and diffuse and urgent measures of safety and consolidation were required. This study is carried out specifically on some selected buildings in Pescara del Tronto (Ascoli Piceno), where a multi-sensor 3D survey was performed in repeated missions in September, October, December, and February.

2. Multi-Sensor Documentation in Emergency Areas

2.1. Methodological Framework

In recent years, we can affirm that the increasing use of Remoted Piloted Aircraft System (RPAS) equipped with cameras have already improved its role in competitiveness and efficiency in surveying operation on the field. The

143

phenomenon is nowadays covering many application domains (e.g., geomatics, geotechnics, archaeology, forestry, structural analysis, etc.). The current aspect is also connected to the diffusion of a large number of operators (not expert people as well) and concurrently to the efficient image-matching algorithm based of SfM photogrammetry even more suitable for 3D information extraction from images and frames videos that are acquired by compact cameras that are embedded in low cost commercial drones. The use of drones equipped with cameras and/or high-definition video devices, along with on-board GPS systems, have thus grown in its recognized role for aerial documentation finalized to metric survey purposes. It is moving in a very profitable way in many contexts due to low-cost and non-hazardous characteristics, and, as is well known in recent research experiences, is very popular for the 3D documentation and monitoring Cultural Heritage sites (Chiabrando et al., 2016a; Ruiz Sabina et al., 2015; Lerma et al., 2012; Remondino et al., 2011).

In particular, photogrammetry by drones is proving increasingly crucial in emergency situations, not only for remote observation of sites and emergency action planning for the collection of qualitative information, but also for production of very large scale metric data (Boccardo et al., 2015; Wang et al., 2013; Hirokawa et al., 2007).

During the last ten years, the use of drones became equally diffuse in urban areas involved in natural disaster for preliminary search and rescue, Building Damage Assessment (BDA) (Meier, 2016; Fernandez et al., 2015). This is sometimes preferred more than the more traditional vertical images from remote sensing data. Due to their scale, their geometric configuration, and ultimately their intrinsic features, satellite imaging does not satisfy the requirements of details and information (Lemonie et al., 2013; Rastiveis et al., 2013; Gerke and Kerle, 2011).

The first data acquisition phase, to be done as quickly as possible after a disaster, is ordinarily carried out in person with many efforts by technicians in damaged sites, and it is a heavy time-consuming operation. Rapidity in acquisition times must be effectively balanced with post processing times, as well as the time-cost ratio must be successful, in favour of low-cost sensors with their top efficiency, as the best possible compromise between timeliness and accuracy (Lemonie et al., 2013). It must maximize, ultimately, the density of data and the metric data extraction from 3D models that are processed ex post, without neglecting productivity in ex ante data acquisition in emergency circumstances, where the practicability of spaces that are connected to the high risk they could cause, can adversely affect the quality of data. Moreover, 3D models already are, and could be increasingly in the near future, an essential platform of dense information for interdisciplinary teamwork on the object of study, which will be analysed in remote, in a second step, for many purposes. They could be, for example: emergency measures of rescue and second aid for civil protection and firefighters, or for damage detection and structural assessment, and for planning a structural strengthening project, or even historical documentation studies, as well as restoration analysis and intervention.

2.2. Regulations on Damage Assessment via Documentation Techniques

A 3D survey for damage sites needs some important considerations in terms of time-costs, as well as human involvement, on balancing the acquisition phase resources and the processing ones. Traditional survey techniques have restrictions in different issues. Ground-based mapping is complex, sometimes dangerous for expert operators and data acquisition is largely limited to terrestrial point of view and façade information. On the other hand, medium range image-based mapping is typically restricted to vertical views for the roof condition, but for collapses, lower levels of damage are much harder to map, because such damage effects are largely expressed along the façades, which are not visible in such imagery (Gerke and Kerle, 2011).

Most operational post-disaster damage mapping, such as the processing of satellite data, acquired through the International Charter *"Space and Major Disasters"* (https://www.disasterscharter.org), remains based on visual interpretation (Voigt et al., 2011; Kerle, 2010). Anyway, also high-resolution satellite images, due to their scale, their geometric configuration, and ultimately their intrinsic features, do not respond comprehensively to the demand of details and information for the scale and complexity of urban context to a clear identification of the damage (Lemonie et al., 2013; Rastiveis et al., 2013; Gerke and Kerle, 2011). Multi perspective oblique imagery, as it is known in literature, seems to be the profitable solution to maximize detail on buildings. For the 3D data processing, many ways can be followed: from a merely visual one for qualitative information, to a manual one, managed by operator on points clouds and 3D models, up to an automatic data extraction. Automatic image analysis techniques for BDA can be broadly grouped into pixel and object-based methods (Fernandez et al., 2015).

For damage assessment, we have thus to chase the solution of integration and/or fusion of nadir oblique cameras where possible, integrated by terrestrial information only if necessary. These multi-sensors models are a kind of complex informative database that must approaching to be a final-use based model: metric and non-metric information define the geometric and conservative characterization. For example, the structural analysis and damage assessment on masonries, in this case, for post-earthquake contexts, establishes the setting up of damage scenarios on the preliminary interpretation and evaluation of visible damages features on the objects. The classification on EMS-98 scale (European Macroseismic Scale) of damage is based on building types. Particularly, for each type of masonry structure, depending on the employed materials and constructive techniques, an empiric vulnerability class is associated (descending scale from A to E) and a level of damages is categorized observing the structure (from 1 to 5) (Figure 1).

Figure 1. Classification of damage to masonry buildings from EMS-98 scale (European Macroseismic Scale) (Grunthal, 1998).

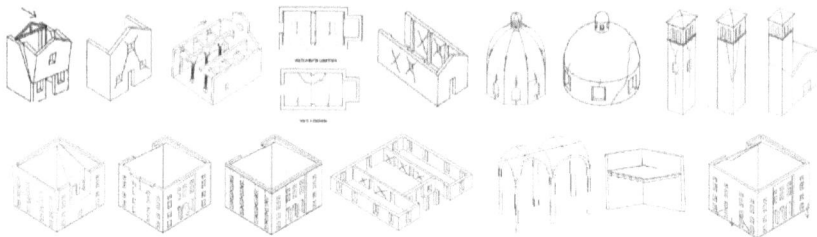

Figure 2. Excerpts from "Abacus of collapse mechanisms" in case of damages assessment of Cultural Heritage, datasheet A-DC "Churches" and B-DP "Buildings" (DPCM 23/02/2006).

The Italian regulation takes into account through several updated regulations over the years the survey finalized to the damage mapping and indexing in a standardized format. The *AeDES* datasheet *(C.N.R.-G.N.D.T. and S.S.N. in Department of Civil Protection),* is the 1st Level datasheet and has been organized in order to help the damage detection and assessment, emergency response, and practicability, for ordinary buildings in post-earthquake emergence. More specifically, it can be possible to approach to monumental buildings with the datasheet for the damage regarding palaces B - DP (2006) and churches A - DC (2011). According to the Ministerial datasheet A-DC and B-DP, the ability to define the structural behavior of the masonries building with its elements is described in the Figure 2 for "Churches" and "Buildings". The potential of metric and non-metric information enclosed in high-definition 3D point clouds and models completed by quick mapping is useful to reach this type of level of detail, and potentialities are many and very remarkable in relation to the classification of potentially recognizable damage to a first visual analysis of artefacts.

3. Testing Rapid Mapping in Pescara Del Tronto after 2016 Earthquake

After the earthquake occurred in August 2016 in the center of Italy, DIRECT Team (Disaster Recovery Team) from Politecnico di Torino in cooperation with the

GEER team (Geotechnical Extreme Events Reconnaissance Association), and with the Remotely Piloted Aircraft Systems (RPAS) group of the Italian Firefighters were involved in several reconnaissance and metric survey missions. In September, October, November 2016, and February 2017, the goals were the rapid mapping of villages and a focused multi-sensors documentation of buildings in urban areas that were deeply repeatedly damaged by shocks. Numerous villages had been involved, such as Pescia, Pescara del Tronto, Cittareale, Accumoli, Norcia, Castelluccio, Amatrice, etc.

The present contribution is focused on the ancient perched village of Pescara del Tronto (Figure 3), in Arquata del Tronto municipality. The techniques that were deployed by the teams have been focused on the documentation post-disaster of the whole village with multi-scale approach and resulting assorted multi-sensor acquisitions. Groups of differently damaged buildings were examined, with the combined use of terrestrial and aerial sensors.

Then, a test area was chosen in the northern part in the specific focus of a stand-alone damaged building, an integrated metric survey has been conducted and was tested in order to evaluate both multi-sensor model information and processing resources. Moreover, after a subsequent earthquake event occurred in the end of October and with the contribution of the mission in December a new 3D model post-event has been available.

Figure 3. Pescara del Tronto in Arquata del Tronto municipality (Ascoli Piceno), is one of the perched villages allocated across the Apennines mountain in Marche region (left, center), center Italy, on the border with regions Abruzzo, Lazio, Umbria. It was destroyed by the reiterated seismic shocks from

It is important to underline that, due to the sequence of earthquakes, the access at many damage sites of interest remains difficult because the sites were (and still are) located in restricted red zones, and are dangerous because many of the structures are unstable and are still prone to collapse. Notwithstanding the site challenges that have made us think about sensor choice and acquisition planning, the higher part of Pescara del Tronto village was accessible. Terrestrial LiDAR technique was excluded regardless, due to the emergency purposes of the mission in those sites. Therefore, for these problematic sites, the preferred approach to investigate their damage involved the use of photogrammetric-based acquisition using UAV, which were integrated with traditional close-range acquisition systems wherever feasible. Next to these ones, a quick SLAM based LiDAR scan was tested.

First of all, in order to define a common reference system for the RPAS and terrestrial acquisition, Global Navigation Satellite System (GNSS) and total station measurements were performed. As Ground Control Points (GCPs), several markers were placed on the area and were then measured using the GNSS in Real Time Kinematic (RTK) mode (Figure 4). Together with the GNSS measurements, some natural points using a total station side shot approach were measured on the façades of the damaged buildings in order to be used during the photogrammetric process. The measured points were georeferenced in a common reference system (UTM-WGS 84 Fuse 33 N ETRF 2000) using the information derived from the Italian Dynamic permanent network controlled by the Italian Geographic Military Institute (IGM, http://www.igmi.org/).

The whole area in Pescara del Tronto involved in the metric survey was covered by almost 40 GCP materialized on ground by targets as Figure 4. Three of these targets where positioned in the neighborhood of the damaged building. Moreover, for each building that was measured and imaged by multi-sensor acquisitions, a set of natural points were detected on buildings. For the buildings blocks points have been measured on the main elements, as the roof edges, the façade, the windows and the door, and the stone elements are useful for both the aerial and the close-range blocks.

Figure 4. Survey operation during the setting up of the topographic measurements. A moment of the aerial markers planning and placing on ground with fire fighters (left). Global Navigation Satellite System (GNSS) measurements on target disseminated in more or less accessible areas in the neighborhood of the damaged building, with the Leica System 1200 GNSS receiver (center left and right), Leica TS06 total station topographic survey (right).

3.1. UAV Platforms for High-Scale Aerial Documentation

The teams incorporated multiple aerial platforms and image-based sensors including COTS (commercial off the shelf) platforms and a customized professional UAV fixed-wing platform (Figure 5). The acquisition strategy for each system varied based on its strengths and capabilities, but provided a wide range of remote sensing data that can be used for subsequent focused processing and analysis. The aim is to test them in different scenarios and setting acquisitions to evaluate their contribution of aerial imaging, by nadir and oblique points of view, in the 3D definition of the building geometry, in terms of details and metric/radiometric information to be extracted too (Chiabrando et al., 2017). Moreover, a georeferencing strategy for the ZEB mobile mapping system point clouds is proposed. It is based on the geometrical features provided by the high-scale aerial point clouds.

Figure 5. The fleet of aircrafts used in Pescara del Tronto by the Disaster Recovery Team (DIRECT) Team in cooperation with Firefighters Remotely Piloted Aircraft Systems (RPAS) corp. The Phantom 4 DJI (top left) flying (third from right image), the Ebee by Sensfly (down left) at the take-off phase (second from right); the customized Inspire 1 by DJI, property of Firefighters (first from right).

3.1.1. A Fixed Wing UAV: eBee by Sensfly

The first flight over the settlement of Pescara del Tronto was performed with an eBee™ small UAV platform, manufactured by Sensefly and commercialized in Italy by Menci Software (http://www.menci.com) (Figure 5). In this case, the UAV was equipped with a digital camera Canon Power Shot S110™, which offers a 1/1.7″ Canon CMOS sensor, 12 MP images, and a focal length of 5.2 mm. The platform is extremely manageable and very useful for rapid map realization in emergency (Boccardo et al., 2015). The eBee system is certified by ENAC as EBM-1539 and it is approved as inoffensive. In order to cover all the damaged areas (almost 83 ha) three flight with the following characteristics were realized: mean flight height 150 m; expected Ground Sample Distance - GSD= 0.05m; Side overlap 60%; Forward overlap 85%.

3.1.2. Multirotor Aircrafts: Phantom 4 and Inspire 1 by DJI

The other flights over Pescara del Tronto were performed using multirotor aircrafts that were deployed for a closer acquisition on focused buildings blocks (Figure 5).

A Phantom 4™ quadrotor small UAV, manufactured by DJI (*Dà-Jiāng Innovations Science and Technology Co., Ltd Technologies*) has been used. The Phantom 4 is equipped with a 4K video camera that has a 1/2.3″ CMOS sensor, 94-degree field of view, 12.4 MP images, and a focal length of 20mm. The Phantom 4 system weighs 1.38 kg, has a maximum flight time of 28 min, and offer the ability to hover and/or collect imagery from vertical faces. Those flights were performed manually with an experienced UAV operator. Imagery from the UAV was transmitted to the operator in real time, ensuring significant image overlap, while maneuvering the UAV to capture the skewed imagery from objects of interest. This approach was also used successfully following the 2014 Iquique earthquake (Franke et al., 2016). The flight with the DJI was performed at an elevation ranging from 10 m to 20m. The data were acquired by UAV (Figure 7), specifically, following the two available approaches: the 4K video recording (and then the frame extraction) and the single shooting set-up, that allow for acquiring nadir and oblique images and videos. A set of 64 nadir images were acquired by DJI camera; moreover, almost 140 frames were extracted from the 08:19 min video with 29 frames/sec (1frame/3.5sec).

Furthermore, a quadcopter INSPIRE1 by DJI, customized by the SAPR team of Italian Firefighters group was used. The weight of about 3kg with (maximum weight at take-off 3.5 kg) and an extreme adaptability, manually piloted or with planned flight, offers a flight range of 20 minutes. It is equipped by the Camera ZENMUSE X5, CMOS sensor, focal length 15mm F/1.7-F/16, field of view 72°, for 4K video and images 16 MP (4608x3456), and offers oblique and nadir images acquisition.

3.2. Hand-Held Mobile Mapping Scanner: ZEB1 by GeoSLAM

In a post-disaster scecatnario and with the need to evaluate the contribution of different methods for the documentation of the damage, a 3D mobile mapping system (Figure 8) was tested. Among the available alternatives, we opted for the hand-held Zeb1 system by Geoslam (http://geoslam.com/).

This device (Figure 8) consists in a two-dimensional (2D) lightweight time-of-flight scanner with 30 m outdoor maximum range (*Hokuyo scanner*) and an Inertial Measurement Unit (IMU), which roughly ensure the altitude. They are both mounted on a spring so that when the operator moves in the environment to be mapped, the device swings freely and randomly determining the 2D scanning plane invests the environment generating a 3D point cloud (Bosse et al, 2012). The mapping system is based on the SLAM technology, which is the mostly suitable for indoor environments since it uses the environment geometric features to update the position of the device (Riisgaard and Blas, 2005). ZEB1 uses the raw trajectory to roughly calculate the surface normals and potential constraints (features recognition) within a single sweep of the scanner. Then, a *cloud-to-cloud* registration of the point cloud profiles generate the 3D cloud using an iterative process, which relies on geometric objects and features within the scans. The importance of the features constrains, essential to align subsequent scans, is visible in Figure 8, which shows how the quality of the recording is less accurate when the operator has moved away from the building. Although the system is provided for outdoor/indoor mapping, the processing presented in the next paragraphs show different accuracies and the level of detail for different portions of mapped objects, as other tests performed (Thomson et al., 2013). The consideration that the cloud was acquired in few minutes shows the remarkable interest of this system.

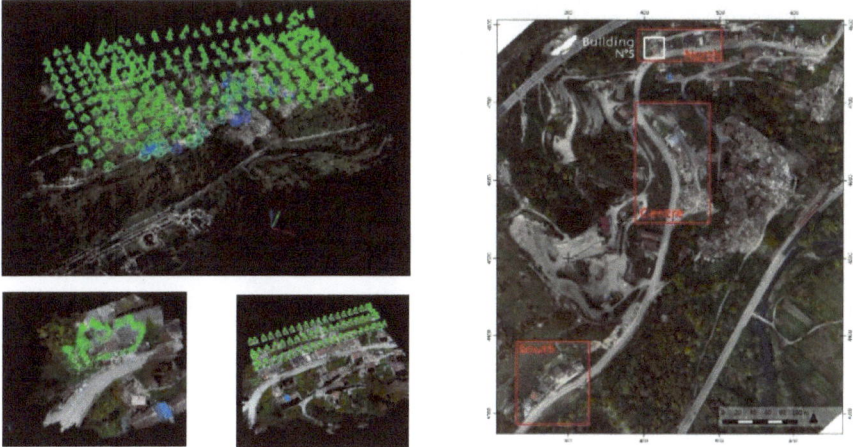

Figure 6. The eBee flight on Pescara del Tronto, organized in three grid blocks: the bigger longitudinal one on the village, and two lateral ones, perpendicular to the mail grid (top left). (down left) An example of oblique (left) and nadir grid (right) acquisition by multirotor DJI. The three test areas are pointed out on the eBee aerial orthoimage, GSD of 5 cm (right), in which the integrated survey by ZEB1 scans and UAV photogrammetry was performed and focused on building n°5 (white square).

Figure 7. Samples datasets of acquired images by flights. North area, Phantom4 (first row) and South area, Inspire1 (second row).

3.3. Point Clouds Data Processing and Integration for Rapid Mapping in Three Test Areas

In order to test a rapid mapping method in this critical area, three areas that were characterized by different geometric conformation characteristics have been chosen (Figure 6). Meanwhile, a post disaster damages survey and assessment, together with structural analysis, were developed by experts.

Due to the extremely difficulty of the site, the topographic and GNSS survey had to be flexible and rapid. Single alone Total Station vertices were set for the detail measurements on spot buildings all over the village of Pescara del Tronto; it were located on aerial targets placed on the surrounding, and measured using the GNSS in RTK mode. So, the whole data processing is directly affected by the precision of the topographic measurements, bound by the precarious area.

For the first processing step each data collection from different sensors were processed in separate blocks:

- *Terrestrial close-range photogrammetry performed walking around the buildings blocks*
- *Fix wing UAV nadir cameras (150 m altitude) generate a 3D photogrammetric model of the whole city.*
- *Multirotor nadir and oblique cameras (variable altitude) focused on the buildings blocks*
- *ZEB1 survey around buildings and blocks.*

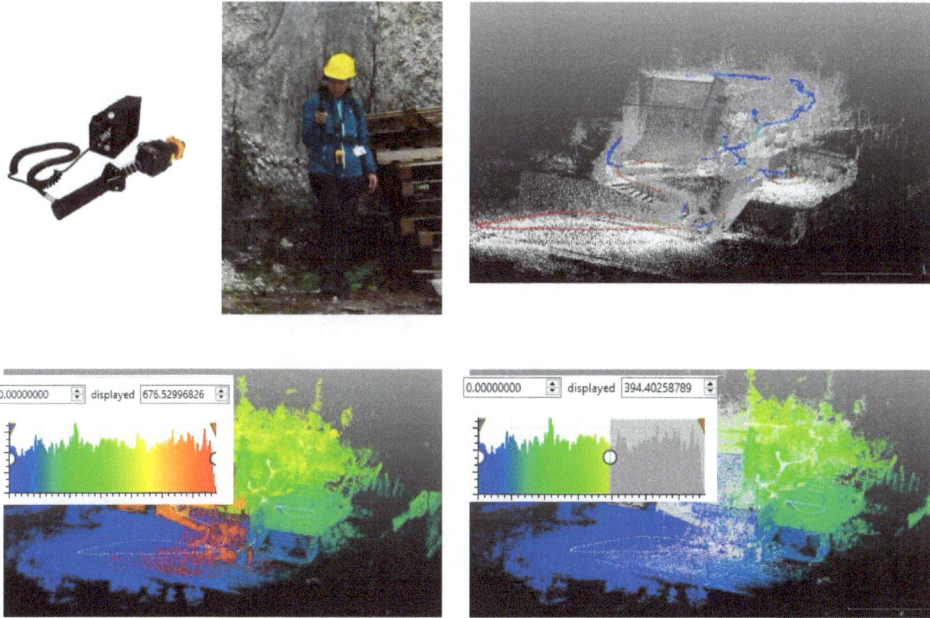

Figure 8. (Top left) The portable ZEB1 system and the operator during an acquisition. (Top right) The raw b/w point cloud and its path, in scalar colour to indicate the quality of registration. (Down left) The point cloud visualization in scalar colour according to "time" variable (seconds unit); in white colour the path. (Down right) A selected portion of point cloud at the beginning of the acquisition: the first six minutes.

For the processing of the photogrammetric blocks three commercial software were tested, with different dense image - matching algorithms: Pix4D (https://pix4d.com) by EPFL Innovation Park, workflow is based on a Structure From Motion approach (Strecha, 2014); Context Capture by Bentley System (https://www.bentley.com), Photoscan Pro by Agisoft (http://www.agisoft.com). For point cloud treatment, optimization, 3D modelling and analysis 3DReshaper (http://www.3dreshaper.com/) by Tecnodigit-Hexagon and open source Cloud Compare (http://www.danielgm.net/cc) were employed. A workstation with high performance hardware was exploited: CPU: Intel(R) Core i7-6800k 3.4 GHz. RAM 128 GB. NVIDIA quadro M2000.

3.3.1. Aerial Photogrammetry Point Clouds

The aerial acquisition, which was performed with different aircrafts, covered two main detail requirements and, as a consequence, two different scales, as it is pinpointed in Table 1. *eBee* complete cameras block was processed to produce a 3D model as an overview of the village in its environment according to a landscape scale (Figure 6). Time of processing of almost three hours provided a high quality point cloud, 3D model, and DSM (15Gb data), as we can see in Table 1, with GSD is almost 5cm. Aerial images, originating by multirotor UAV, from both acquisition by cameras positions and frames extracted from HQ video, have been oriented with GCPs, and the achieved models (almost 10 hours) offer an accuracy of a mean value of 0.095 m on Ground Control Points (GCPs) and an average value of 0.025 m on Check Points (CP)s.

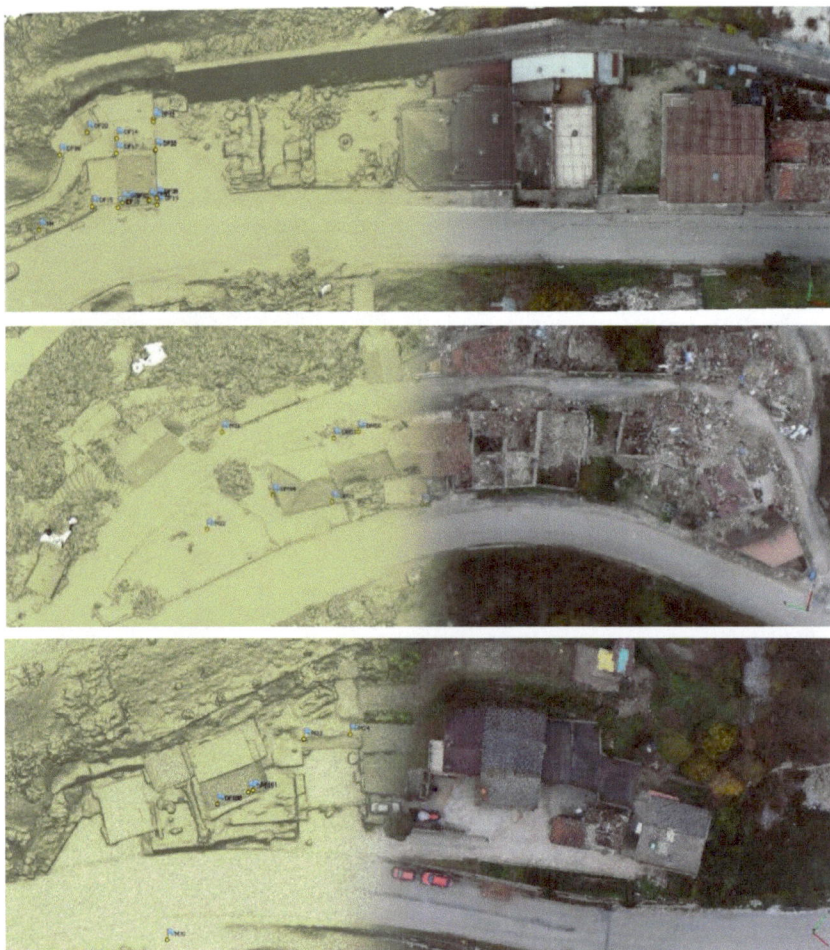

Figure 9. The polygonal surface model of the dense point cloud generated with Photoscan workflow from the UAV images, without and with RGB information. (From the top) the Northern, the Central, the Southern blocks.

High scale aerial models, with a mean GSD of 1.3 cm/pixel, have been calculated (Figures 6 and 9):

- NORTHERN AREA (*Buildings n°5 and n°6*), a moderately flat area with stand-alone buildings facing the street. A huge rock wall stands behind the buildings block, on which the viaduct is set.
- CENTRAL BLOCK (*n°4*), a big block bounded between two semi-parallel streets connecting north and south village areas with a difference in height of about 2m.
- SOUTHERN AREA (*Buildings n° 1, n° 2 and n° 3*), a significantly complex context, with a group of buildings arrayed on a difference in height of about 7 m. Behind a very impassable space with thick vegetation.

Table 1. Synthesis table of aerial datasets, processing, Root Mean Square Error (RMSE) on Ground Control Points and Check Points.

		Data Processing Results about Aerial Photogrammetric Blocks		
	EBEE	**PHANTOM 4 DJI**	**INSPIRE 1 DJI**	
		North	*Center*	*South*
Camera config.	nadir	nadir+oblique	nadir+oblique	nadir+oblique
Area Covered (m²)	830,000	6,390	17,100	13,100
Flight altitude (m)	150	35	95	55
Cameras n°	354	380	256	291
Image resolution	3000x4000			
av. GSD (cm/px)	5	1.11	1.65	1.18
tie points	1,505,299	264,452	402,924	313,981
Dense point cloud	104,733,709	27,802,455	28,853,157	21,219,411
RMSE on GCPs(m)	**0.026**	**0.0198**	**0.0194**	**0.0202**
X	0.022	0.0227	0.0208	0.0196
Y	0.024	0.0151	0.0136	0.0174
Z	0.032	0.0216	0.0237	0.0235
RMSE on CPs (m)	**0.036**	**0.0428**	**0.0174**	**0.0190**
X	0.031	0.0248	0.0215	0.0164
Y	0.016	0.0135	0.0062	0.0100
Z	0.058	0.0903	0.0245	0.0306

3.3.2. ZEB1 Point Clouds

In Pescara del Tronto, ZEB1 instrument was tested in many areas with groups of buildings, with different configurations in the acquisition trajectory. In these test areas, different point clouds (Table 2) were processed by automatic SLAM Cloud-to-cloud registration in GeoSLAM *pay-as-you-go* cloud processing. The crucial point of this technology is the control of the trajectory during the movement, which is estimated and corrected because of the 3D cloud that was acquired using a variation of traditional ICP (iterative closest point) scan-matching (Bosse et al., 2009). The development of the system has taken advantage of the opportunity to help the correction by the execution of closed loop trajectories during the mapping path, which also leads to better assess to the overall quality of the final 3D cloud.

The marketed system guarantees an absolute accuracy of position variable between 3 and 40 cm depending on the type of environment that is mapped (http://geoslam.com/).

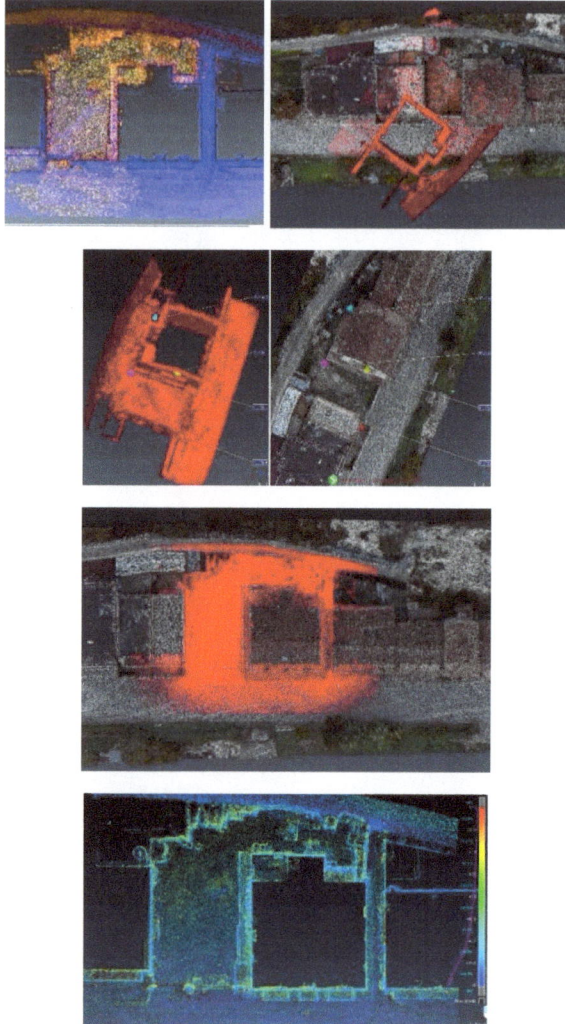

Figure 10. The process of registration and georeferencing of the ZEB point cloud through the fitting algorithms with the aerial point cloud. The co-registration of two ZEB clouds (orange, blue) around a building and its courtyard in Northern area (top, left). The cloud union (red) not georeferenced (right). The preliminary alignment with corresponding points coordinates (down, left) and the georeferenced result after *cloud-to-cloud* fitting (right).

In Pescara del Tronto, due to the critical area conformation and the contingent performance of the path, the procedure of cloud registration ended with a residual error of up to 30 cm. Another advantage of the system is that the results are offered as a series of structured datasets, relative to the cloud and the trajectory, offering the ability to segment the cloud using the time function (Figure 8). After a point cloud processing according to the time (time data embedded in the 3D data), alignment with other ZEB clouds (Figure 10), segmentation, post registration, georeferencing, and optimization, the point cloud was ready to be analyzed with other sensors results, with the lack of radiometric information.

Due to the experimental use of these clouds, some problems about point clouds SLAM registration related to context complexity occurred. Different processing strategies are preliminary proposed in order to georeference and empirically improve the alignment of row ZEB1 points characterized by acquisition time:

-   Georeferencing ZEB cloud attributing coordinates on identifiable homologous points,
-   Georeferencing ZEB cloud with points and optimizing with fitting algorithms on *aerial point cloud,* and
-   Segmentation and georeferencing ZEB cloud if any there would be errors during the acquired trajectory and bad quality clouds registration; then, optimization with fitting algorithms on the aerial point cloud.

According to these first experimental tests (Table 2), the *Central area,* in which a closed semi-circular path was performed, without retracing in the return the outbound track, reports the worst results. The need of quite close geometrical features around which the scan sustains the SLAM processing of progressive registration is thus very crucial. Changes of elevation with extensive spaces in these complex contexts have proven to be no help for a recording with a controlled error lower than 10 cm.

Above, Figure 12, a sample of three profile sections extracted on the integrated point clouds, aerial and ZEB scan (Figure 11), in which their different contribution to the geometry definition is pinpointed. It is important to point out, thanks to this section typology, the relevant contribution of the ZEB1 scans for the comprehensive definition of the building geometry. Below, alignment errors that are identified in Table 2 and are visible in a comparison map in range color on the sample in Figure 13, about the Building n°1. Statistical result of mean errors in clouds fitting results are very influenced first of all by the noise of the ZEB point clouds, and on the other hand, by the different accuracy reached by the aerial point cloud triangulation about the geometric definition in narrow spaces on ground. Excluding this point cloud noise and considering the significant areas of the buildings, the errors values in the comparisons below by clouds fitting algorithm can be reasonably and limited under 10-15 cm, which satisfy the expected accuracy of the technological solution aims of rapid mapping and damage documentation.

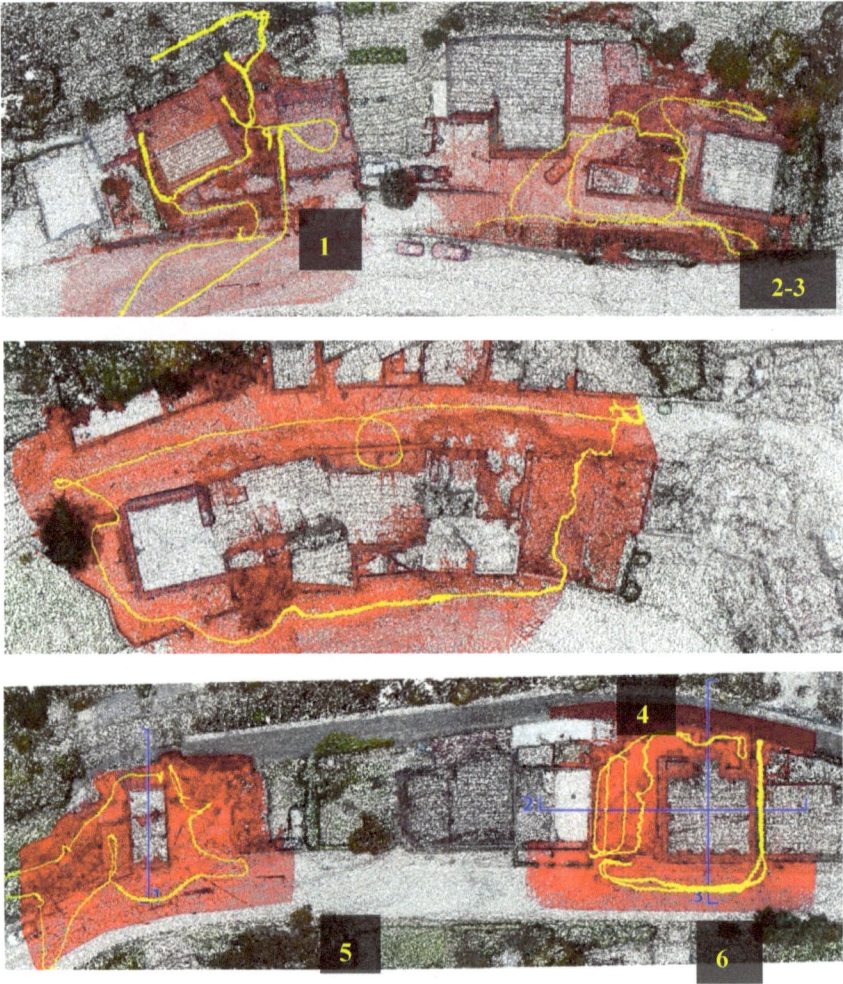

Figure 11. The aerial RGB point clouds and the ZEB1 georeferenced scans (red) with indicated the trajectory of acquisition (yellow), in order South, Center, North. Marked in blue, the sections profiles showed of Figure 12. The walkway and the path closure mode (Tracked trajectory in a closed ring or in a roundtrip through to the same path), influenced Simultaneous Localization And Mapping (SLAM) registration of the ZEB1 clouds, as visible here and in Table 2.

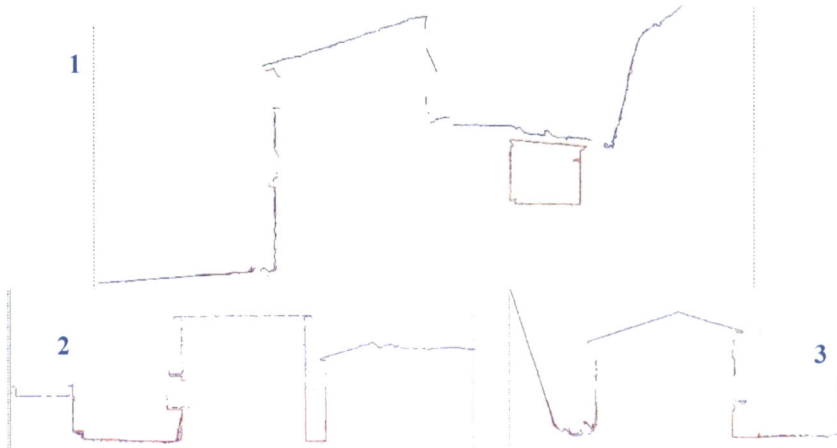

Figure 12. Three section profiles on northern block regarding the integration of point clouds. In blue colour the aerial contribution and in red the ZEB1 scan contribution.

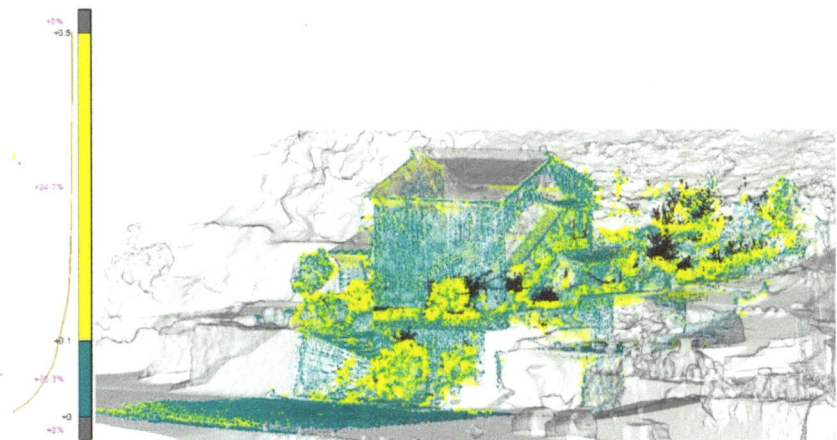

Figure 13. An example of cloud comparison, between ZEB and aerial photogrammetric one. Points statistically differ by 0.00 < **74.5%** < 0.10m (green)/0.01 < **24.7%** < 0.6m (yellow) (Details available in Table 1).

Table 2. Synthesis table of data processing results for the ZEB point clouds (Density values are expressed by n° of points in a spherical volume, with r = 1 cm).

Area	Building	N° Points	Density (pt/V. Sphere r = 0.01m)	Density St.dev	Alignment Mean Error (m)		
					0.00 < x < 0.10	0.10 < x < 0.30	0.03 < x < 0.60
South	B1	8.210.000	4.44	5.03	74.5%	20.8%	4.7%
	B2-3	14.500.000	3.35	3.80	68.4%	27.3%	4.2%
Center	B4	16.100.00	2.26	2.09	44.2%	50.2%	5,6%
North	B5	9.600.000	3.33	4.94	81.8%	12.5%	5.7%
	B6 (a+b)	15.800.000	3.73	3.90	69.4%	28.1%	2.4%

4. Deriving Metric and Non-Metric Data from 3D Model: A Test-Case

Starting from the subsequent reconnaissance missions in the area, the different datasets have been organized and combined, in order to evaluate and classify the level of detail and usability of the different data, and for which applicative context. The preliminary data processing of the Northern dataset was limited to a selected focus object: this building (n°5, Figure 11) is one of the been kept under the observation of the structural team.

A pertinence area around the test building have been chosen too (Figure 14). A large amount of information have been extracted from aerial and terrestrial acquisition and classified from images and 2D/3D production, as in following examples.

Figure 14. An images of the building n°5 in October 2016 and relative orthoimage by multi-rotor drone (aerial images of both shooting and frames extraction from 4k video with DJI drone).

4.1. Data Acquisition and Processing

Different datasets are trying to be integrated in the test area. In Figure 15, the datasets are listed: the aerial acquisition (distinguished by camera configuration), and the photogrammetric model that is produced with both the camera configurations; the terrestrial close-range acquisition, and the fusion model using the complete dataset.

As it is said before, the topographic and GNSS survey were affected by site conditions and constraints, and the difficulties of freely positioning markers for measurements, together with GNSS signal problems from permanent stations. Here, the processed datasets referred to the Building n°5. Estimated errors on CP for each processing by RMSE analysis (table and graph in Figure 15) allow for us to choose the last one, the data fusion, as the ground-truth for further analysis. According to the information extraction strategy, 3D data must be processed for filtering by noise reduction and optimized to finalize a 3D model or 3D information as in next paragraphs.

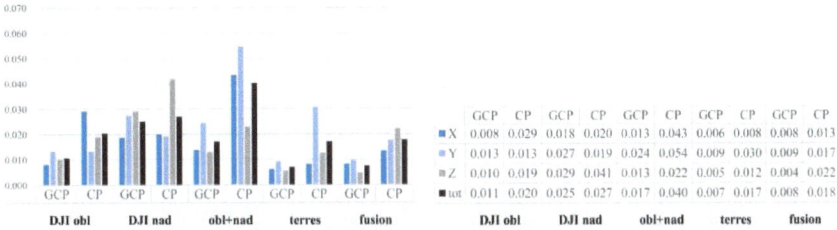

	GCP	CP	GCP	CP	GCP	CP	GCP	CP	GCP	CP
X	0.008	0.029	0.018	0.020	0.013	0.043	0.006	0.008	0.008	0.013
Y	0.013	0.013	0.027	0.019	0.024	0.054	0.009	0.030	0.009	0.017
Z	0.010	0.019	0.029	0.041	0.013	0.022	0.005	0.012	0.004	0.022
tot	0.011	0.020	0.025	0.027	0.017	0.040	0.007	0.017	0.008	0.018

	DJI obl	DJI nad	obl+nad	terres	fusion

Figure 15. Accuracy on ground control points (GCPs) and Check Points (CPs) RMSE (m): DJI flights, fusion of aerial data, close-range acquisition, and global integration of photogrammetric datasets. We can consider this last one as the best verified result so that it can be used as ground-truth.

4.2. Evaluation on Multi-Sensor Data Acquisition and Information Interpretation, Classification and Extraction

Operating camera configuration and sensor integration, it is possible to make the 3D model accessible to different needs, and expertise information that is embedded in 3D data had not to be simply geometric. As we affirmed before, the great part of first building damage assessment (BDA) is still based on visual evaluation, and then the metric assessment of the metric structural condition of buildings, finalized to accessibility constraint.

4.2.1. Orthoimages and DSM

A photogrammetric model by commercial DJI drone, as in Orthoimage, is highly competitive in terms of quality of RGB information. The standard of comparison is a well-known terrestrial close-range photogrammetry, a solution that is less viable in these contexts of timing and accessibility to spaces. Most damages and creeps on all of the façades are clearly visible even in the aerial model (as in example in Figure 17), better than in the fusion one. More reflections that are interesting to be carried out are on the typical aerial representation of the site surface. For a DSM, data extraction could usually concern, for example, isolines and elevation points. For the 3D documentation of the building and its surrounding areas at environmental scale, ZEB1 point cloud has provided good results, as in Figure 16. For planimetric information at architectonical scale, a HQ fusion model offers the best level of detail of the building and around (Figure 12).

No RGB info

Figure 16a. Comparison, on a small area, between used platforms and sensors: eBee, average GSD of 5 cm/pixel (left); DJI Phantom, average GSD of 2.18 cm/pixel (center left); fusion model, average GSD of 0.92 cm/pixel (center right); no RGB data from Zeb1.

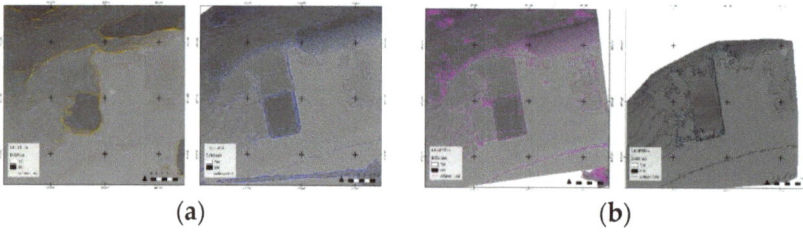

(a)   (b)

Figure 16b. DSM comparison. eBee (left); DJI (center right). Fusion model (center left); ZEB1 (right).

Figure 17. Orthoimage of the East side. The aerial model by integrated oblique and nadir cameras (right), can be comparable in terms of available detail and the accessible information to the terrestrial photogrammetric one (left).

4.2.2. Geometric Detail and Section Profiles

We identify some crucial points in the structure in which it could be very useful for a structural reconnaissance after a damage event. The main geometrical and structural building condition could be evaluated first of all with a series of horizontal and vertical sections that could better clarify any geometrical behavior of the volume, the façade, the roof, and the other main walls. Here, the transversal section, intersecting the point of weakness of the chimney axe, could show possible misalignments and out of plumb of the lateral masonries; in longitudinal section (Figure 18, Table 3) it is possible to analyse the openings and possible damages of the façade.

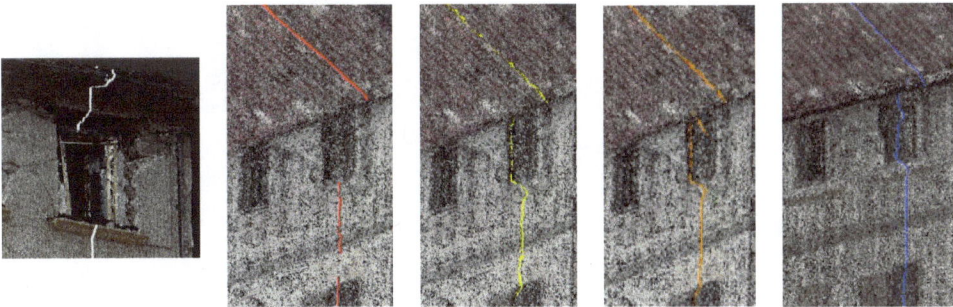

Figure 18. A zoomed portion of the longitudinal section superimposed to the fusion model showing the contribution to the geometric definition. From left: terrestrial photogrammetric model (white); nadir cameras (red); oblique cameras (yellow), aerial acquisitions model (orange), HQ data fusion model (blue).

Density indexes (Table 3) are here significantly for differentiate the contribution of each sensor for the geometrical definition. Despite the close-range one is the richer in dense points on the façade, is the most non-uniform (St. Dev), together with the ZEB1 one. As expected, the fusion model presents a uniform distribution of measurements of the whole building in comparison to the aerial one. We can verify that the fusion of models contributes to the geometric definition, not in terms of average density, but for the standardization of distribution.

We can affirm that the extensiveness data from fusion of terrestrial and aerial images can be integrated with a union of the LiDAR ZEB1 data (Figure 19) that have been georeferenced, in order to produce a complete profile of the building with hits geometry in relationship with the rear rock face.

163

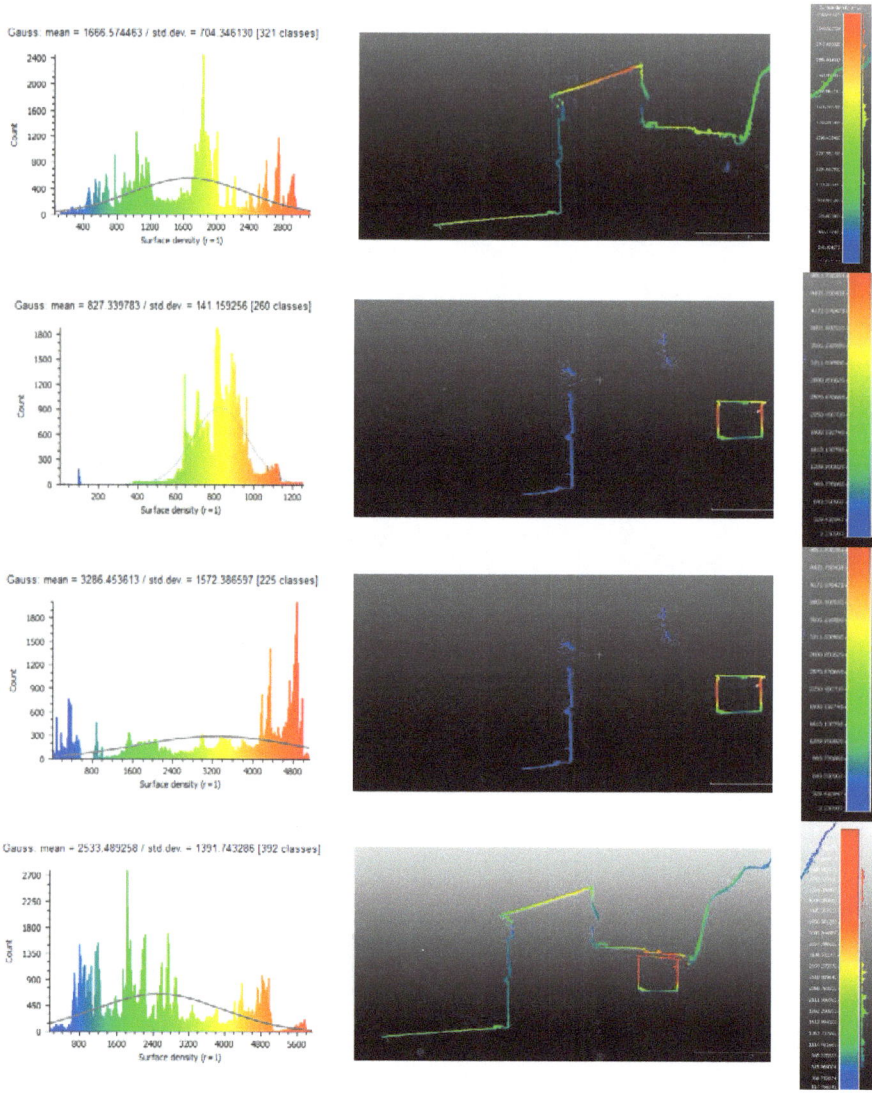

Figure 19. Density for aerial point cloud (1), data fusion point cloud (2), ZEB point cloud (3), and integrated Light Detection and Ranging (LiDAR) and aerial point cloud (4).

Table 3. Synthesis table of statistical values on longitudinal section (20 cm thickness).

| | | | Longitudinal Section | | | |
| | | | Density (n° points) | | | |
		n° points	Average Density (point/m²)	st.dev.	Min	Max
	ZEB	50,444	3,286	1,572	9	5100
	NIKON	high	370,716	27,300	7,200	84	37,220
	EBEE	806	12	3	1.9	18
DJI	obl	95,119	1,578	658	1	2,970
	nad	8,667	128	35	16	197
	fusion	102,493	1,666	704	47	3,100
FUSION	low	1,400	18	4	0.6	32
	med	12,111	147	31	0.9	228
	high	67,400	827	141	1	1,256
INTEGRATION AERIAL+ZEB POINTS CLOUD		152,937	2,533	1,392	117	5,936

5. Conclusions and Future Perspectives

In this contribution, a work-in-progress test approach for rapid mapping is presented, which is finalized to testing and refine the methodological approach, as well as improve the operational procedure with a view to a standardization, in the Geomatics techniques scenario, focused on rapid mapping in emergency contexts. A step-by-step analysis about procedures to acquire 3D data for rapid mapping by quick terrestrial scanning (ZEB1 portable scan) and aerial photogrammetry by COTS UAV equipment and sensors have been displayed. For an environmental scale in Pescara del Tronto, different areas according to blocks geometry and topographical conformation have been chosen and modelled with the contribution of aerial and terrestrial point clouds, and they give interesting results in terms of application according to the typological and topographical geometry of the object, as well as the surrounding. Therefore, it is also interesting to analyse the better practice of SLAM-based point clouds manipulation on outdoor environments (despite the implementation of this technique to be employed in indoor scenarios or otherwise in narrow locations). The aims of the proposed documentation approach is first of all the optimization of points clouds processing from a multi-sensor acquisition at different scale, in order to balance competitiveness of resources (human and technical) and the effectiveness of metric and non-metric information.

Afterward, in order to exploit the richness of metric and radiometric information of the 3D model optimization, a test building in a higher scale has been deepened. In this case, we can reflect that a high-quality model integration of models is preferable instead of a data fusion processing: extreme density on radiometric and metric information by close-range photogrammetry can be fulfil by oblique image processing for upper parts. An indoor mapping, if practicable, can complete the reconnaissance multi-sensor model. In terms of accuracy, the realized model by data fusion finally has the lower discrepancy on CPs and better fits with the terrestrial measurements (Figure 15). Unless very accurate information is necessary for specific focus analysis on the building, an aerial model could be suitable; the integration of aerial and terrestrial does not add a significant improvement in level of geometric definition for a first-step damage documentation and assessment (Figure 14), and in any case, the detail of terrestrial imaging could not be reached. Nevertheless, according to the achieved results and preliminary analysis, is possible to underline that a UAV nadir-oblique close-range acquisition, open to improvement in the use of high resolution cameras, could obtain a very strategic level of information. It is not so much the density that rises using the aerial camera merging, but, of course, is the homogeneousness (st. dev values variation) cloud density, which allows for extracting less fractionated sections (Figure 18).

Acknowledgments: The author would like to thank especially Team Direct (https://www.facebook.com/Team-Direct-461829537253316) and all people involved in the missions, including Antonia Spanò, Filiberto Chiabrando, Andrea Lingua, Lorenzo Teppati Losè, Paolo Maschio, Vincenzo Di Pietra, Paolo Dabove from Politecnico di Torino. During the in-site campaign in Pescara del Tronto, Accumoli, Amatrice, the Fire Fighters SAPR group accompanied the Direct Team with dedication, assisted the fieldwork with people and instruments. Thanks above all to Politecnico di Torino for promoting and financing the emergency mission in center Italy areas affected by earthquake (Prof. Sebastiano Foti for the task force organization), and also GEER team (Geotechnical Extreme Events Reconnaissance Association). Thanks to Alessandra Sperafico for his dedication and help in data processing. A mention also to MESA s.r.l. for ZEB1 and sincerely thank Cristina Bonfanti e Nadia Guardini.

References

1. Aicardi, I.; Chiabrando, F.; Grasso, N.; Lingua, A.M.; Noardo, F.; Spanò, A. UAV photogrammetry with oblique images: First analysis on data acquisition and processing. *Int. Arch. Photogramm. Remote Sens. Spat. Inf. Sci.* **2016**, *XLI-B1*, 835–842.
2. Boccardo, P.; Chiabrando, F.; Dutto, F.; Tonolo, F.G.; Lingua, A. UAV Deployment Exercise for Mapping Purposes: Evaluation of Emergency Response Applications. *Sensors* **2015**, *15*, 15717–15737.
3. Bosse, M.; Zlot, R. Continuous 3D Scan Matching with a Spinning 2D Laser. In Proceedings of the 2009 ICRA'09 IEEE International Conference on Robotics and Automation, Kobe, Japan, 12–17 May 2009; pp. 4244–4251.
4. Bosse, M.; Zlot, R.; Flick, P. Zebedee: Design of a Spring-Mounted 3D Range Sensor with Application to Mobile Mapping. *IEEE Trans. Robot.* **2012**, *28*, 1104–1119.

5. Chiabrando, F.; Sammartano, G.; Spanò, A. A comparison among different optimization levels in 3D multi-sensor models. A test case in emergency context: 2016 Italian earthquake. *Int. Arch. Photogramm. Remote Sens. Spat. Inf. Sci.* **2017**, *XLII-2/W3*, 155–162.

6. Chiabrando, F.; D'Andria, F.; Sammartano, G.; Spanò, A. 3D Modelling from UAV Data in Hierapolis of Phrygia (TK). In Proceedings of the 8th International Congress on Archaeology, Computer Graphics, Cultural Heritage and Innovation 'ARQUEOLÓGICA 2.0', Valencia, Spain, 5–7 September 2016; pp. 347–349.

7. Chiabrando, F.; Di Pietra, V.; Lingua, A.; Maschio, P.; Noardo, F.; Sammartano, G.; Spanò, A. TLS models generation assisted by UAV survey. *Int. Arch. Photogramm. Remote. Sens. Spat. Inf. Sci.* **2016**, *XLI-B5*, 413–420.

8. Fernandez Galarreta, J.; Kerle, N.; Gerke, M. UAV-based urban structural damage assessment using object-based image analysis and semantic reasoning. *Nat. Hazards Earth Syst. Sci.* **2015**, *15*, 1087–1101.

9. Gerke, M.; Kerle, N. Automatic structural seismic damage assessment with airborne oblique Pictometry© imagery. *Photogramm. Eng. Remote Sens.* **2011**, *77*, 885–898.

10. Kerle, N. Satellite-based damage mapping following the 2006 Indonesia earthquake— How accurate was it? *Int. J. Appl. Earth Obs. Geoinf.* **2010**, *12*, 466–476, doi:10.1016/j.jag.2010.07.004.

11. Lemoine, G.; Corbane, C.; Louvrier, C.; Kauffmann, M. Intercomparison and validation of building damage assessments based on post-Haiti2010-earthquake imagery using multi-source reference data. *Nat. Hazards Earth Syst. Sci. Discuss.* **2013**, *1*, 1445–1486.

12. Lerma, J.L.; Seguí, A.E.; Cabrelles, M.; Haddad, N.; Navarro, S.; Akasheh, T. Integration of Laser Scanning and Imagery for Photorealistic 3D Architectural Documentation. In *Laser Scanning, Theory and Applications*; Wang, C.-C., Ed.; InTech Open: Rijeka, Croatia, 2011; doi:10.5772/14534.

13. Maier, P. Assessing Disaster Damage: How Close Do You Need to Be? 2016. Available online: https://irevolutions.org/2016/02/09/how-close/ (accessed on 28 July 2017).

14. Rathje, E.M.; Franke, K. Remote sensing for geotechnical earthquake reconnaissance. *Soil Dyn. Earthq. Eng.* **2016**, *91*, 304–316.

15. Rastiveis, H.; Samadzadegan, F.; Reinartz, P. A fuzzy decision making system for building damage map creation using high resolution satellite imagery. *Nat. Hazards Earth Syst. Sci.* **2013**, *13*, 455–472, doi:10.5194/nhess-13-455-2013.

16. Bosse, M.; Zlot, R.; Flick, P. Zebedee: Design of a Spring-Mounted 3D Range Sensor with Application to Mobile Mapping. *IEEE Trans. Robot.* **2012**, *28*, 1104–1119.

17. Remondino, F.; Barazzetti, L.; Nex, F.; Scaioni, M.; Sarazzi, D. UAV photogrammetry for mapping and 3D modeling–current status and future perspectives. *Int. Arch. Photogramm. Remote Sens. Spat. Inf. Sci.* **2011**, *38*, C22, doi:10.5194/isprsarchives-XXXVIII-1-C22-25-2011.

18. Riisgaard, S.; Blas, M. Slam for Dummies. A Tutorial Approach to Simultaneous Localization and Mapping, 2005. Available online: http://ocw.mit.edu/NR/rdonlyres/ Aeronautics-and-Astronautics (accessed on 28 July 2017).

19. Ruiz Sabina, J.; Gallego Valle, D.; Peña Ruiz, C.; Molero García, J.; Gómez Laguna, A. 2015. Aerial Photogrammetry by drone in archaeological sites with large structures. *Virtual Archaeol. Rev.* **2015**, *6*, 5–19, doi:10.4995/var.2015.4366.

20. Strecha, C. The rayCloud—A Vision Beyond the Point Cloud. In Proceedings of the FIG Congress 2014, Engaging the Challenges—Enhancing the Relevance, Kuala Lumpur, Malaysia, 16–21 June 2014.

21. Thomson, C.; Apostolopoulos, G.; Backes, D.; Boehm, J. Mobile laser scanning for indoor modelling. *Int. Arch. Photogramm. Remote. Sens. Spat. Inf. Sci.* **2013**, *II-5/W2*, doi:10.5194/isprsannals-II-5-W2-289-2013.

22. Voigt, S.; Scheneiderhan, T.; Twele, A.; Gahler, M.; Stein, E.; Mehl, H. Rapid damage assessment and situation mapping: Learning from the 2010 Haiti earthquake, *Photogramm. Eng. Remote Sens.* **2011**, *77*, 923–931.

Giulia Sammartano. Optimization of Three-Dimensional (3D) Multi-Sensor Models For Damage Assessment in Emergency Context: Rapid Mapping Experiences in the 2016 Italian Earthquake. In *Latest Developments in Reality-Based 3D Surveying and Modelling*; Remondino, F.; Georgopoulos, A.; González-Aguilera, D.; Agrafiotis, P.; Eds.; MDPI: Basel, Switzerland, 2018; pp. 141–168.

Chapter 3
3D Modelling and VR/AR

Integration of Pipelines and Open Issues in Heritage Digitization

Luca Cipriani [a] and Filippo Fantini [a]

Dept. of Architecture, Alma Mater Studiorum University of Bologna, Ravenna Campus, Bologna, Italy; (luca.cipriani, filippo.fantini2)@unibo.it

Abstract: In this paper, a series of topics, concerning acquisition, communication, and the analysis of buildings will be discussed. In particular, a strategy aimed at exploiting the visual potentials of reality-based models, will be described inside render engines, real-time applications, as well as their role for the production of conventional drawing for documenting and studying built heritage. This methodology is inspired by pipelines coming from computer-generated imagery (CGI) and the video game industry: areas of research that are progressively showing an increasing interest towards the world of photogrammetry remote sensing, in particular for the new possibilities offered by Structure from Motion (SfM)/Multi View Stereo (MVS) in terms of automatic texturing of complex shapes. Quad-dominant remeshing, displaced subdivision surfaces, and accurate interactive/ automatic parametrization are the main points of this workflow, whose final goal is the achievement of multipurpose models, which are capable to comply with graphic codes of traditional survey, as well as semantic enrichment, and, last but not least, data compression/portability and texture reliability under different lighting conditions.

Keywords: photogrammetry; terrestrial laser scanner; normal mapping; displaced subdivision surfaces; archaeology

---

1. Introduction

Digitization is a fundamental part of the current scenario of Cultural Heritage documentation and intervention; it grew up and become a protagonist in the script of preservation and valorisation of heritage at risk (Grün et al., 2004), enabling new reading keys to formerly in-depth researched monuments (Bianchini, 2013; Cipriani et al., 2016, Adembri et al., 2014). Nevertheless, several basic issues concerning three-dimensional (3D) model true exploitation are still source of debate, here, in particular, two aspects of these problems will be discussed: the first concerns the pairing of data reliability and the needs of dissemination/interaction, and the second is about the use of digital models as bases for the production of traditional drawings.

The higher the achievable resolution and accuracy, the higher the amount of hardware resources, as well as the software efficiency needed: for this reason, several investigation lines—centred on data portability (Hoppe, 1996, Lee et al., 2000, Gobetti et al., 2012)—3D online streaming and offline visualization of huge 3D models (Potenziani et al., 2015) have been carried out for two decades. After years of applications and progressive implementation inside commercial software belonging to the field of computer-generated imagery (CGI), the duo subdivision surfaces–displacement maps was generally adopted in the field of visual effects as a standard for Level of Detail (LoD) models, due to their scalability, portability, and efficiency. Also, inside game engines, such as Unity, the LoD feature allowed by DirectX 11 opened the possibility to use the tessellation of subdivision surfaces derived from highly detailed meshes from active and passive sensors (Merlo et al., 2013). For these reasons, some years ago, an investigation line started that was focused on the development of specific pipelines on how to convert in the more accurate, automatic and reliable way data from sensors into "displaced subdivision surfaces" (in accord to the terminology from Lee et al., 2000). In this sense, a series of experiments from the world of Cultural Heritage have been conducted (Fantini, 2012; Merlo et al., 2013, Guidi et al., 2016). Other aspects concerning these methods made them increasingly more flexible and reliable: on the one hand, the introduction of OpenEXR (http://www.openexr.com/) as displacement image format, on the other, the development of automatic quad-dominant remeshing tools (Jakob et al., 2015), and last but not least, the increasing automatization and accuracy of mesh parameterization tools (Cipriani et al., 2014).

The ease with which low-resolution quadrilateral meshes can be "sealed" in order to complete models enabling the achievement of constructive solid geometries, CSG, is also undeniable (Fantini, 2009): this issue is fundamental for the true exploitation of digital models as bases for the creation of conventional drawings. In fact, watertight meshes can undergo Boolean operations without problems, allowing for the production of sections on both convex and concave objects, without the need to perform a true slice or modification of the mesh thanks to interactive Booleans that are included in several modelling applications (Maxon Cinema 4D, The Foundry Modo, etc.).

Another issue is the one concerning the role of simplified meshes of "quads" in relation to the semantics that are applied to complex architectural and archaeological surveys. Can a structured mesh—formed by connected and regular sequences of edges in accord to features and curvature flow—facilitate segmentation/partition of a model?

The last topic has to do with the true use of render engines and is deeply connected to the control of the operator on the (u,v) parameter space, in particular it deals with the conversion of apparent colour texture into diffuse colour texture (for Lambertian surfaces), or into a set of textures that are aimed at supplying render engines with proper information concerning complex optical behaviours (for instance mosaic tiles, translucent materials, etc.).

2. Aims of the Study

The purpose of this research on archaeologic and architectural representation is to define a comprehensive pipeline that is capable of converting heavy meshes from active and passive sensors into 3D digital assets that are capable to dialogue with applications aimed at visual realism from the word of CGI/ Visual Effects (VFX). The strategy that is proposed here (Figure 1) focuses on how to produce optimized models: meaning models that are flexible enough to be rendered under different lighting and environmental conditions, which is reliable in relation to physical simulation of light, as well as topologically correct in order to produce two-dimensional (2D) drawing (plans, sections, elevations, etc.). This last aspect is still a matter under investigation, since it requires time-consuming and complex strategies (Martos and Cacheroa, 2015; Tryfona and Georgopoulos, 2016) in order to fit virtual simulacra with standard graphic codes and long-time tradition of archaeological drawings from XVIII and XIX century architects (Jacques and Bonfait 2002). Those examples are still used since they supply scholars and restorers with relevant information about masonries before XX century interventions; however, the charm of those watercolours and engravings require great effort and time to be reproduced by means of the current technological framework, and it could also be considered a sort of empty stylistic exercise (i.e. using non-photorealistic rendering techniques). The outline of work set out in this paper develops a pipeline that integrates the severe use of orthographic projections (architectural drawing codes), and the possibilities that are supplied by render engines (the updated version of manual rendering) in order to achieve true and full exploitation of different surveying techniques.

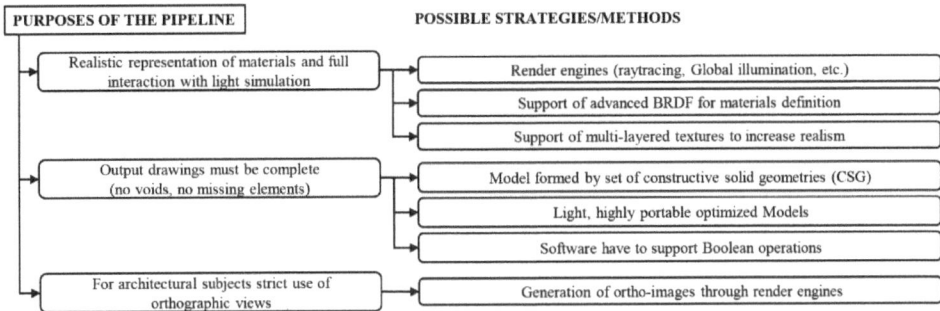

Figure 1. Summary of the purposes and possible strategies.

3. Optimization of Models and Textures

In the field of Visual Effects (VFX) and Computer-generated imagery (CGI), meshes that are formed by triangles are considered very poor in terms of rendering and animation. The main reason is the fact that animation/morphing of virtual characters has to be LoD representations, based on subdivision surfaces (subD), rather than blocked resolution meshes that are made of triangles, or Non-uniform rational B-spline (NURBS) models that present severe topological restrictions (De Rose et al., 1998). Thanks to the achievements in both the fields of mesh parameterization and compression by means of displaced subdivision surfaces, or DSS (Lee et al., 2000), new standards have been achieved in the first decade of this century through the progressive implementation of this variable-level-of-detail representation in the main geometric modelling applications that are addressed to entertainment. The main advantages of the displaced subD are: compression of morphologic detail, high portability, adaptive tessellation, back-face culling (Bunnell, 2005). A strong limitation to an extensive use—in the field of reality-based models for documentation—is that the main subdivision criteria/rule that is implemented in commercial applications, the Catmull-Clark (Catmull and Clark, 1978; De Rose et al., 1998), works more efficiently with quad-dominant meshes, that are far from being a standard in the field of geomatics. The basic idea of subdivision surface is that a mesh M^0 (a polyhedron roughly approximating a shape) can be iteratively subdivided using a refinement scheme, producing a smooth surface that is, in the case of Catmull-Clark approximating scheme, a bi-cubic uniform B-spline. Lee et al. (2000) developed a research line that "merged" the possibility to increase the geometric resolution of a mesh (Figure 2) with the opportunity to modify the position of the vertices belonging to progressively higher resolution mesh using a displacement map (Blinn, 1978).

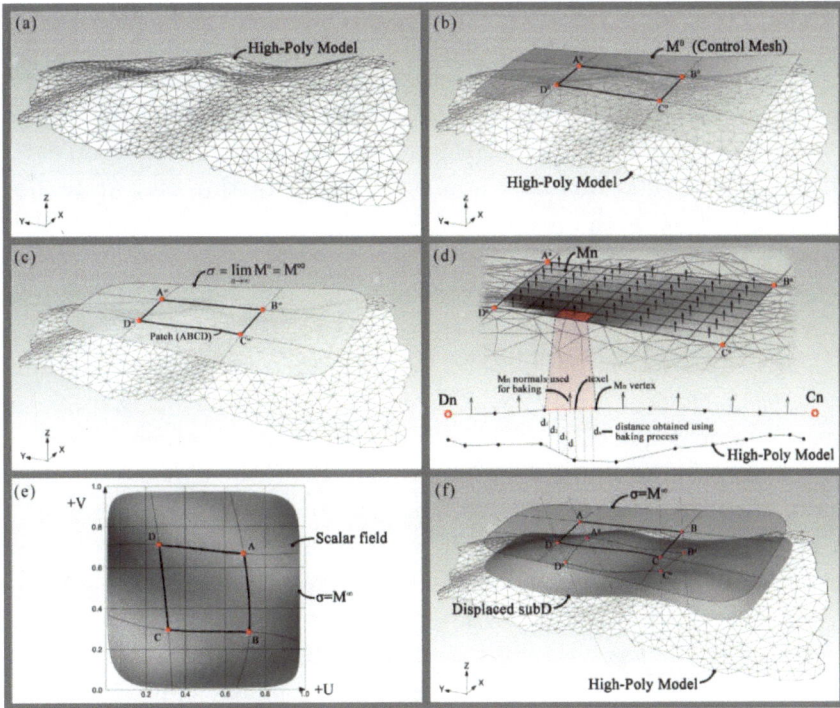

Figure 2. (**a**) high-poly model from active/passive sensor; (**b**) the mesh covered by quadrilateral polygons, M^0; (**c**) M^0 converted into a subdivision surface; (**d**) baking process performed in order to calculate the distances between subdivision surface and high-poly model into a greyscale bitmap; (**e**) parameterization and corresponding scalar field obtained through baking; and, (**f**) the final output is a Level of Detail (LoD) model that fits with the original high-poly mesh.

The last decade has seen further developments and implementation of this techniques, in particular, once High Dynamic Range (HDR) file formats, as the OpenEXR, extended the possibility to store data inside a single displacement map: from 256 "steps" of a 8-bit grayscale to the wider range provided by 16-bit floating-point, 32-bit floating-point, and 32-bit integer pixels (http://www.openexr.com/). Time-consuming manual operations called "re-topology"—in the CGI jargon— have been necessary for years in order to "cover" high-poly meshes that are formed by triangles with quadrilateral polygons, and make them fit with the requirements of subdivision surfaces. Among them, in addition to quad-dominancy, the problem of valence should be mentioned: the continuity of first and second derivatives (C^2) is verified only in points of valence equal to four, namely all of the vertices in which four edges converge. For all the other cases, called "singularities", C^2 cannot be verified and shading discontinuities may

175

appear. This approach can be considered as to be equivalent to reverse-modelling pipelines based on coating, by means of NURBS patches, high level of detail meshes from sensors (Figure 3). The need to escape, or limit, manual "re-topology" (it also requires skilled and experienced polygonal modellers) led to the development of several researches based on the automatic and semi-automatic detection of features and curvature lines on non-structured triangular meshes with the aim to use these morphological hints to obtain quad-dominant meshes (Lai et al., 2008; Jacob et al., 2015). Several entertainment applications, focused on character modelling and organic modelling through mesh sculpting—namely anthropomorphic and phytomorphic assets with smooth flowing shapes and absence of sharp creases—progressively implemented automatic remeshing with quads (e.g., Pixologic ZBrush has Zremesher, The Foundry Modo implements Automatic Retopology Tool, Pillgrim 3D Coat has Auto-Retopo, etc.), other research projects led to stand-alone applications that were aimed at these kind of remeshing (Instant Meshes). In general, the output quad-mesh from these automatic solutions tends to be isotropic, and, in the case of complex shapes (characterized by multi-connected elements), several singularities are produced; in order to get rid of these problems, several tools and parameters are included in such applications (interactive guides, definition of areas with higher detail, general smoothing to be applied before remeshing, etc.), but a final manual completion is, in general, needed (Figure 4). The higher the number of polygons of these final meshes, the higher the time needed to complete them by means of bridges, hole filling, etc.

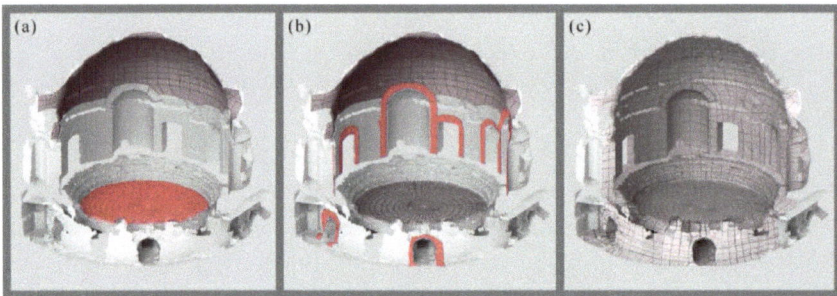

Figure 3. Manual quad-dominant remeshing applied to Heliocaminus's Baths at Hadrian's Villa: (**a**) primary surfaces; (**b**) secondary surfaces; (**c**) the final model obtained by making bridges among bands of polygons.

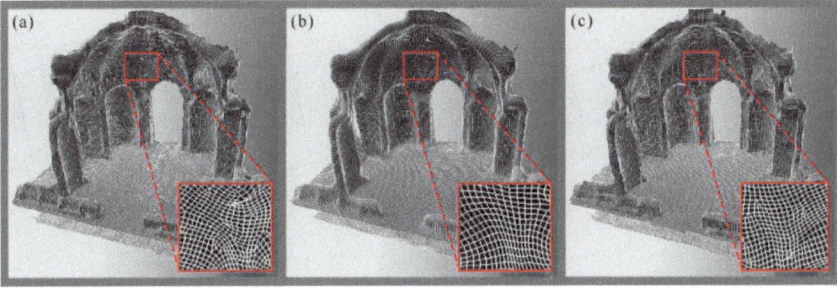

Figure 4. Automatic re-topology tests on the Vestibule of Piazza d'Oro in Tivoli:
(a) The Foundry Modo; (b) Pixologic ZBrush; (c) Instant Meshes (ETH, CNR-ISTI,
Università dell'Insubria).

In order to establish a correct relation between a quad-mesh and the bitmap that is responsible for the displacement, an adequate parameterization is needed: also, in this case, increasingly more accurate automatic and semi-automatic methods have become widespread in the field of CGI applications (Cipriani et al., 2014). The current state of the art of parameterization includes isometric (length preserving), conformal (angle preserving), and equi-areal (area preserving) solutions to be applied to open or watertight meshes. In this last case, the manual selection of a connected tree of edges needs to be used for braking the connectivity of the mesh inside the parameter space. The possibility to mix different parameterization criteria inside modelling applications allows for the full use of the parameter space, in which colour and other information are encoded (normals, displacement, global illumination solution, occlusion, etc.). A fundamental aspect of the pipeline explained here deals with avoiding automatic parameterization of heavy triangular meshes that leads to poor (u,v) layouts and a general waste of pixels, as well as an uncontrolled number of islands (or charts) in the parameter space: a typical feature when working on dense textured meshes from SfM/MVS. Optimized and easy-to perceive 2D versions of quad-meshes, with a correct texel density (with respect to the needed output), is the fundamental step in order to obtain a reliable displaced subdivision surface (Figures 5 and 6). It is safe to say that Structure from Motion (SfM)/ Multi View Syereo (MVS) applications have been inspired by some (u,v) checking tools included inside entertainment software to perceive the magnitude of the polygonal stretching amount occurred during the 3D-to-two-dimensional (2D) transformation, fill ratio and overlap ratio. A crucial factor for a correct result, once the displaced subD is completed, is a correct parametrization, as well as a correct resolution of the OpenEXR displacement map, with respect to the original high-density mesh that is made of triangles. The process that leads to this bitmap is called "baking", namely a render-to-texture solution, whose basic functioning is illustrated in Figure 2d: it should be noted that the theoretical subdivision surface σ—the limit surface that is obtained applying infinite times the subdivision criteria—never enters into play during this process. In fact, the calculation of each distance from subD and high-poly model is de facto the deviation from a blocked resolution subD and the high resolution mesh from scanner/photogrammetry. Before launching the calculation (baking), it is fundamental to balance the following factors: the resolution of the original mesh should be matched by the resolution of M^n; the resolution in pixels of the displacement map should match the desired number of polygons (e.g., 4096x4096 pixels can theoretically supply the subD with 16777216 height displacement values). The metric accuracy of displaced subD in the frame of Cultural Heritage applications has already been tested (Guidi and Angheleddu, 2016; Merlo et al., 2013) using deviation tools inside the specific applications. In Figure 7, the results of a test that is conducted on a niche from Piazza d'Oro at Hadrian's Villa are shown, in which the subdivision criteria that are adopted were the proprietary one from the Foundry Modo, and not the classic Catmull-Clark. Then, the role of parameterization is fundamental for a reliable result, but the interest towards optimized and easy-to-perceive 2D versions of meshes provide further opportunities concerning both semantics and the photo-realism of 3D digital assets. If the "pseudo-development" provided by parametrization tools allow an easy interaction between the user and the texture (Cipriani et al., 2015), also an easier and more accurate 2D bitmap editing is possible; for instance, materials like conductors (e.g., gold tesserae of mosaic tiles) need special alpha images in order to specify different bidirectional reflectance distribution function (BRDF) behaviours upon the same mesh (Figure 8).

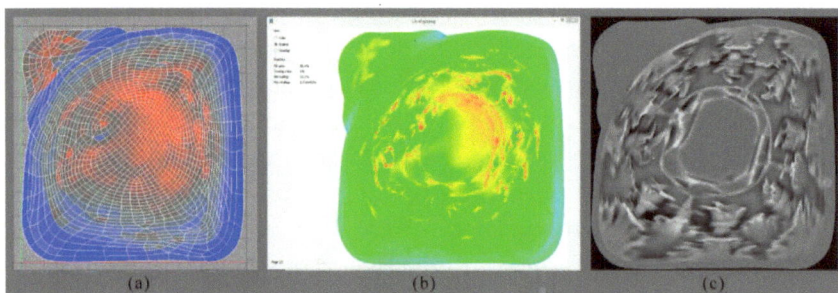

(a)   (b)   (c)

Figure 5. A parametrized subdivision surfaces (subD): (**a**) visualization of parameterization distortions inside a computer-generated imagery (CGI) application; (**b**) an analogous visualization recently included inside a SfM/MVS application; and, (**c**) displacement map storing distances between subD and high-poly mesh: excessive stretching leads to less reliable displaced subD.

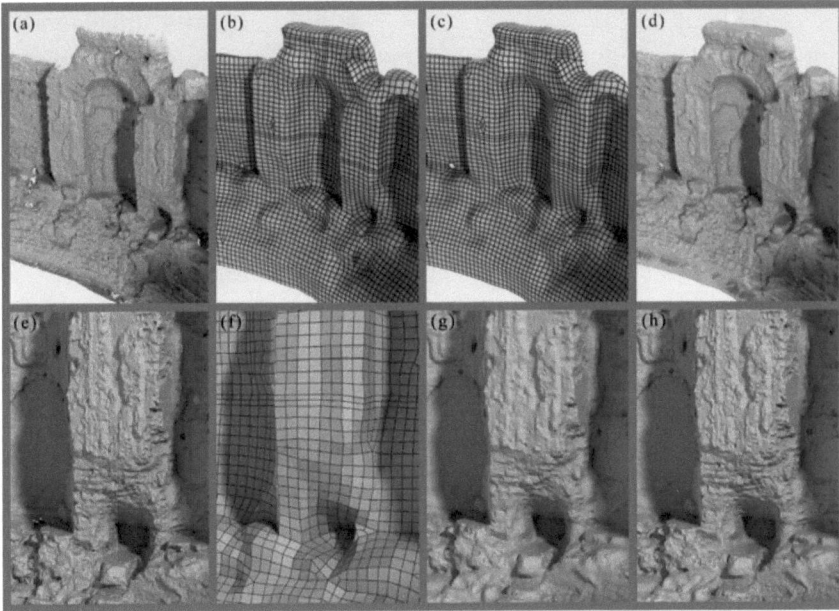

Figure 6. (a) high-poly mesh, 42.8 millions of polygons; (b) quad-dominant mesh M^0, 40,956 polygons; (c) M^n, LoD; (d) displaced subD, LoD; (e) detail of the masonry (high-poly model); (f) detail of the coarse mesh, or M^0; (g) increasing of the detail: 78,4330 polygons; and, (h) increasing of the detail: 20.8 millions of polygons.

In order to use the full power of render engines—in particular global illumination—on Lambertian surfaces, a shadow attenuation/removal on apparent colour textures is advisable. In fact, once a 3D digital model and the corresponding texture are created, a number of problems arise during the rendering: it is the so-called double shadow effect (DSE), which is due to the excessive pieces of information encoded in colour texture from SfM applications. For this reason, in CGI, they distinguish between apparent colour and diffuse colour: the first is obtained once the frames are re-projected on the model, while the second is the final result of an editing process aimed at removing shadows and other chromatic alterations caused by the interaction with other objects and environment. Focusing on the problem of BRDF diffusive component, several commercial applications facilitate the conversion of apparent colour into diffuse colour (e.g., CrazyBump Software CrazyBump or Rendering Systems ShaderMap). However, those solutions do not always fit with the chromatic reliability that is needed for Cultural Heritage. Neither, academic studies on sharp shadow removal on single pics (Fredembach and Finlayson, 2005; Kumar and Kaur, 2010; Guo et al., 2011) seem to be fitting with the typical problems to solve in architectural representation. In fact, we always try to get rid of sharp shadows in favour of photographic campaigns that are carried out during overcast sky (Martos and Cachero, 2015). Sophisticated shadow removal procedures are based on the knowledge of a more detailed ensemble: the object shape (including its position and orientation), natural, and artificial lighting sources, the exact timing of the photographic campaign and environmental lighting (Debevec et al., 2004); in this way, it is possible to achieve the spatially varying diffuse surface reflectance of a complex scene, with diffuse surfaces that are lit by natural outdoor illumination. The shadow removal technique (Cipriani et al., 2015b) applied in the case studies from Hadrian's Villa are based on two main points: the first is the possibility to have at one's disposal both three-dimensional geometry and the illumination of the scene. This includes the sensor/cameras location, the light source direction, and the observed object geometry, from which a priori knowledge of shadows projection is derived. Both direct and indirect illumination are "baked" and stored inside a grey scale image, saved in OpenEXR format, subjected to specific exposure correction and HDR toning.

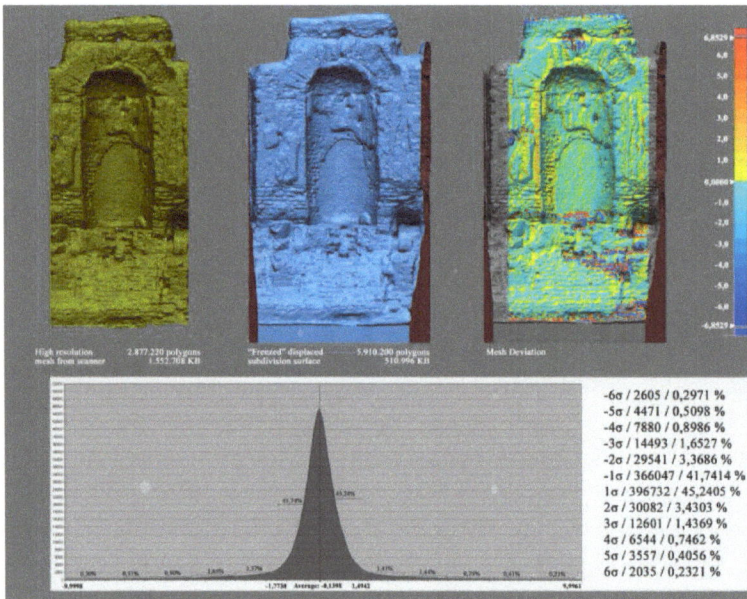

Figure 7. Deviation between the original high-density model from scanner and the corresponding displaced subD model obtained through automatic quad re-meshing (with some minor manual interventions).

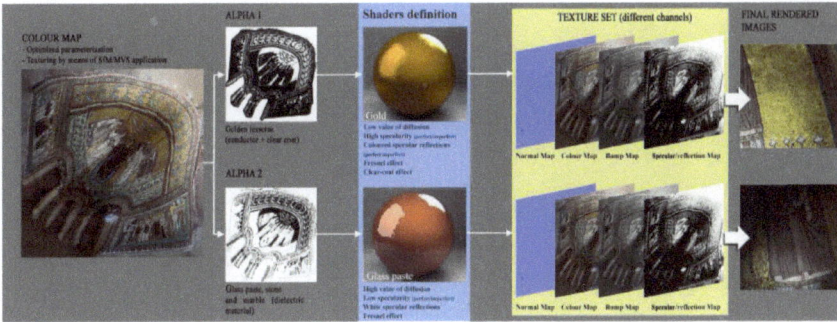

Figure 8. Texturing scheme for San Vitale's apse (Ravenna). The control on mesh parameterization enables the possibility to define different shading behaviours upon the same mesh. In case of poor (u,v) layouts characterized by high number of islands such possibility is precluded.

This grey scale image, once converted into a conventional one (8-bit per channel), can be used as a mask that is able to "blend" two different textures inside a bitmap-editing application. The two textures are obtained from the same set of images undergone to two different colour calibration using different "hero shots" (Gaiani et al., 2016). The first X-Rite ColorChecker is captured once it is placed on a reference area for bright zones, and the other chart is placed and acquired as a sample for darker areas. The first texture is then correctly exposed for the majority of the subject, while the second is over-exposed, except for deeper shadowed zones. Thanks to the blending provided by the above mentioned HDR map, the diffuse colour texture results are "flat" because it is deprived of shadows. The model in Figure 9a,c with such textures applied do not present too dark areas (DSE) once re-illuminated if compared to those obtained through standard colour processing techniques. On the contrary, other strategies based on tone mapping from HDR images (in this case from RAW image files) tend to produce over-exposed areas on brighter parts of the building, also loosing colour consistency (Figure 9b,d).

181

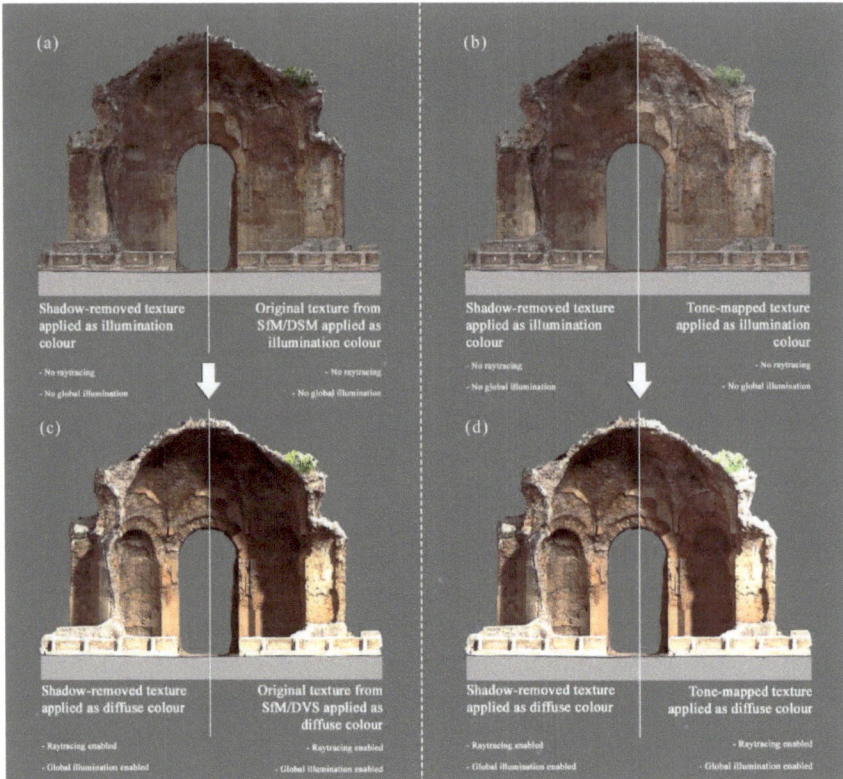

Figure 9. Vestibule of the Piazza d'Oro at Hadrian's Villa. Comparison among textures obtained by means of different pipelines: shadow-removal technique, colour calibration based on chromatic reference (X-Rite Colourchecker, tone mapping.

4. Case Studies

The tests carried out by the research unit are centred on two Italian UNESCO sites: Hadrian's Villa and late Antiquity and Byzantine monuments from Ravenna.

Cases from the pavilions of Hadrian's Villa are presented in order to underline different issues ranging from digitization of optically non-cooperating materials, to the texturing of approximately Lambertian surfaces, and, finally, semantic characterization of the following constructive elements:

- Main masonry: *opus mixtum.*
- Entablatures: travertine, Carrara marble, concrete.
- Columns: serpentine marble, concrete.
- Floor/groundworks: *opus caementicium* traces with footprints of removed elements in opus sectile.

182

In the case of Maritime Theatre—the Hadrian's personal domus—and the great southern nymphaeum of Piazza d'Oro (a vast multifunctional tricliniar area) archaeologists and architects working on it needed "multi-purpose" 3D model that could be used for the production of drawings (orthographic projections, perspectives, axonometric views), to be compared to archival documentation and historical drawings.

Models had to be specifically designed in order to store and facilitate the visual outputs underlining man-made alterations that occurred during the last centuries (ranging from plundering of archaeological finds to bad or non-intended interventions), and, last but not least, interactive models for enabling virtual anastylosis.

The achievement of multi-scale 3D models, namely LoD representations to be used as tools for virtual anastylosis become than a fundamental issue of the research unit working on both the pavilions of Hadrian's Villa, and on scattered elements once belonging to those buildings, in particular marble entablatures (stored in museums and *antiquaria*).

For the monuments from Ravenna (Arian Baptistery and the Basilica of San Vitale), the main focus was the achievement of a methodology enabling a more efficient BRDF definition for digital models of mosaic surfaces obtained through SfM-DSM pipelines. For these last cases, the problem of topology, as well as image and mesh segmentation, was crucial in order to supply render engines with a detailed modulation of reflectance values.

5. The Proposed Approach

With the aim to solve the problem of how to achieve reliable, multi-purpose models in compliance with the focus of this research, a five-step methodology was developed:

- A first draft point cloud segmentation is carried out manually in order to export toward mesh-processing applications single structural elements of the comprehensive point cloud. Then they are converted into high resolution meshes slightly bigger than the architectural element they represent. The set of independent high definition watertight meshes are achieved through global remeshing (in 3D Systems Geomagic design X) in order to get rid of both topologic and geometric defects (Cipriani et al., 2014). In case of Maritime Theatre or Piazza d'Oro, the list of categories is the following: columns (C), walls (M), entablatures (T) and sectors of floor split into original floor (OF), and simple soil (F). These meshes are aimed at forming a semantic partition on which the optimization and integration process will be carried out through techniques coming from entertainment applications (displaced subD, normal, displacement mapping, etc.).

183

- After an interdisciplinary discussion phase (with archaeologists, restorers, experts of history of art) a priority list is defined: each part of the complex is evaluated in order to understand its relevance and reliability. Semantics, in case of complex archaeological site/buildings, has to be considered as an all-encompassing tool to deepen the understanding of a building: the model, as well as its parameter space (namely, the way the model is split into (u,v) reference space), have to encode and make explicit characters/issues that otherwise could be misinterpreted on the basis of classical treatises. In other terms: strictly formal readings, based on classic orders (*genus*), may be quite far from being complete and reliable. Moreover, elements like floors, as well as portions of masonries, still include relevant formal features, enabling further interpretations, in particular, for removed elements.

- In order to improve the model's portability, reliable and less reliable/erroneous elements are optimized in two different ways. On the one hand, highly detailed models of a building's specific parts (flat elements as floors without specific decorations, ground, restored columns, etc.) have been converted into "low-poly" models, then they were parametrized, and finally, in order to apparently re-establish the original detail, normal maps have been calculated using the typical render-to-texture solutions called baking (Cohen et al., 1998). On the other hand, entablatures, columns, marble elements that are characterized by refined decorations and bass-reliefs, *opus caementicium*, and earthenware regular trails on the floor undergone a quad-dominant re-meshing in order to convert them into displaced subD with OpenEXR texture applied (Lee et al., 2000, Fantini 2012, Guidi and Angheleddu, 2016). Both quad-dominant and low-resolution triangle meshes are converted into Constructive Solid Geometries (CSG). In some cases, depending on special needs of the final model, it can be more advisable to carry out a semantic partition (Cipriani et al., 2014): in that case, it is the summation of a set of meshes that leads to a watertight mesh thanks to specific modelling procedures aimed at supplying continuity (Figures 10 and 11).

- All of the models were then textured using SfM applications since laser scanner and photogrammetric campaigns were carried out at once using a common network formed by ringed automatically detected (RAD) coded targets (Cipriani et al., 2015a).

- For approximately Lambertian surfaces a shadow removal process is carried out (masonry, bricks, mortar); for mosaic tiles an additional segmentation is performed in 2D, using alpha channels in the parameter space in order to apply different shaders to glass paste, golden tiles and stones (Figure 8).

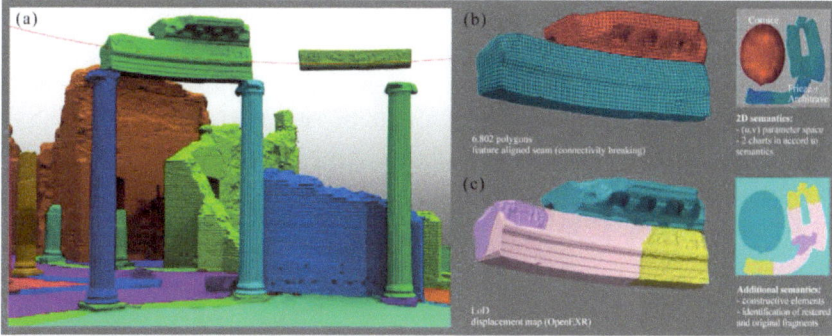

Figure 10. the Maritime Theatre at Hadrian's Villa: (**a**) view of the model showing a detail of the entablature, with a fragment of frieze located in the supposed original position: each colour is associated to an individual mesh and it represents a first draft segmentation; (**b**) the entablature, as well as other parts forming the whole model, are displaced subD made of quads: parametrization can be used as a second segmentation since the two islands are in accord to structural nature of the piece; (**c**) additional segmentation stages can be carried out in order to underline the nature of each fragment included inside a unique piece.

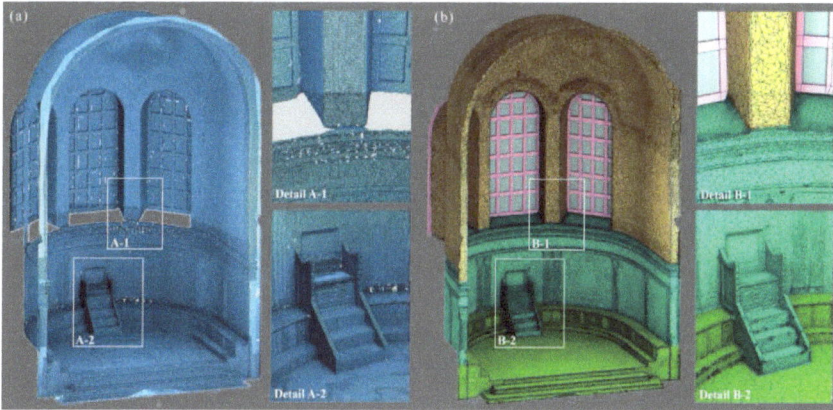

Figure 11. Semantic partition on the San Vitale apse: (**a**) High-poly mesh made of triangles and detail underlining the presence of voids; (**b**) a low-poly meshes made after the semantic partition. A complex task, once working with triangles is the one of following the curvature flow when breaking the connectivity among faces. In these cases quad-dominant meshes are more advisable.

6. Conclusions

The pipeline that is summarized in this paper leads to the achievement of highly compressed meshes, while keeping all of the acquired morphologic detail

185

gathered with sensors of a different nature. The strategy adopted here, underlines the role of displaced subD as one of the most interesting strategies in terms of visual reliability and data compression, even if this representation technique is currently far from being a standard in the field of remote sensing/surveying.

With the only exception of some high-end program belonging to design/mechanical engineering, subdivision surfaces cannot "dialogue" with the current scenario of applications for geomatics, so they have to be "frozen", i.e., blocked at a proper resolution and exported in other formats. At the same time, displaced subD are widely spread among entertainment applications, and from this point of view, their implementation inside mesh processing and reverse modelling applications may facilitate the dialogue with experts of dissemination by means of animations, rendering, as well as game developers who feel more comfortable with such models.

To understand why low resolution quad-dominant meshes are a standard inside entertainment software is food for thought for the community of developers in the field of commercial and open-source applications dedicated to survey.

It will be interesting to see whether or not state of the art applications in surveying will introduce polygonal "quads" in the next few years; moreover, due to the high interest shown by VFX software houses toward automatic quads-remeshing, it is safe to say that in the years ahead, we shall see robust implementations of these algorithms in several applications. In commercial terms, a software house in the field of SfM/MVS, which will introduce special tools that are aimed at facilitating the workflow for the production of 3D digital assets for gaming industry (quad re-meshing, integrated parametrization tools, baking of normals, displacement, etc.) could gain high market shares (avoiding the current practice to import and export data, jumping from an application to another).

A 2D (u,v) reference system that is visually consistent with the corresponding 3D model in (x,y,z) can also open a door for a deeper exploitation of parameter space, seen as an auxiliary to semantic enrichment.

The final models comply with the requirements stated: the visual outputs in suitable for both animations, rendering, and graphic outputs that are useful for documentation, and also in the frame of restoration and maintenance practices (Figures 12–14).

Figure 12. The Maritime theatre digital model obtained through the pipeline exposed in the paper. It can be efficiently used as the base for virtual anastylosis and other studies. Every part of the model is in the form of Constructive Solid Geometries (CSG) and allows sections as well as the production of reliable high-resolution ortho-image.

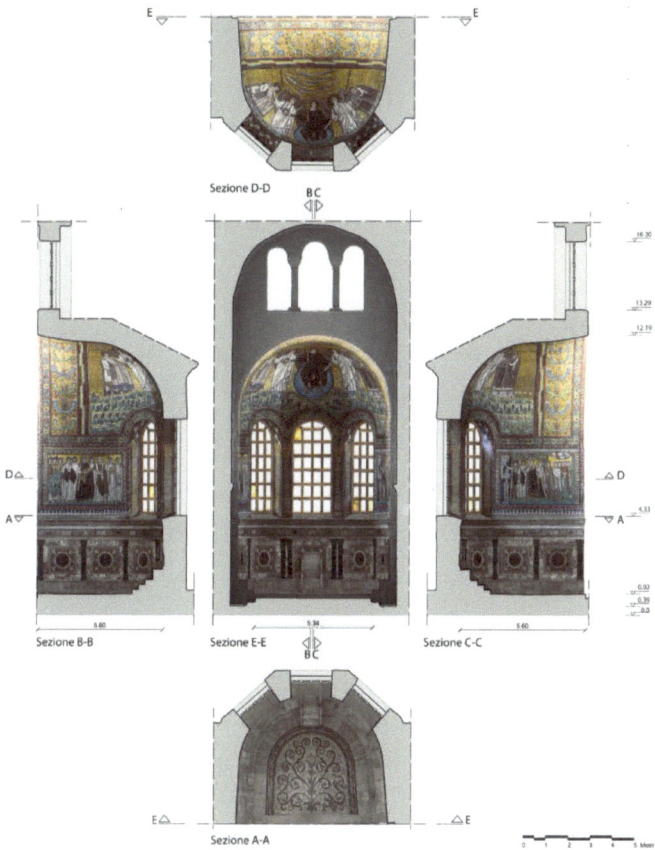

Figure 13. Architectural drawing obtained using the digital model of the apse. The model is also suitable for accurate simulation of light and materials.

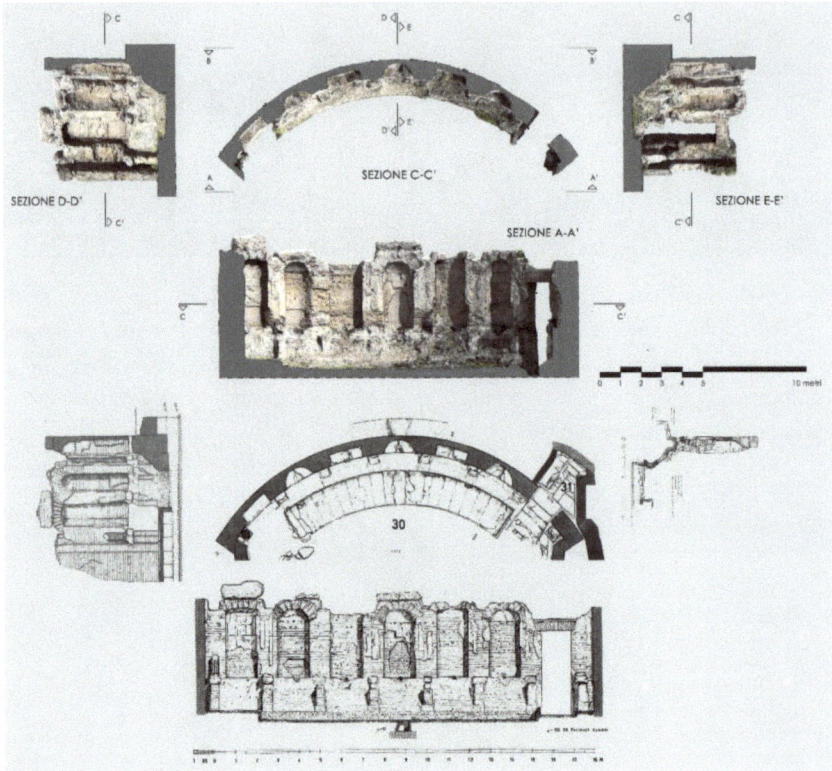

Figure 14. Comparison between orthographic representations of the Southern Nynphaeum at Piazza d'Oro. Displaced subdivision surface model (modelling and texturing by Giacomo Mussoni), traditional survey by Friedrich Ludwig Rakob (1967).

Acknowledgments: Authors would like to thank Benedetta Adembri of the Istituto "Villa Adriana e Villa d'Este" (formerly, Soprintendenza Archeologia del Lazio e dell'Etruria Meridionale) for the scientific coordination of all the archeological aspects of the research on mixtilinear plan-design of Hadrian's Villa and Sergio Di Tondo (MicroGeo s.r.l.) for technical advice during all the surveying campaign in these years. Thanks to the kind cooperation of other collaborators/tutors that during the last years carried out several surveying campaigns: Silvia Bertacchi, Gianna Bertacchi, Simone Vianello, Simone Rostellato, and Luca Grossi. Authors wish to thank all the collaborators due to their efforts during surveying campaigns and during time-consuming tests and modelling operations. In particular: Vincenza Carollo for semantics/meshing of the Maritime Theatre, Giacomo Mussoni for testing several quad-dominant remeshing applications and for the editing of the southern nymphaeum of Piazza d'Oro, Francesca Gagliardi for modelling and texturing San Vitale's apse.

References

1. Adembri, B.; Di Tondo, S.; Fantini, F.; Ristori, F. Nuove prospettive di ricerca su Piazza d'Oro e gli ambienti mistilinei a pianta centrale: confronti tipologici e ipotesi ricostruttive. In *Adriano e la Grecia, Villa Adriana fra Classicità ed Ellenismo*; Calandra, E., Adembri, B., Eds.; Studi e Ricerche. Electa: Milan, Italy, 2014; pp. 81–90.

2. Bianchini, C. (Ed.). La documentazione dei teatri antichi del Mediterraneo. In *Le Attività del Progetto ATHENA a Mérida*; Gangemi: Rome, Italy, 2013.

3. Blinn J. Simulation of Wrinkled Surfaces. *ACM SIGGRAPH Comput. Graph.* **1978**, *12*, 286–292.

4. Bunnell, M. Adaptive Tessellation of Subdivision Surfaces with Displacement Mapping. In *GPU Gems 2: Programming Techniques for High-Performance Graphics and General-Purpose Computation*; Pharr, M., Randima, F., Eds.; Addison-Wesley Professional: Boston, MA, USA, 2005; Chapter 7, pp. 109–122.

5. Catmull, E.; Clark, J.; Recursively generated B-spline surfaces on arbitrary topological meshes. *Comput.-Aided Des.* **1978**, *10*, 350–355.

6. Cipriani, L.; Fantini, F.; Bertacchi, S. 3D models mapping optimization through an integrated parameterization approach: cases studies from Ravenna. *Int. Arch. Photogramm. Remote Sens. Spat. Inf. Sci.* **2014**, *XL-5*, 173–180.

7. Cipriani, L.; Fantini, F. Modelli digitali da Structure from Motion per la costruzione di un sistema conoscitivo dei portici di Bologna. In *Disegnare Idee Immagini n° 50*; Gangemi: Roma, Italy, 2015; pp. 70–81.

8. Cipriani, L.; Fantini, F.; Bertacchi, S. El color en las piedras y en los mosaicos de Rávena: nuevas imágenes de los monumentos antiguos a través de la fotogrametría no convencional de última generación. In *EGA. Revista de Expresión Gráfica Arquitectónica, No. 26*; Departamento de Expresión Gráfica Arquitectónica: Valencia, Spain, 2015, pp. 190–201.

9. Cipriani, L.; Fantini, F.; Bertacchi, S. 3D Digital Models for Scientific Purpose: Between Archaeological Heritage and Reverse Modelling. In *Handbook of Research on Emerging Technologies for Architectural and Archaeological Heritage*; Ippolito, A., Ed.; IGI Global: Hershey, PA, USA, 2016; Chapter 10, pp. 291–321.

10. Cohen, J.; Olano, M.; Manocha, D. Appearance-Preserving Simplification. In Proceedings of the 25th Annual Conference on Computer Graphics and Interactive Techniques, Orlando, FL, USA. 19–24 July 1998; pp. 115–122.

11. Debevec, P.; Tchou, C.; Gardner, A.; Hawkins, T.; Poullis, C.; Stumpfel, J.; Jones, A.; Yun N.; Einarsson, P.; Lundgren, T.; Fajardo, M.; Martinez, P. *Estimating Surface Reflectance Properties of a Complex Scene under Captured Natural Illumination*; USC ICT Technical Report ICT-TR-06.2004; University of Southern California Institute for Creative Technologies: Los Angeles, CA, USA, 2004; pp. 1–12.

12. De Rose, T.; Kass, M.; Truong, T. Subdivision surfaces in character animation. In Proceedings of the 25th Annual Conference on Computer Graphics and Interactive Techniques (SIGGRAPH '98), Orlando, FL, USA. 19–24 July 1998; pp. 85–94.

13. Fantini, F. Capitolo n°2: Teorie e tecniche della rappresentazione numerica o poligonale. In *Geometria descrittiva, Vol. II, Tecniche e Applicazioni*; Migliari, R., Ed.; Città Studi edizioni-De Agostini Scuola: Novara, Italy, 2009; pp. 60–94.

14. Fantini, F. Modelos con nivel de detalle variable realizados mediante un levantamiento digital aplicados a la arqueología. In *EGA. Revista de Expresión Gráfica Arquitectónica*; Departamento de Expresión Gráfica Arquitectónica: Valencia, Spain, 2012; Volume 19, pp. 306–317.

15. Fredembach, C.; Finlayson G. Hamiltonian path based shadow removal. In Proceedings of the 16th British Machine Vision Conference (BMVC), Oxford, UK, 5–8 September 2005; Clocksin W., Fitzgibbon Andrew, W., Torr Philip H.S., Eds.; British Machine Vision Association (BMVA Press): Los Alamitos, CA, USA, 2005; pp. 502–511.

16. Gaiani, M.; Remondino, F.; Apollonio, F.I.; Ballabeni, A. An Advanced Pre-Processing Pipeline to Improve Automated Photogrammetric Reconstructions of Architectural Scenes. *Remote Sens.* **2016**, *8*, 1–27.

17. Gobbetti, E.; Marton, F.; Rodriguez, M.B.; Ganovelli, F.; Di Benedetto, M. Adaptive quad patches: an adaptive regular structure for web distribution and adaptive rendering of 3d models. In Proceedings of the 17th International Conference on 3D web Technology, Los Angeles, CA, USA, 4–5 August 2012; pp. 9–16.

18. Grün, A.; Remondino, F.; Zhang, L. Photogrammetric reconstruction of the Great Buddha of Bamiyan, Afghanistan. *Photogramm. Rec.* **2004**, *19*, 177–199.

19. Guo, R.; Dai, Q.; Houiem, D. Single-image shadow detection and removal using paired regions. In Proceedings of the IEEE Conference on Computer Vision and Pattern Recognition (CVPR), Colorado Springs, CO, USA, 20–25 June 2011; Volume 1, pp. 2033–2040.

20. Guidi, G.; Angheleddu, D. Displacement mapping as a metric tool for optimizing mesh models originated by 3D digitization. *ACM J. Comput. Cult. Herit.* **2016**, *9*, 9–23.

21. Jacques, A.; Bonfait, O. (Eds.) *Italia Antiqua. Envois Degli Architetti Francesi (1811–1950). Italia e Area Mediterranea", Catalogo Della Mostra*; École Nationale Supérieure des Beaux-Arts: Paris, France, 2002.

22. Kumar, S.; Kaur A. Shadow detection and removal in colour images using Matlab. *Int. J. Eng. Sci. Technol.* **2010**, *2*, 4482–4486.

23. Hoppe, H. Progressive meshes. In Proceedings of the 23rd annual conference on computer graphics and interactive techniques (SIGGRAPH'96), New Orleans, LA, USA, 4–9 August 1996; ACM Press: New York, NY, USA, 1996; pp. 99–108.

24. Lai, Y.K.; Kobbelt, L.; Hu, S.M. An incremental approach to feature aligned quad dominant remeshing. In Proceedings of the SPM '08 ACM symposium on Solid and Physical Modelling, Stony Brook, NY, USA, 2–4 June 2008; pp. 137–145.

25. Lee, A.; Moreton, H.; Hoppe, H. Displaced subdivision surfaces. In Proceedings of the 27th Annual Conference on Computer Graphics and Interactive Techniques, New Orleans, LA, USA, 23–28 July 2000; ACM Press/Addison-Wesley Publishing Co.: New York, NY, USA, 2000; pp. 85–94.

26. Jakob, W.; Tarini, M.; Panozzo, D.; Sorkine-Hornung, O. Instant field-aligned meshes. *ACM Trans. Graph.* **2015**, *34*, 1–15.

27. Merlo, A.; Vendrell-Vidal, E.; Fantini, F.; Sanchez-Belenguer, C.; Aliperta, A. 3D model visualization enhancement in real time game engines. *Int. Arch. Photogramm. Remote Sens. Spat. Inf. Sci.* **2013**, *XL-5/W1*, 181–188.
28. Martos, A.; Cacheroa, R. Acquisition and reproduction of surface appearance in architectural orthoimages. *Int. Arch. Photogramm. Remote Sens. Spat. Inf. Sci.* **2015**, *XL-5/W4*, 139–146.
29. Potenziani, M.; Callieri, M.; Dellepiane, M.; Corsini, M.; Ponchio, F.; Scopigno, R. 3DHOP: 3D Heritage Online Presenter. *Comput. Graph.* **2015**, *52*, 129–141.
30. Tryfona, M.S.; Georgopoulos, A. 3D image based geometric documentation of the Tower of Winds. *Int. Arch. Photogramm. Remote Sens. Spat. Inf. Sci.* **2016**, *XLI-B5*, 969–975.

Cipriani, L.; Filippo Fantini, F. Integration of Pipelines and Open Issues in Heritage Digitization. In *Latest Developments in Reality-Based 3D Surveying and Modelling*; Remondino, F., Georgopoulos, A., González-Aguilera, D., Agrafiotis, P., Eds.; MDPI: Basel, Switzerland, 2018; pp. 171–191.

191

Step into Virtual Reality—Visiting Past Monuments in Video Sequences and as Immersive Experiences

Thomas P. Kersten [a], **Felix Tschirschwitz** [a], **Simon Deggim** [a] and **Maren Lindstaedt** [a]

HafenCity University Hamburg, Photogrammetry & Laser Scanning Lab, Überseeallee 16, D-20457 Hamburg, Germany; (Thomas.Kersten, Felix.Tschirschwitz, Simon.Deggim, Maren.Lindstaedt)@hcu-hamburg.de

Abstract: Recent advances in contemporary Virtual Reality (VR) technologies are going to have a significant impact on everyday life. Through VR, it is possible to virtually explore a computer-generated environment as a different reality, and to immerse oneself into the past or in a virtual museum without leaving the current real-life situation. Cultural heritage monuments are ideally suited both for thorough multi-dimensional geometric documentation and for realistic interactive visualisation in immersive VR applications. In this contribution, the generation of virtual 3D models of past monuments in Bad Segeberg, Germany, and its processing for data integration into the game engine Unreal is presented. The workflow from data acquisition via 3D modelling to VR visualisation using the VR system HTC Vive, including the necessary programming for navigation, is described. Furthermore, the use (including simultaneous multiple-user environments) of such a VR visualisation for Cultural Heritage (CH) monuments is discussed in this contribution.

Keywords: 3D; 4D; cultural heritage; HTC Vive; monuments; reconstruction; virtual museum

1. Introduction

1.1. Virtual Reality

Virtual Reality (VR) will change our future life. Although VR is not very new, it is currently at the beginning of a technological transition. The term Virtual Reality was introduced by the author Damien Broderick in his science fiction novel *The Judas Mandala* published in 1982. As early as 1962, Morton Heilig built the Sensorama, a machine that is one of the earliest known examples of immersive, multi-sensory (now known as multimodal) technology and which could be named as the first VR system (Rheingold, 1991). Nevertheless, it took another thirty years until the first Head-Mounted Display (HMD) for the mass market was released with the VFX 1 from Forte (Cochrane, 1994). Unfortunately, this headset caused nausea and disturbances of equilibrium for users due to the long latency when visually updating the display. After almost another twenty years, the new HMD Oculus Rift was announced by Palmer Luckey, which really worked and which rang in the new era (Desai et al., 2014).

It is already obvious that this new technology will offer great opportunities for many applications such as medicine, engineering, computer sciences, architecture, cultural heritage and virtual restoration. VR typically refers to computer technologies that use software to generate the realistic images, sounds and interactions that replicate a real environment, and simulate a user's physical presence in this environment. Furthermore, VR has been defined as a realistic and immersive simulation of a three-dimensional environment, created using interactive software and hardware, and experienced or controlled by movement of the user's body or as an immersive, interactive experience generated by a computer. VR offers an attractive opportunity to visit objects in the past (Gaitatzes et al., 2001) or places, which are not easily accessible, often from positions which are not possible in real life. Moreover, these fundamental options are increasingly being implemented today through so-called "serious games", which embed information in a virtual world and create an entertaining experience (edutainment) through the flow of and interaction with the game (Anderson et al., 2010; Mortara et al., 2014). The first virtual museum using the VR system HTC Vive as a HMD for immersive experiences was introduced by Kersten et al. (2017).

1.2. Virtual Museum

The function of a museum is to aid non-specialists in understanding information and context via an interaction of short duration. Ideally, museums should also deepen visitors' interest in the subjects that they present. In accordance with their educational mission, museums must constantly present and re-present complex issues in ways that are both informative and entertaining, thus providing access to a wide target audience. Visitors with prerequisite knowledge, prior experiences, as well as associated individual interests and objectives tend to take a more active role in engaging with museums (Reussner, 2007). Today, these fundamental ideas are transferred to the digital world through so-called "serious games", which embed information in a virtual world and create an entertaining experience through the flow of and interaction with the game (Mortara et al., 2014).

For the museum field, the consolidation and implementation of culture and information technology is often called Virtual Museum (VM). The definition of a Virtual Museum is, however, not fixed. Since the 1990s, many different definitions for a VM have been published with significant differences depending on the contemporary status of information and communication technology (ICT)

(Shaw, 1991; Schweibenz, 1998; Jones and Christal, 2002; Petridis et al., 2005; Ivarsson, 2009; Styliani et al., 2009). According to V-MusT (2011), "a virtual museum is a digital entity that draws on the characteristics of a museum, in order to complement, enhance, or augment the museum experience through personalization, interactivity and richness of content. Virtual museums can perform as the digital footprint of a physical museum, or can act independently [...]". Pujol and Lorente (2013) use the term VM to refer to a digital spatial environment, located in the WWW or in the exhibition, which reconstructs a real place and/or acts as a knowledge metaphor, and in which visitors can communicate, explore and modify spaces and digital or digitalized objects. Pescarin et al. (2013) evaluated VMs. They found that the impact of interactive applications on the user seems to depend on the capability of the technology to be "invisible" and to allow a range of possibilities for accessing content. To achieve this, VMs need a more integrated approach between cultural content, interfaces, and social and behavioural studies. Content presentation ranges from text, images, sound and videos to interactive techniques such as animated 3D models, to act as a central platform for the informative supplement of the real museum visit (Samida, 2002). The design of the VM varies from simple Web pages (Bauer, 2001) to panorama-based virtual tours (Kersten and Lindstaedt, 2012) to interactive apps for smartphones or tablets (Gütt, 2010). A good example for a VM is AfricanFossils.org, which presents a spectacular digital collection of fossils and artefacts found mostly at Lake Turkana in East Africa in a three-dimensional virtual lab space on a website (http://africanfossils.org/). The digital collection of animals, human ancestors, and ancient stone tools offers a unique tool for scholars and enthusiasts to explore and interact with the collection online. Another example for a digital collection of exhibits is Smithsonian X 3D (https://3d.si.edu/), for which various 3D capture methods are applied to digitize iconic collection objects. The idea of Smithsonian X 3D is to promote the use of 3D data for many different applications for professional and laymen users alike.

A VM that is retrievable on the Internet would offer the possibility of making a time- and location-independent virtual visit to the museum. It would also facilitate preparation for and evaluation of an actual museum visit, as this medium stimulates the attention of the visitor while also providing further information. The great strength of a VM is the ability to utilise current ICT to supplement conventional exhibition techniques via the presentation and integration of content into the real exhibition, thus significantly contributing to a visitor's understanding.

1.3. Past and Preserved Monuments

Cultural Heritage (CH) monuments are significant testimonies of the human past. These memorials are in danger around the world today due to increasing devastation by war, terrorism and vandalism, as well as by creeping weathering.

For many CH objects, it is already too late, i.e., they were destroyed in the past and they are probably already forgotten. Examples of such meaningless destruction are, among many others, the Great Buddha statues from Bamiyan in Afghanistan (Gruen et al., 2002), blown up in March 2001 by the Taliban, the Minaret of the Umayyad Mosque of the UNESCO heritage site of the ancient city of Aleppo (Fangi and Wahbeh, 2013), which was destroyed during the current Syrian civil war and the Great temple of Bel in the archaeological momentous site Palmyra in Syria (Wahbeh et al., 2016), one of the most important Syrian heritage monuments, which was destroyed in September 2015 by the so-called "Islamic State". These terrible acts of destruction demonstrate clearly that metric documentation of (important) ancient monuments is an essential and very useful reference for a future reconstruction or restoration of those CH objects. However, a metric documentation is very often not available due to missing appropriate recording techniques and systems in the past. Photography and photogrammetry have been used for the documentation of buildings and monuments since the late 19th century (Meydenbauer, 1867). Wiedemann et al. (2000) used, for example, photographic glass plates from the Meydenbauer archives for reconstructing the historical commandant's office building in Berlin, which was destroyed in WW II. Another example for the documentation of historic buildings such as North-German castles by digital photogrammetry is given by Kersten et al. (2004). The advantages of another technique, terrestrial laser scanning for the documentation of huge historic buildings such as the Imperial Cathedral (Kaiserdom) of Königslutter, Germany, are described in Kersten and Lindstaedt (2012).

However, such modern metric documentation techniques were not available for lost places such as the Siegesburg, which is also known as Segeberg castle (Figure 1), an ancient monument from the Dark Ages and the Early Modern Age (from a time more than 500 years ago), which was destroyed in the 17th century and demolished during the following centuries. In this contribution, the return of the Siegesburg by 3D virtual reconstruction is presented. The laboratory for Photogrammetry & Laser Scanning of the HafenCity University Hamburg carried out this reconstruction in co-operation with the museum Old-Segeberg Town House (Alt-Segeberger Bürgerhaus) using historic sources such as paintings, engravings and isometric maps to reerect the important monument. The reconstruction has been conducted in three different phases: (a) reconstruction of the Kalkberg, (b) reconstruction of the castle in AutoCAD and (c) texture mapping and visualisation of the castle with Lumion 3D. A detailed workflow of this project is described in Deggim (2015).

Additionally, a VM has been developed for the museum of Old-Segeberg town house, example of preserved monuments representing the past, as an interactive tour for a Windows-based computer system and as a virtual reality application in 3D using the Virtual Reality System HTC Vive. Based on this concrete example, this contribution provides examples of how museums can fulfil the technological and media requirements in the 21st century using detailed geo data and appropriate ICT.

2. The Cultural Heritage Monuments

2.1. The Siegesburg in Segeberg — The Lost Monument

The first castle on the Kalkberg was built in the first half of the 12th century (1128) by the Danish Duke Knud Lavard, but it was already destroyed in 1130 by the Schauenburg count Adolf I, who felt threatened by the fortress. The Roman-German Emperor Lothar of Supplinburg (Lothar III) ordered to build a new castle on the Kalkberg following the advice of the missionary Vicelinus in 1134. The castle should serve as a base for the Christianisation at the edge of the border to the Slavic tribes. In the course of various armed conflicts in the following centuries, the castle was besieged several times by enemy troops, sometimes destroyed, and then rebuilt. However, the centrality of the Siegesburg increasingly strengthened its suitability as a centre of political power and as a residence. After the owners of the castle, the Schauenburg family, died out in the 15th century, the castle and the County of Holstein passed on to the Danish King Christian I in 1459. However, the castle was so badly damaged in 1534 by a devastating city fire that the Danish Governor and bailiff Heinrich Rantzau ordered the restoration of the dilapidated Siegesburg as an extensive building complex as is shown in the engraving by Braun-Hogenberg from 1588 and the contemporary illustration from 1595 (Figure 1). After the outbreak of the Thirty Years' War, the castle underwent various occupations. From 1627 until 1629, imperial troops of Wallenstein occupied Segeberg and took the unfortified castle on the Kalkberg without a fight. In retaliation, the open castle was burned down at the end of the Thirty Years' War in 1644 by the Swedish troops. After centuries of intensive mining of the Kalkberg, only the lower half of the sole castle fountain of Northern Germany, approximately 42 meters in depth, reflects the once-powerful Siegesburg. It was driven into the rock and is today located in a steep reduction wall. Between 1934 and 1937, the Reich Labor Service built the Kalkberg Stadium in the pit created by gypsum mining. This is an outdoor stage with about 7800 seats and standing room for 12,000. The Karl May Festival (open-air theatre festival about Karl May's adventure novels about the Wild West) has been held here every year since 1952.

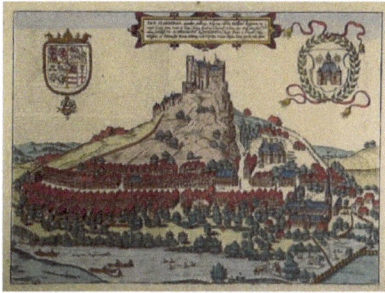

Figure 1. Isometric map of Segeberg including the castle Siegesburg and the Old-Segeberg Town House in the year 1588 from Braun-Hogenberg (view from North to South, top left) and a contemporary illustration of the Siegesburg on top of the Kalkberg from 1595 (view from southeast, top right). Photo ((bottom left) and synthetic view (bottom right) of the front façade of the Old-Segeberg Town House.

2.2. The Old-Segeberg Town House — The Preserved Monument

Even at the end of the 19th century, the Old-Segeberg Town House (Figure 1 bottom), located in the city of Bad Segeberg 40 km northeast of Hamburg, was already known as the oldest house of the city. Today, it is one of only a few well-preserved, small urban town houses from the beginning of the early modern period in the federate state of Schleswig-Holstein. In the newly installed council book from 1539, the building was already included in the historic rent listing. After Segeberg was almost completely destroyed in June 1534 during the Count feud from 1533–1536/37, the town house was re-established in 1541. First, as a simple

197

hall building with a single-storey, in-frame construction with brick-bracing to the property that is today's Lübecker Straße No. 15. The method of construction was poor and building materials from neighbouring ruined properties (e.g., in the roof framing) were partly recycled. However, the basic structure of the framework construction was established from fresh wood (oak). It is presumed that the cellar with walls, constructed from boulders, hailed from the medieval predecessor building (Reimers and Hinrichsen, 2015). Later, in the following centuries, the house was extended and converted several times.

With the support of a historian and based on historical sources (Reimers and Hinrichsen, 2015), six construction phases of the building could be identified. These were each modelled in AutoCAD and are presented in chronological order in Figures 2 and 3, from left to right: (1) construction work (1541), (2) the first extension (around 1587), (3) Stall addition at the south front (before 1805), (4) extension of the living space (from 1814), (5) renovation and conversion of the front façade (ca. 1890), and (6) refurbishment of the building and conversion to a museum (1963/64). A detailed description of the six construction phases of the building is presented in Kersten et al. (2014).

After the refurbishment in 1963–1964, the building contained the local museum of the City Bad Segeberg. For the next few decades, exhibits from the petty-bourgeois living and working environment of the 19th and 20th centuries were shown in its historic rooms. After the adult education centre (Volkshochschule), Bad Segeberg took over sponsorship of the museum in 2012; it was renamed "Museum Alt-Segeberger Bürgerhaus" and successive permanent exhibitions on the topics "500 years development of civic culture in the mirror of a 470-year-old house" and "800 years history of the city of Segeberg - from the medieval castle settlement to the modern resort" were hosted.

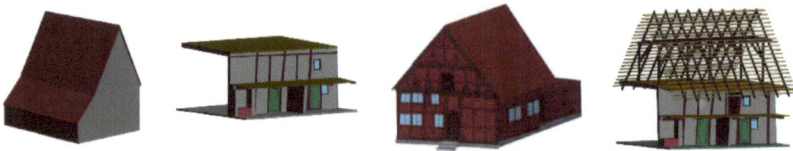

Figure 2. 3D model of Old-Segeberg town house. From left to right: 1541 outside, 1541 inside, 1584–1588 outside and 1584–1588 inside (Kersten et al., 2014).

Figure 3. 3D model of Old-Segeberg town house. From left to right: 1814 outside, 1814/1890 inside, 1890 outside and 1964 outside/inside (Kersten et al., 2014).

3. Reconstruction and Modelling

3.1. The Kalkberg and the Siegesburg

The initial step of the project was the reconstruction of the historic Kalkberg. First, the modelling of the Kalkberg was carried out with butter in order to derive just the shape of the mountain without any scale using historic sources such as paintings, isometric maps and descriptions. Secondly, today's existing landscape has been modelled by meshing of airborne laser scanning (ALS) data from the national survey of Schleswig-Holstein. Figure 4 (left) illustrates the ALS data as a mesh including the current Kalkberg and the open-air stage of the Karl May Festival in Bad Segeberg. From this DEM, sections with a spacing of 10 m were generated and plotted in a scale of 1:330. These sections were transferred to cardboards, cut out and then fixed upright on a planar platform (Figure 4 centre) for further modelling with gypsum (Figure 4 right). Third, the shape of the modelled butter mountain coming out of the fridge was visually transferred to the physical 3D model by manual modelling. The finished physical 3D model of the historic Kalkberg (Figure 5 left) was photographed with a Nikon D800 for digitisation using dense image matching. The final meshed model is illustrated in Figure 5 (centre), which was combined with the meshed model from ALS data after scaling (Figure 5 right). This combined meshed 3D model was the base for the modelling of the Siegesburg.

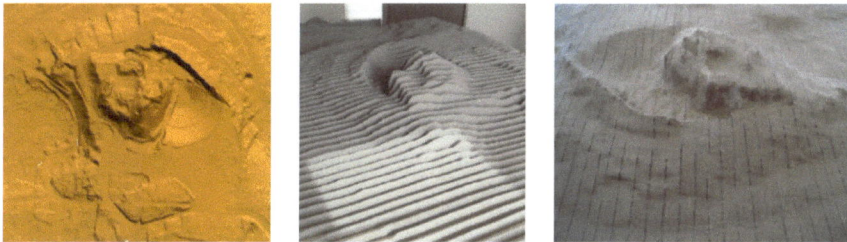

Figure 4. The reconstruction of the Kalkberg from airborne laser scanning data: digital elevation model (original data, left), printed scaled sections of the DEM (centre) and physical 3D model of today's existing Kalkberg (right). Photos centre and right: Nils Hinrichsen.

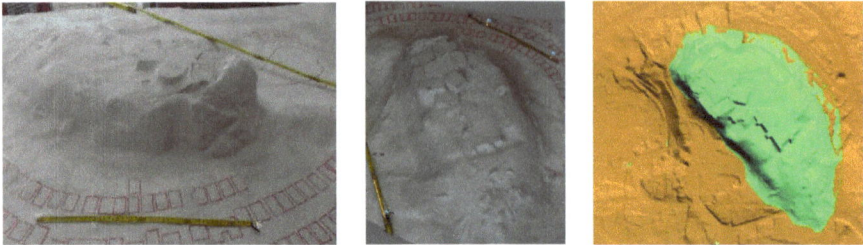

Figure 5. The historic physical 3D model of the Kalkberg (left), the historic digital 3D model from dense image matching (centre) and the combination of two DEMs from both today (orange) and the past (green) (right).

The first virtual 3D model of the Kalkberg and the Siegesburg (Figure 6 left) was created in Cinema 4D by Dipl.-Ing. Uwe Oswald, lecturer for 3D design at HTK – Academy of Design in Hamburg, some years ago. However, this virtual object includes some small false assumptions about the shape of the castle and it was not reconstructed with great detail. Nevertheless, a second virtual model was created and adapted to the reconstructed digital terrain model of the Kalkberg in Cinema 4D using the expertise of a historian (Figure 4 centre). This was used by a model builder to create a physical castle model on top of the Kalkberg for the museum exhibition (Figure 6 right).

Figure 6. The first virtual 3D model of the Kalkberg and the Siegesburg (left) created by Dipl.-Ing. Uwe Oswald in Cinema 4D, and the final physical castle model in the museum exhibition (right).

The initial CAD model shows the rough position and size of the castle, but it lacks any details which are necessary for the generation of a video sequence allowing close-up shots from the castle to be focused on. Thus, the basic CAD model (Figure 6 left), the physical model in the museum and the available historic illustrations were used as the base for the new model to be constructed. The first step was to research typical elements of Northern European castles in the Early Modern Age and to collect example images. The Siegesburg had existed already for over 450 years in the year 1600 AD and had often been attacked, partly destroyed, rebuilt, renovated and expanded. Thus, the castle should not have a continuous style but should consist of various architectural elements.

The 3D modelling has been carried out with Autodesk AutoCAD. Due to the size of the castle, it was split into different object parts, constructed individually and was later merged in the visualisation software Lumion 3D. The balance between details and data reduction should be considered, since a high number of polygons results in performance problems of the computer system during the work process (Deggim, 2016). This is a very significant factor to be considered, especially if the models are reused for real-time virtual reality applications as used later in an ongoing project.

The modelled buildings are based on the image collection. They combine elements from various sources or are based on single images. The shape of each building, given by the initial CAD model, was the starting point for the construction and correction of each object to be modelled. Afterwards, doors, windows, roofs and other details such as a stepped gable or timber framework were added to the building models (Figure 5 and Figure 6 top).

To increase the degree of reality, the former use of each building was taken into account. Tower platforms and the defence galleries are reachable via ladders or staircases or they can be accessed from adjacent buildings. Walls without a defence gallery were modelled with arrow slits and machicolations were placed at reasonable locations. The former use of each individual building of the castle is not known in general. However, a typical storehouse scene with a crane was added to the main castle to make the scene lively. For additional variety, small objects such as chests, buckets, fences, boxes, torches, tables, benches, cups, weapons, wheels and many other small constructions such as a wooden shelter and a cart were modelled and placed in the scene (Figure 12 bottom). In total, 21 buildings and 31 scenery objects were modelled and later integrated into the castle scenery.

Figure 7. Comparison between the initial and final model for the keep (left), a building at the main castle (centre) and the drawbridge (right).

3.2. The Old-Segeberg Town House

The entire museum Old-Segeberg Town House was modelled in 3D so that visitors could virtually explore the exterior and interior of the building in close

201

relationship to the various exhibits in the museum. Special focus was given to developing visitors' understanding of the complex history of the building via an interactive visualisation of the nearly 500-year-old museum building's extensive construction history.

The 3D object recording was conducted over two separate acquisition campaigns, on 21 April and 2 August 2011, using the IMAGER 5006h terrestrial laser scanner and two digital SLR cameras, Nikon D40 and D90, for the exterior and interior areas, respectively. The base for the development of the virtual museum was this detailed 3D recording of the exterior and interior of the building using digital photogrammetry and terrestrial laser scanning. Intensive 3D CAD modelling, using coloured point clouds from laser scanning and manually-measured photogrammetric 3D points, represented the second stage of activity. For the 3D modelling and visualisation of huge point clouds in AutoCAD, the plug-in PointCloud (today Faro PointSense), from the company Kubit in Dresden, Germany, was used. Using this plug-in, CAD elements, e.g., surfaces such as the half-timbered bars, could be directly digitised in the point cloud. In the oriented images, each object point was measured manually in at least four photos from different camera stations. After the manual image point measurements were completed for one object element, for example a window, the computed 3D object points for this element were transferred to AutoCAD. There, polylines were generated from these points, and these were later used to generate surfaces. Simple object parts were constructed using geometrical primitives (e.g., cuboid, pyramid, cylinder, cone, sphere, torus, etc.), while some more complex object parts were created with the Boolean operators in CAD (union, subtraction and intersection). Kersten et al. (2014) give a detailed description of the data acquisition and modelling of the town house.

Based on this reconstructed 3D model, further 3D modelling was carried out to fulfil the requirements for the development of the virtual museum. To bring the interior to life, the most important exhibits, information panels and furnishings were also modelled and placed in their appropriate places within the building. Additionally, the six different historical construction stages of the Old-Segeberg Town House were modelled, in collaboration with the historian Nils Hinrichsen (Director of the Museum Old-Segeberg town house). The appearance of the building, in particular for the early construction phases, is only ensured for building parts based on historical scientific evidence collected in recent years. For example, a dendrochronological analysis of individual timbers was conducted, which determines the age of various parts of the building and which could be assigned to the corresponding construction phases (Reimers and Hinrichsen, 2015). For data reduction, the six construction phases were modelled together, i.e., objects, which occur in several construction phases, were created only once and stored in a database, thus allowing utilisation by the program in multiple phases.

Figure 3 shows the first four construction stages and their most distinctive changes from the same perspective. Finally, based on terrestrial photos and Google Earth data, the environment of the building was also reconstructed to ensure that this historic building was embedded in its current urban environment. As a stylistic device, the surrounding buildings were coloured grey to emphasize the museum in the visualisation.

The texture mapping of the model was carried out using the software Autodesk 3ds Max. The photos used for texturing were mainly images taken in situ. However, textures that were freely available online were also integrated after appropriate editing. Furthermore, bump and alpha textures were used to improve the depth effect and the appearance of details. In total, 239 textures were used for visualisation.

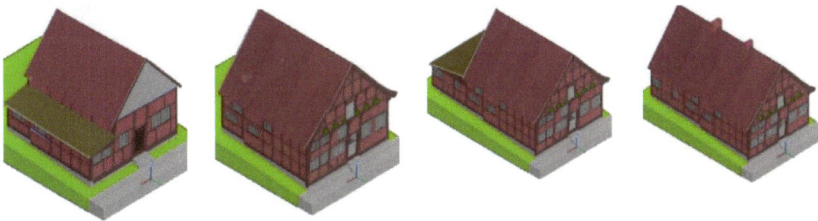

Figure 8. Advanced 3D models of the different construction stages of the town house. From left to right: 1541–1585, 1585–ca. 1805, ca. 1805–1814 and 1814–ca. 1890.

4. Visualisation

4.1. The Generation of a Video Sequence for the SIEGESBURG

4.1.1. Software

The software package "Lumion 3D" is a visualisation software. Its main target audience is architects and designers. It comes with a material library, foliage items, landscape tools and a high-speed renderer. Compared to established software packages such as 3ds Max, Cinema 4D and many others, it has only minor texture settings explicitly well-prepared for architectural visualisation and a complete lack of export functionality. It is, on the other hand, an easy-to-learn software with a fast workflow and a high number of video and graphic settings. This software was used for texture mapping and all following processing steps such as environment modelling, animation and video production for this project.

4.1.2. Texture Mapping

The material library of Lumion mainly consists of modern architectural textures. Consequently, the textures had to be edited in several ways to fit the old, rough appearance of a post-medieval castle: A) up-scaling of most of the textures, especially the ones including stones, B) creating and integrating normal maps to provide reliefs, which supply an overall more realistic look, especially when light sources (e.g., sun) are included in the scene and C) colour correction. Figure 9 shows the non-textured (left/centre) and textured (right) models of the same building. Further aging of the objects could be achieved by placing small fauna objects on and at the bottom of a wall, e.g., stones and ivy (Figure 10).

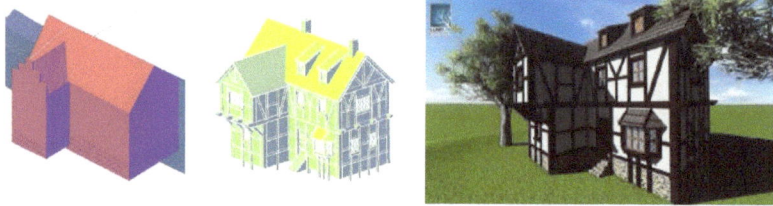

Figure 9. Development of a building in the outer ward: Initial CAD model (left), detailed model (centre), textured model in Lumion 3D (right).

4.1.3. Environment

Landscape: The landscape underneath and around the castle was modelled with the Lumion landscape tools using the 3D reconstruction of the Kalkberg as a basis. Steep parts were mapped with rock textures, while shallow parts were designed as a combination between grass and dirt (Figure 10). An additional 3D effect was applied to the grass parts, showing individual blades of grass when the camera moves closer to the object.

Figure 10. Landscape with rock, grass and trees beneath the west side of the castle (left) and view into the main street of the town Segeberg (right).

The surrounding landscape was modelled afterwards based on both today's and the historic appearance of the area. As the past 400 years extensively changed the landscape, only the rough pattern of the present look can be used. The historic sources, on the other hand, are known to be inaccurate at some points, e.g., the distribution of hills around the village Segeberg. The shape and position of both lakes near the castle were derived from current aerial photographs. The environment around the urban area was modelled with typical agricultural elements, such as farms and grazing land with fences between them. Small forests were added where they are shown in historic sources. Over 6000 trees and plants were placed in the whole scenery. To block the visual axis in the distance, a slight fog was added, not only providing a realistic feeling of distance but also limiting the area, which had to be filled with landscape elements and 3D objects. Figure 11 (left) shows the whole scenery of the castle and its environment from an aerial perspective.

Figure 11. Perspective view of the castle including the village Segeberg (left) and the rural area around the castle (right).

Urban area: The village of Segeberg (the "Bad" (bath) was not added until the end of the 19th century) itself consisted of approximately a hundred buildings at that time, which were located mainly at the north side of the Kalkberg along the main road from east to west (Lübecker Straße), following the curvature of the terrain. The positions of the buildings are shown in high detail on the isometric map of Braun-Hogenberg (Figure 1). This source was used for the placement of all buildings for the physical model. The resulting ground plan was the basis for the creation of the digital model (Figure 12). Although the focus of this project was on the castle, the surrounding area should nevertheless be historically correct and provide a consistent impression of the background scenery. To facilitate the workload for this reconstruction part, only one building was modelled, having four slightly different sides, which allows for many different appearances due to rotation and combination of two or more copies of the same model. Furthermore, the model was imported twice into the object library using two different textures to allow for a higher variety. Additionally, some of the smaller castle buildings were reused as city buildings.

Figure 12. Map of the present city (green) with the state of 1600 as overlay (red, not complete) (left) and as a complete overlay (white/red) in Google Earth (right) with reverse map orientation (top = north facing). Figures: Nils Hinrichsen.

4.1.4. Video with Lumion

Depending on the potential target group, there are several possibilities to generate a video sequence. The camera path, use of special effects, lighting, cutting and choice of background music may influence both the atmosphere and the message of the video. The aim of this project was a trailer-like sequence, which could be used both for advertisement of the Siegesburg as well as for information purposes in a museum context. The following concepts and effects were used to create the video using the software Lumion.

1) Camera: To capture the Siegesburg both as an entire ensemble in the context of the urban surrounding as well as the details inside the castle, a carefully planned camera path was essential. During the whole sequence, the viewer should not become lost, i.e., he should always be aware of where the current camera position is located in relation to earlier shots. Therefore, several long tracking shots were created and each following shot starts at a position which has already been shown in an earlier shot. Thus, different areas of the castle can be shown without having the location of each area established after each cut.

The video can be partitioned in different sections: A) introduction sequence, with some written information and short cuts with glimpses on some details of the castle to create a sense of excitement; B) overview shot to establish the whole scene with a wide shot, approaching the urban and surrounding area of the castle once; C) first part of the castle tour including the change of the camera angle to eye-level in front of the main gate and keeping the human observer perspective for most of the following sequences. This creates an immersive feeling for the viewer while the camera shows some details in the outer ward; D) tour around the main castle showing the camera view mostly from below to have a more imposing view of the buildings. For a fluent tracking shot, the camera follows objects in the foreground while moving around, e.g., the rope of a little crane or the handrails of a staircase; E) night scenes consisting of short shots for a similar atmosphere as for the introduction; F) closing credits showing the castle in the background. The visualisation is based on the drawing of the scene in the foreground (Figure 13). All scenes were created with the video editor in Lumion, within which all camera views were defined with snapshots of the scene. The software then smoothly interpolates between those frames to render each frame in between.

Figure 13. Drawing of the castle as shown in the closing credits of the video sequence.

2) Visual Effects: The raw version of the video sequence of the 3D environment has been manipulated in the video editor to enhance the graphical quality of specific scenes. The video editor offers a library of effects. Each effect can be applied for specific parts of the video and is variable in length and intensity. In total, this video used 17 different types of effects, the most important of which are described below: (A) Clouds and fog were used to create the impression of a long distance towards the border of the scene in order to hide the skyline; (B) depth of field to blur objects very distant or very close to the camera resulting in a more realistic look. Incidentally, this effect can be used to guide the viewers' attention to a certain point in the video; (C) titles for implementing writing of the opening titles and the closing credits; (D) vignette effect for creating a soft black frame at the border of the image to restrict the field of view, letting the viewer focus on the shown part in the centre; (E) editing of the position and intensity of the sun for simulating the course of the day; (F) colour correction allowing the editing of the RGB-values of the images. This was used to apply a red shade to opening titles and the closing credits sequences to create the atmosphere of morning/evening; (G) volume sunlight and lens flare effects for editing the sunlight at particular angles.

Both visual effects and animations are managed by placing markers on a timeline to specify fixed conditions between which the software interpolates. Figure 14 shows an example of the changes in the same image, created by visual effects.

Figure 14. A frame of the opening sequence without (left) and with (right) volumetric sunlight and colour correction. Both images already have the vignette effect (borders fade to black).

3) Rendering: The rendering time depends on the following criteria: the number of effects, processing demands of these effects, length of the video sequence, frames per second (fps) rate, resolution and content of each frame. The video was rendered in 720p (1280 × 720 pixel) using 30 frames per second. The generated video consists of 11,370 frames, which corresponds to a length of 6.19 min, resulting in a rendering time of about 36 hours. In addition to the 2D version, a 3D video version was generated in Lumion. Therefore, the library of effects provides an effect with stereoscopic options. With the specification of eye distance and focal length, the software renders two camera parts slightly off centre to the original one. Since there is only one calculation for lighting and shadows for the stereo images, the rendering time for the 3D version takes only eight hours longer than the 2D version providing MP4 file formats for both versions. Consequently, the 3D video version needs a 3D screen for correct display. The two final Lumion videos (2D/3D) have a length of 6:19 min and a file size of ca. 700Mb (2D) and ca. 950 Mb (3D) using the file format MP4. The 2D video version with the title "Siegesburg Segeberg 1588" is available for free viewing on the YouTube channel "HCUHamburgGeomatics" of the Hafencity University Hamburg at the following link: https://www.youtube.com/watch?v=M1pQnAhvQ4w. Additionally, some single high-resolution still images were rendered for additional use in the museum exhibition (Figure 15).

4.1.5. Audio

The background music of the video is a very important tool to influence the spirit of the video and to support or highlight certain elements. Since no appropriate, publicly available soundtrack for this Lumion video could be found on the Internet, the music integrated in the video sequence was self-composed. This allowed for fine-tuning between audio and video, so that both media form a harmonic unit. Visual atmospheres are supported by a corresponding use of instruments, rhythms and sounds. Thus, camera path and movements and even individual cuts could be set in the appropriate musical theme.

4.1.6. Video with 3ds Max

As already mentioned, a second modelling and visualisation project ran in parallel using 3ds Max as visualisation software and focussing on a more realistic Brick Gothic Texture for the castle. The resulting video is also available for free viewing on the same YouTube channel at the following link: https://www.youtube.com/watch?v=V-hESSCwOe0. A detailed description of the 3D reconstruction and visualisation of the Siegesburg using AutoCAD und 3ds Max is presented in Herzberg (2015). Two perspective views of the Siegesburg with Brick Gothic textures are illustrated in Figure 16.

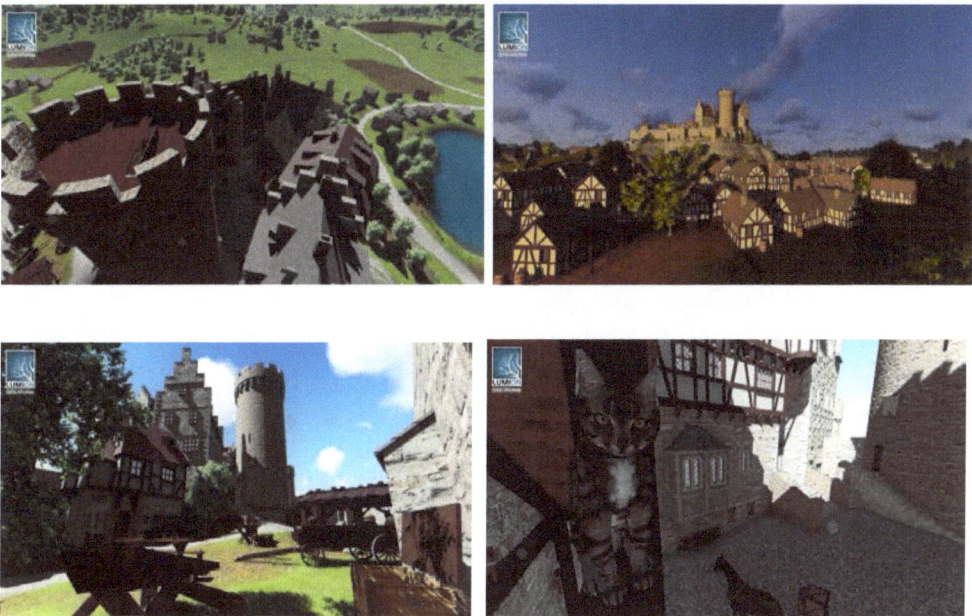

Figure 15. Views from the castle keep (top left), from the north (top right) and from the castle yards (bottom) generated with Lumion.

209

Figure 16. Two perspective views of the Siegesburg with realistic Brick Gothic Texture on top of the Kalkberg rendered with 3ds Max (Herzberg 2015).

4.2. The Old-Segeberg Town House in Virtual Reality Application

4.2.1. Game Engine Unreal

A game engine is a software framework designed for the creation and development of video games for consoles, mobile devices and personal computers. The core functionality typically provided by a game engine includes a rendering engine for 2D or 3D graphics to display textured 3D models (spatial data), a physics engine or collision detection (and collision response) for the interaction of objects, an audio system to emit sound, scripting, animation, artificial intelligence, networking, streaming, memory management, threading, localisation support, scene graph, and may include video support for cinematics. A game engine controls the course of the game and is responsible for the visual appearance of the game rules. For the development of a virtual museum, game engines offer many necessary concepts with much functionality so that users can interact with the VM.

In the past, the development of game engines was mostly based on the development of a specific game with paid licensing to external game developers. In recent years, however, most of the large engine providers have focused more on the advancement of engines and additionally offer free access for developers. Examples of game engines with free potential use are the engine Unity from Unity Technologies, the CryEngine of the German development studio Crytek, and the engine Unreal from Epic Games (www.epicgames.com). A current overview and comparison of different game engines can be found, e.g., in O'Flanagan (2014) and Lawson (2016). The selection of the appropriate engine for a project is based on the provided components mentioned above, the adaptability in the existing work processes as well as special preferences of the (game) developer. In the framework of this project, the game engine Unreal was selected due to the opportunity to develop application and interaction logics using a visual programming language, the so-called Blueprints. Visual programming with Blueprints does not require the

210

writing of machine-compliant source code. Thus, it provides opportunities for non-computer scientists to program all functions for a VM using graphic elements. The saving in time associated with this method of software development allows for the generation of additional scenarios and for more intensive user testing. Game engines are, therefore, very well-suited to the development of virtual museums.

4.2.2. Virtual Reality System HTC Vive

HTC Vive (www.vive.com) is a virtual reality headset (with a weight of 555 grams, Figure 17) for room-scale virtual reality. It was developed by HTC and by Valve Corporation. It was released on April 5, 2016, and it is currently available on the market for EUR 899. Basic components are the headset for the immersive experience, two controllers for user interactions and two "Lighthouse" base stations for tracking the user's movement.

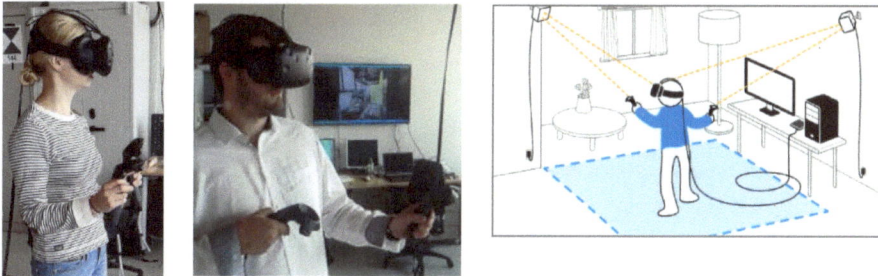

Figure 17. The virtual reality system HTC Vive in use (left). The screen in the background shows the same sequence as it appears to the user in the VR glasses (centre). The setup of the Virtual Reality System HTC Vive (right, HTC Corporation, 2017).

The technical specifications of the HTC Vive are summarized in the following: (a) two screens with a field of view of approximately 110 degrees, one per eye, each having a display resolution of 1080x1200 with a refresh rate of 90 Hz; (b) more than 70 sensors including a MEMS (Microelectromechanical systems) gyroscope, accelerometer and laser position sensors; (c) 4.6 by 4.6 m tracking space for user operation using two "Lighthouse" base stations for tracking the user's movement with sub-millimetre precision by emitting pulsed IR lasers; (d) SteamVR running on Microsoft Windows as the platform/operating system; (e) Controller input by SteamVR wireless motion tracked controllers; and (f) a system for overlaying the real room's bounds for static objects in the real world called Chaperone and (g) a front-facing camera for looking around in the real world to identify any moving objects in a room as part of a safety system.

The following technical specification is required as the minimum for the computer to be used: processor Intel™ Core™ i5-4590 or AMD FX™ 8350, graphic card NVIDIA GeForce™ GTX 1060 or AMD Radeon™ RX 480, 4 GB RAM, video output 1x HDMI 1.4-connection or DisplayPort 1.2 or newer, 1x USB 2.0-connection or newer, operating system Windows™ 7 SP1, Windows™ 8.1 or more up to date or Windows™ 10.

The device uses a gyroscope, accelerometer, and laser position sensor to track the head's movements as precisely as one-tenth of a degree. Wireless controllers in each hand, with precise SteamVR-tracking, enable the user to freely explore virtual objects, people and environments, and to interact with them. The VIVE-controller is specifically designed for VR with intuitive control and realistic haptic feedback. The Lighthouse system uses simple photo sensors on any object that needs to be captured. To avoid occlusion problems, this is combined with two lighthouse stations that sweep structured light lasers within a space.

4.2.3. Windows-Based VM Implementation

The major part of the work dealt with the programming of the user movement in the museum, with information queries and the corresponding animations in the game engine "Unreal Engine".

The intuitive handling of the program was an essential prerequisite, allowing also easy access to and use of the VMs for inexperienced PC users (Figure 18 left).

Figure 18. Graphical User Interface for the virtual museum tour (left) and graphical user interface for the building history with an animated view into the interior of a construction phase (right).

The control of the software is exclusively available via mouse interaction and is based on many well-distributed positions throughout the building, which can be directly selected by clicking on a map or approached using a defined camera path within the 3D environment. In these positions, the users can freely look around, as in a 360^0 panorama. Users can also zoom in and out, and click on the available information button. As a special highlight of the VM, the visualisation of the building history was realized with a "model in the model" (Figure 18 right and Figure 19 left).

At one station of the virtual tour, which is located in front of the model, the user can open this model to display and animate, upon request, each of the building states including all related information. The user can look at the building model from all sides using virtual rotation. In addition, with a mouse click, the roof can be removed, the individual floors can be driven apart by animation, and the appearance of the building interior in the respective building phase (Figure 8) will be shown. Each room can now be selected to display information about the development of building use. Furthermore, it is possible to start an animated transition from the activated construction phase, which shows and describes the structural changes to the next state for each construction phase. These animations required a subdivision of all 3D models into 387 smaller objects, to precisely control the movement of the objects for each animation. In the animation, the user is guided by predefined camera movements to appropriate viewpoints.

Figure 19. Impressions from the 3D environment of the virtual museum Old-Segeberg Town House including the model of the building (left), part of the exhibition (centre) and a view at the attic (right).

Menus and information boards, which can be opened during the individual tour using the info button placed next to the selected objects, were created for the exhibits. These menus include brief explanations and mostly a figure that can be enlarged via a mouse click. Some information is directly imparted using detailed point-of-view shots in the 3D environment. In such cases, cameras were distributed in the whole Museum at appropriate points. These can also be selected using the info button.

Finally, comfort functions were created such as tool tips, an overview map and a help menu. For quality assurance, the VM program has been tested by several people with different PC experience to subsequently customize the software details.

4.2.4. Virtual Reality Application with HTC Vive

Based on the modelled and textured 3D data in the game engine, an immersive virtual reality visit was developed utilizing the new Virtual Reality System HTC Vive. The visit offers the possibility of experiencing the museum and the history of the building from a real person's point of view and interaction scheme. For this purpose, the controlling positions have been replaced by the possibility of free movement by the user. To bridge long distances in the virtual object, a teleportation function is available for the navigation of the user (Figure 20 left). The users' hands can interact with various components of the virtual world to control the building's presentation. The object selection and menu operation are enabled via a "laser beam", which is controlled by the motion controller (Figure 20 centre). The highlight of the virtual museum visit is the animation of the architectural building history, which is vividly represented by the HTC Vive glasses directly in front of the visitor's eyes on the basis of the 3D models (Figure 18 right and Figure 20 right). Different historically-confirmed construction phases are visibly demonstrated in 3D and changes are illustrated by transition animations.

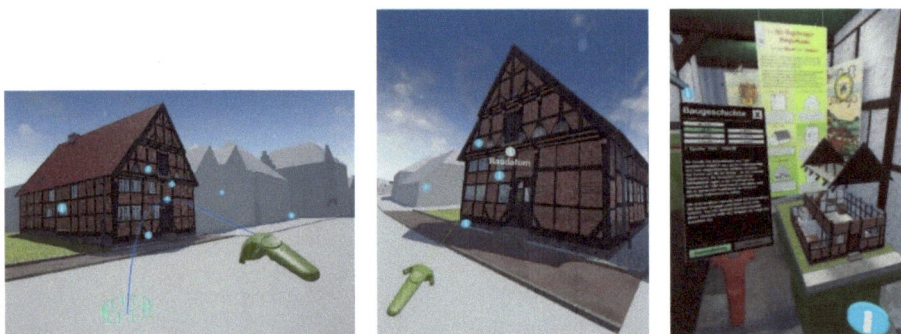

Figure 20. The virtual museum Old-Segeberg Town House in the Virtual Reality System HTC Vive as an intense experience: navigation in the virtual museum using the developed teleportation function (left), menu navigation with green motion controller by "laser beam" (centre) and second red controller for the information menu (right).

5. Conclusions and Outlook

This contribution has presented the virtual 3D reconstruction of the destroyed cultural heritage monument Siegesburg including the Kalkberg in Segeberg, and of the preserved monument Old-Segeberg town house, today the oldest building in Bad Segeberg and probably in the state Schleswig-Holstein, Germany. Using isometric maps such as the one from Braun-Hogenberg, historic sources and expert knowledge, the castle and the appertaining Kalkberg were successfully reconstructed in 3D for the generation of video sequences in Lumion 3D, representing a half-timbered architecture from Central Germany (Figure 15), while a second video generated with 3ds Max represents the castle in the northern European Brick Gothic (Figure 16). These video sequences symbolise the virtual return of the Siegesburg, while the physical return was already realised using the physical castle model in the museum exhibition. Furthermore, a stereoscopic 3D video has also been generated using Lumion 3D.

In the second part of this project, the VR visualisation of the Old-Segeberg Town House has been presented. Therefore, the exterior and the interior of the historic building including the six construction phases were modelled in CAD using data from photogrammetry and terrestrial laser scanning. Using the textured 3D models of the building and its interior, a Virtual Museum was successfully developed and implemented for the museum Old-Segeberg Town House with two options: a) interactive software application for windows-based computer systems and b) virtual reality application for the VR system HTC Vive. Many visitors and participants have tried and tested the VM by using both a windows-based computer system and the VR system HTC Vive. The Old-Segeberg Town House can look back on 475 years of architectural development, identified and explained in the form of animations, which are the highlights of the virtual museum. The developed computer program contains 13 guided viewpoints distributed at important positions in the museum and 52 info menus with detailed information for visiting the virtual museum. The program has a size of 500 MByte and is executable as a standalone program on Windows operating systems. It is developed in the game engine Unreal, which offers not only complex visualisations of 3D objects, but also provides every programming tool necessary for creating extensive interactions between the user and the environment. It is planned to make this program accessible for visitors using a PC-terminal in the museum. It allows, in addition to the current exhibition, a multimedia interaction with the history of the city and the building. Thus, based on this building, it makes an important educational contribution for the urban development of Bad Segeberg and about 500 years of housing tradition in Schleswig-Holstein. Developed entirely in 3D, the VM is unique in this form in Germany as an informative component of a museum.

The current follow-up project "Segeberg 1600" is concerned with the historically correct reconstruction of the complete urban and rural area of the two villages Segeberg and Gieschenhagen in the year 1600 (Figure 21). Some of the modelled objects of the Siegesburg will be reused as revised and improved versions for this project, especially concerning textures in the Brick Gothic and details. The idea behind this project is the implementation of the entire 3D data of the village Segeberg in a game engine for a virtual reality application using the Virtual Reality System HTC Vive as it is already realised for the museum Old-Segeberg Town House (Kersten et al., 2017).

The 2D version of the history of the Siegesburg and the village is shown regularly in the exhibition of the museum Old-Segeberg town house. The combination of the physical model, the video and the interactive VR visualisation helps the visitors to visualise the situation of Segeberg and especially the castle

215

more than 400 years ago. Since the castle and the complete urban environment have changed drastically since then, very little is known about parts of the history of Segeberg—to the point that many people are not even aware that there once was a castle on top of the Kalkberg. This project helps to increase the historical awareness of the local community and the tourists interested in such historic events. This 3D reconstruction of these cultural heritage monuments also demonstrates the potential in the documentation, visualisation and preservation of current or already destroyed cultural heritage objects.

A realised VR application using the HTC Vive is a very immersive experience. It provides the opportunity to walk through a virtual reality environment such as the museum of the Old-Segeberg Town House or the virtual villages of Segeberg and Gieschenhagen in the past, to collect all of the provided information and to see all of the different animations, which explain the topics more than in a physical exhibition, e.g., in serious games.

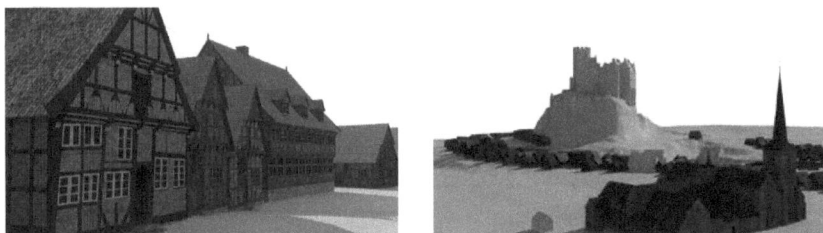

Figure 21. Textured 3D building models of the villages Segeberg and Gieschenhagen, Germany, in the Early Modern Age at 1600: Old-Segeberg Town House and town hall (first and forth building, left) and an overview of already modelled buildings for the VR project "Segeberg 1600" (right) with the St. Mary's church and the monastery in the front and the Siegesburg in the background.

Acknowledgments: The significant support of the historian Nils Hinrichsen (Head of the museum Alt-Segeberger Bürgerhaus) is gratefully acknowledged. The results of this contribution are significantly influenced by his important historic information and expertise.

References

1. Bauer, T. Museen und Internet. *Museol. Online* **2001**, *3*, 112–161.
2. Deggim, S. Visualisierung der historischen Siegesburg in Bad Segeberg. Unpublished project report, Master study program Geomatics; HafenCity University, Hamburg, Germany, 2015; p. 36.
3. Deggim, S. Entwicklung Eines Virtuellen Museums für ein Historisches Gebäude am Beispiel des Alt-Segeberger Bürger-Hauses. Unpublished Master thesis, Master study program Geomatics, HafenCity University, Hamburg, Germany, 2016; p. 82.

4. Deggim, S.; Kersten, T.; Lindstaedt, M.; Hinrichsen, N. The Return of the Siegesburg - 3D-Reconstruction of a Disappeared and Forgotten Monument. In *3D Virtual Reconstruction and Visualization of Complex Architectures*, Proceedings of the International Archives of the Photogrammetry, Remote Sensing and Spatial Information Sciences Conference, Nafplio, Greece, 1–3 March 2017; D. Aguilera, A. Georgopoulos, T. Kersten, F. Remondino, E. Stathopoulou, Eds.; Vol. XLII-2/W3; ISPRS: Nafplio, Greece, 2017, pp. 209–215.

5. Fangi, G.; Wahbeh, W. The destroyed Minaret of the Umayyad Mosque of Aleppo, the Survey of the Original State. *Eur. Sci. J.* **2013**, *4*, 403–409.

6. Gruen, A.; Remondino, F.; Zhang, L. Reconstruction of the great Buddha of Bamiyan, Afghanistan. *Int. Arch. Photogramm. Remote Sens. Spat. Inf. Sci.* **2002**, *34*, 363–368.

7. Gütt, I. Smartphone-Applikationen im Museumsbereich. Bachelor Thesis, Studiengang Museumskunde, Fachbereich Gestaltung, HTW Berlin, Berlin, Germany, 2010; p. 74.

8. Herzberg, A.-C. Visualisierung der Siegesburg von Bad Segeberg mit 3Ds Max Design. Unpublished project report, Master study program Geomatics, HafenCity University, Hamburg, Germany, 2015; p. 28.

9. HTC Corporation. Vive PRE User Guide. Available online: http://www.htc.com/managed-assets/shared/desktop/vive/Vive_PRE_User_Guide.pdf (accessed on 31 May 2017).

10. Ivarsson, E. Definition and Prospects of the Virtual Museum. Master Thesis, Uppsala University. Available online: http://www.elinivarsson.com/docs/virtual_museums.pdf (accessed on 31 May 2017).

11. Jones, G.; Christal, M. The future of virtual museums: On-line, immersive 3-D environments. *Created Real. Group* **2002**, *4*, 1–12.

12. Kersten, T.; Acevedo Pardo, C.; Lindstaedt, M. 3D Acquisition, Modelling and Visualization of north German Castles by Digital Architectural Photogrammetry. *Int. Arch. Photogramm. Remote Sens. Spat. Inf. Sci.* **2004**, *35*, 126–132.

13. Kersten, T.; Lindstaedt, M. Virtual Architectural 3D Model of the Imperial Cathedral (Kaiserdom) of Königslutter, Germany through Terrestrial Laser Scanning. In *EuroMed 2012—International Conference on Cultural Heritage*; Lecture Notes in Computer Science (LNCS), Ioannides, M., Fritsch, D., Leissner, J., Davies, R., Remondino, F., Caffo, R., Eds.; Springer-Verlag: Berlin/Heidelberg, Germany, 2012; Volume 7616, pp. 201–210.

14. Kersten, T.; Hinrichsen, N.; Lindstaedt, M.; Weber, C.; Schreyer, K.; Tschirschwitz, F. Architectural Historical 4D Documentation of the Old-Segeberg Town House by Photogrammetry, Terrestrial Laser Scanning and Historical Analysis. In *Cultural Heritage. Documentation, Preservation, and Protection*, Proceedings of the 5th International Conference, EuroMed Progress, Limassol, Cyprus, 3–8 November 2014; Ioannides, M., Magnenat-Thalmann, N., Fink, E., Zarnic, R., Yen, A.-Y., Quak, E., Eds.; Lecture Notes in Computer Science (LNCS) 8740; Springer International Publishing: Cham, Switzerland, 2014, pp. 35–47.

15. Kersten, T.; Tschirschwitz, F.; Deggim, S. Development of a Virtual Museum including a 4D Presentation of Building History in Virtual Reality. In *3D Virtual Reconstruction and Visualization of Complex Architectures*, Proceedings of the International Archives of the Photogrammetry, Remote Sensing and Spatial Information Sciences Conference, Nafplio, Greece, 1–3 March 2017; D. Aguilera, A. Georgopoulos, T. Kersten, F. Remondino, E. Stathopoulou, Eds.; Vol. XLII-2/W3; ISPRS: Nafplio, Greece, 2017, pp. 361–367.

16. Lawson, E. Game Engine Analysis, 2016. Available online: https://www.gamesparks.com/blog/game-engine-analysis/ (accessed on 12 January 2017).

17. Meydenbauer, A. Ueber die Anwendung der Photographie zur Architektur- und Terrain-Aufnahme. *Zeitschrift für Bauwesen* **1867**, *17*, 61–70.

18. Mortara, M.; Catalano, C.E.; Bellotti, F.; Fiucci, G.; Houry-Panchetti, M.; Petridis, P. Learning cultural heritage by serious games. *J. Cult. Herit.* **2014**, *15*, 318–325.

19. O'Flanagan, J. Game Engine Analysis and Comparison, 2014. Available online: https://www.gamesparks.com/blog/game-engine-analysis-and-comparison/ (accessed on 11 January 2017).

20. Pescarin, S.; Pagano, A.; Wallergård, M.; Hupperetz, W.; Ray, C. Evaluating virtual museums: Archeovirtual case study. In *Archaeology in the Digital Era*, Proceedings of the 40th Annual Conference of Computer Applications and Quantitative Methods in Archaeology (CAA), Southampton, 26–29 March 2012; Earl, E., Sly, T., Chrysanthi, A., Murrieta-Flores, P., Papadopoulos, C., Romanowska, I., Wheatley, D., Eds.; Amsterdam University Press: Amsterdam, The Netherlands, 2013; pp. 74–82.

21. Petridis, P.; White, M.; Mourkousis, N.; Liarokapis, F.; Sifniotis, M.; Basu, A.; Gatzidis, C. Exploring and interacting with virtual museums. In Proceedings of the 33rd Annual Conference of Computer Applications and Quantitative Methods in Archaeology (CAA), Tomar, Portugal, March 2005; pp. 73–82.

22. Pujol, L.; Lorente, A. The Virtual Museum: a Quest for the Standard Definition. Archaeology in the Digital Era. In Proceedings of the 40th Annual Conference of Computer Applications and Quantitative Methods in Archaeology (CAA), Southampton, UK, 26-29 March 2012; pp. 40–48.

23. Reimers, H.; Hinrichsen, N. Das Alt-Segeberger Bürgerhaus. Vom Stadtwohnhaus des 16. Jahrhunderts zum Stadtmuseum im 21. Jahrhundert. Lutherstadt Wittenberg, Torgau und der Hausbau im 16. Jahrhundert; Goer, M., Furrer, B., Klein, U., Stiewe, H. and Weidlich, A., Eds.; Jahrbuch für Hausforschung 62; Jonas Verlag: Marburg, Germany, 2015; pp. 341–355.

24. Reussner, E. Wissensvermittlung im Museum – ein überholtes Konzept? *Kultur und Management im Dialog* **2007**, *5*, 20–23.

25. Samida, S. Überlegungen zu Begriff und Funktion des „virtuellen Museums": Das archäologische Museum im Internet. *Museol. Online*, **2002**, *4*, 1–58.

26. Schweibenz, W. The "Virtual Museum": New Perspectives for Museums to Present Objects and Information Using the Internet as a Knowledge Base and Communication System. In *Knowledge Management und Kommunikationssysteme, Workflow Management, Multimedia, Knowledge Transfer*, Proceedings of the 6th International Symposium for Information Science (ISI 1998), Prague, 3–7 November 1998; Zimmermann, H.H.; Schramm, V., Eds.; UVK Verlagsgesellschaft mbH: Constance, Germany, 1998; 185–200.

27. Shaw, J. The Virtual Museum. In *Installation at Ars Electrónica*, ZKM, Karlsruhe: Linz, Austria, 1991.

28. Styliani, S.; Fotis, L.; Kostas, K.; Petros, P. Virtual museums, a survey and some issues for consideration. *J. Cult. Herit.* **2009**, *10*, 520–528.

29. V-MusT. What is a Virtual Museum? Virtual Museum Transnational Network, 2011. Available online: http://www.v-must.net/virtual-museums/what-virtual-museum (accessed on 13 January 2017).

30. Wahbeh, W.; Nebiker, S.; Fangi, G. Combining Public Domain and Professional Panoramic Imagery for the Accurate and Dense 3D Reconstruction of the Destroyed Bel Temple in Palmyra. *ISPRS Ann. Photogramm. Remote Sens. Spat. Inf. Sci.* **2016**, *III-5*, 81–88.

31. Wiedemann, A.; Hemmleb, M.; Albertz, J. Reconstruction of historical buildings based on images from the Meydenbauer archives. *Int. Arch. Photogramm. Remote Sens. Spat. Inf. Sci.* **2000**, *33*, 887–893.

Kersten, T.P; Tschirschwitz, F.; Deggim, S.; Lindstaedt, M.. Step into Virtual Reality—Visiting Past Monuments in Video Sequences and as Immersive Experiences. In *Latest Developments in Reality-Based 3D Surveying and Modelling*; Remondino, F., Georgopoulos, A., González-Aguilera, D., Agrafiotis, P., Eds.; MDPI: Basel, Switzerland, 2018; pp. 192–219.

Virtual Reality Technologies for the Exploitation of Underwater Cultural Heritage

Fabio Bruno [a], Antonio Lagudi [b], Loris Barbieri [a],
Maurizio Muzzupappa [a], Marino Mangeruga [a], Marco Cozza [b], Alessandro Cozza [b],
Gerardo Ritacco [b] and Raffaele Peluso [b]

[a] Department of Mechanical, Energy and Management Engineering (DIMEG) University of Calabria, P. Bucci, 46C, 87036 Rende (CS), Italy; (f.bruno, loris.barbieri, m.muzzupappa, m.mangeruga)@unical.it

[b] 3D Research Srl, P. Bucci, 45C, 87036 Rende (CS), Italy; (a.lagudi, m.cozza, a.cozza, g.ritacco, r.peluso)@3dresearch.it

Abstract: The Underwater Cultural Heritage (UCH) represents a resource with huge, but yet largely unexploited, potentials for the maritime and coastal tourism. In the last 10 years, national and international government authorities are supporting and strengthening research activities and development strategies, plans and policies to make underwater archaeological sites' exploitation more sustainable and accessible to large-scale tourism. To this end, the paper presents an outcome of the Virtual and augmented exploitation of Submerged Archaeological Sites (VISAS) project that allows users to explore underwater archaeological sites. In particular, the virtual diving system combines the advantages offered by Virtual Reality (VR) technologies and the latest three-dimensional (3D) reconstruction techniques to create virtual tours for the exploitation of the UCH. User studies' results demonstrate that the proposed VR system is able to provide a playful learning experience, with a high emotional impact, and it has been well appreciated by a large variety of audiences, even by younger and inexperienced users.

Keywords: virtual reality; virtual diving system; underwater cultural heritage; semi-immersive VR environment; immersive VR environment.

1. Introduction

Underwater archaeological assets represent a relevant part of the world cultural heritage and a particularly important element in the history of people, nations, and their relationship with each other concerning their common heritage, especially in the European context. Sunken cities, ancient shipwrecks, remains of ancient fishing installations, and ports represent a vast historical and scientific resource of highest importance for understanding the development of human civilization, and, therefore, a fundamental and essential source for the upbringing and education of the next human generations.

In addition to this historical, cultural, and social role, the Underwater Cultural Heritage (UCH) also stands as an interesting opportunity for the tourism development of the coastal areas. To this end, considerable improvements are going on since the last decades especially thanks to an articulated implementation of the UNESCO conventions 1970 (prevention of illicit traffic), 1972 (World Cultural and National Heritage), and 2001 (Underwater Cultural Heritage). In particular, the 2001 UNESCO Convention on the Protection of the Underwater Cultural Heritage (UNESCO, 2001) has led to the definition of basic principles and recommendations for the protection and in-situ conservation of the submerged archaeological sites. The European Commission too has issued an EU's Blue Growth strategy (EU Commission, 2014), aiming for the cooperation among research institutes, museums, tourism companies, and other stakeholders for the development of innovative and sustainable solutions and products, with a maximum use of information technology, which responds to visitors' expectations.

These recommendations have encouraged the international collaboration among universities, research centers, and companies that has led to the running of a plethora of projects for the development of innovative methods and technologies for in-situ conservation (Bacpoles, 2005; Gregory, 2012; Gündoğdu et al., 2015; Bruno et al., 2016a), but also, as a positive side-effect, to the strengthening of research activities and development strategies to make the underwater archaeological sites' exploitation more sustainable and accessible to large-scale tourism (Chapman et al., 2008; Haydar et al., 2011; Varinlioğlu, 2011). One of them is the VISAS project (Virtual and augmented exploitation of Submerged Archaeological Sites—http://www.visas-project.eu) (Bruno et al., 2016b), which is a collaborative research project that is funded by MIUR (Italian Ministry of Education, University and Research); it has started on 1st April 2014 and ended on 30th September 2016. The project scope was to enhance the cultural and tourist offer that was related to underwater archaeology through innovation of the modes of experience, both on site and remote, of the underwater environments of archaeological interest.

This paper presents an outcome of the VISAS project that combines the advantages that are offered by the recent advances in Virtual Reality (VR) with the newest three-dimensional (3D) reconstruction techniques to create virtual tours for the exploitation of the UCH. The proposed virtual diving system indeed represents an innovative solution that overcomes the limits that are imposed by the underwater environment, and offer to the general public a playful and educational experience, by diving into faithful and realistic reconstructions of submerged archaeological sites. This VR system has been implemented for the testbeds of the

VISAS project, i.e. two underwater archaeological sites of South Italy: Capo Colonna and Cala Minnola. The first one is located in the Ionian Sea, on the East coast of Calabria and 10 km away from Crotone, where raw and semi-finished marble products, transported by a Roman cargo ship, lay on the seabed at a depth of seven meters. The second one is located in the Tyrrhenian Sea, a few miles away from the west coast of Sicily, on the island of Levanzo (Aegadian Islands), where a wreck of a Roman cargo ship lies on the seabed at a depth of 27–30 meters.

2. VR Applications in the UCH

VR technologies have proven their effectiveness in increasing the value of cultural heritage (Barcelo and Forte, 2000; Roussou, 2002; Vote et al., 2002; Lepouras and Vassilakis, 2005; Pavlidis et al., 2007; Bruno et al., 2010), but the possible applications for underwater archaeology have not been investigated to a fair extent. For a few years now, different frameworks for the collection and visualization of the UCH by means of VR technologies have been investigated, but these systems limit the exploitation to a single underwater archaeological remain (Varinlioğlu, 2011), or are more oriented to the digitization for scientific purposes (Katsouri et al., 2015), rather than focusing on edutainment for general audiences. An example of digital repositories of underwater remains is the Big Anchor (Hunter, 2009) and the VENUS (Virtual exploration of underwater site) project (Chapman et al., 2006). In particular, digital models of underwater sites realized in the VENUS project have been used in a couple of VR tools for interactive and immersive visualization (Chapman et al., 2008), allowing for archaeologists to study the virtual site from within. These VR based demonstrators have also been readapted for the general public, but the virtual exhibit is more oriented to the presentation and visualization of archaeological data (Haydar et al., 2011), rather than to edutainment purposes.

There are very few examples of virtual heritage demonstrations that are applied to the underwater environment in which user's engagement is provided by means of an edutainment approach. Stone (Stone et al., 2009) has applied serious games to the underwater environment in order to increase the cultural awareness about marine biology rather than underwater archaeology. On the contrary, the immersive underwater VR environment of the Mazotos shipwreck site (Liarokapis et al., 2017) has been designed for raising users' archaeological knowledge by exploring underwater archaeological shipwrecks, but the virtual environment is not a precise and faithful reconstruction of the real site because the placement of the archaeological objects, rocks, and vegetation is generated procedurally using a stochastic approach.

Differently from the above-mentioned works, the proposed virtual diving system has been designed to raise users' archaeological knowledge and cultural awareness by providing them with a faithful and realistic virtual replica of real underwater archaeological sites that can be explored by means of an edutainment approach. In fact, the VR tour combines an educational purpose with ludic activities, allowing its users to enjoy the virtual environment by simulating a real diving session from the point of view of a scuba diver, other than learning archaeological and historical information in a playful manner.

3. VR System Overview

The VR system is a virtual diving exhibit that allows users to explore the 3D reconstruction of the underwater site and receive historical and archaeological information about the submerged exhibits and structures of the site, but flora and fauna are also described, with a particular attention on their interaction with the submerged artifacts. The VR system can be also be used by diver tourists because of its capability to make a detailed planning of the operations, and of the itinerary to carry out in the underwater archaeological site. The system indeed represents a reliable instrument to plan and simulate the tourist itinerary that is performed at a later time in the real submerged environment. The VR system presents two different versions, each one characterized by the type of devices, the provided levels of immersion, interaction, and presence: the VR semi-immersive and VR immersive experiences. In the first one, users can perform a semi-immersive visualization by means of a full HD monitor that is based on passive 3D technology. Users interact with the system by means of a multi-touch screen tablet, featuring a user-interface that provides all of the input functionalities that are needed to explore the 3D environment and get access to the multimedia data. The visualization can be performed also in an immersive environment by means of Head Mounted Display (HMD) technology.

3.1. Software Architecture

The software architecture of the VR system is depicted in Figure 1. It consists of five main elements: a database, a web service, a scene editor module, a visualization module, and the controller module.

In particular, the SQL database manages all of the data of the virtual scene. The web service provides a bilateral communication between the database and the other modules by means of a HTTP protocol. The scene editor, visualization, and controller modules have been implemented by means of the cross-platform game engine, Unity. Thanks to the adoption of the Unity framework, these modules can be used directly via web, and can communicate by means of the web service software, with the database for data uploading and downloading. The scene editor module allows to compose the virtual scene by integrating 3D objects and multimedia information stored in the database. Once the scene is created, the interaction module is adopted to implement the logics of the virtual scenario defining the physics and behaviours of the elements that belong to the virtual environment. Furthermore, it loads from the database the graphical assets of the submerged, terrestrial, and aerial environments, such as refractions, fog, caustics of the particulate, etc. The interaction module allows to perform the exploration within the virtual scenario according to the user input by means of the controller.

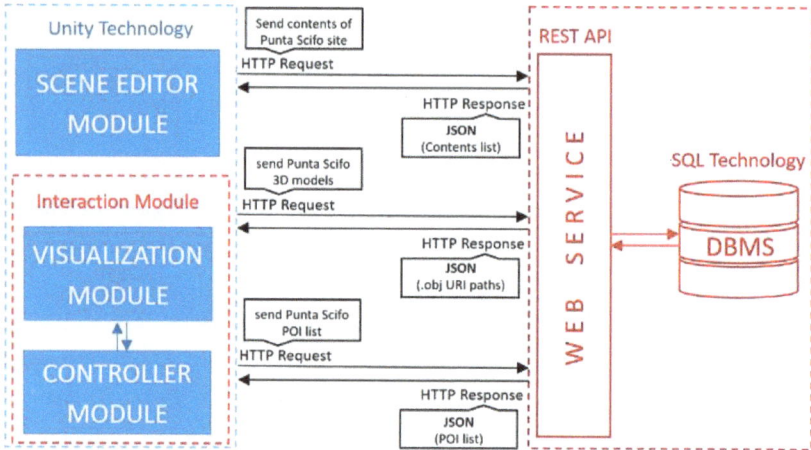

Figure 1. Software architecture of the Virtual Reality (VR) system.

3.2. Virtual Scene Creation

The 3D reconstruction of underwater archaeological assets requires attention because of the long and articulated process. One of the main objectives of the VISAS project concerned the integration of optical and acoustic techniques for the generation of multi-resolution textured 3D models of underwater archaeological sites. The proposed method exploits the high-resolution data obtained from photogrammetric techniques (Bruno et al., 2011; Bianco et al., 2013) and the latest techniques for the construction of acoustic micro bathymetric maps (De Alteriis et al., 2003; Passaro et al., 2013) to build 3D representations that combine the resolution of optical sensors with the precision of acoustic bathymetric surveying techniques. The method allows for obtaining a complete representation of the underwater scene, and to geo-localize the optical 3D model using the acoustic bathymetric map as a reference. For the scope of the project a high frequency multibeam equipment has been adopted to obtain an acoustic bathymetry of the seabed with a sample spacing of about 3cm; while, photogrammetric techniques have been used to build a textured 3D model of the archaeological remains with a sample spacing of about 3mm. In particular, after a first inspection of the site, the photogrammetric acquisition process is performed according to standard aerial photography layouts that consist of overlapping straight lines and also cross lines with oblique poses to minimize the occluded areas. Opto-acoustic markers, which are placed on the seabed and whose number depends by the extension of the site, are used to accurately compute the registration between the optical and acoustic point clouds. While, a set of triangular targets are adopted to scale the 3D model. The last steps of the process consist of meshing and texturing the opto-acoustic point cloud of the underwater archaeological site. The meshing step is carried out using a dedicated software, which has the ability to create a mesh by using an efficient multi-resolution algorithm and to perform further refinements of the model by using the point cloud as reference, so that the model reconstruction is performed in a coarse-to-fine fashion.

During the optical and acoustic survey activities, various useful and interesting locations are also defined and geolocated in order to be subsequently implemented as Points of Interests (POIs) in the virtual reproduction of the underwater archaeological site. In fact, once the creation of the textured 3D model of the underwater archaeological area is achieved, it is adopted as a starting point by the web editor module to build and assemble the immersive environment (Figure 2).

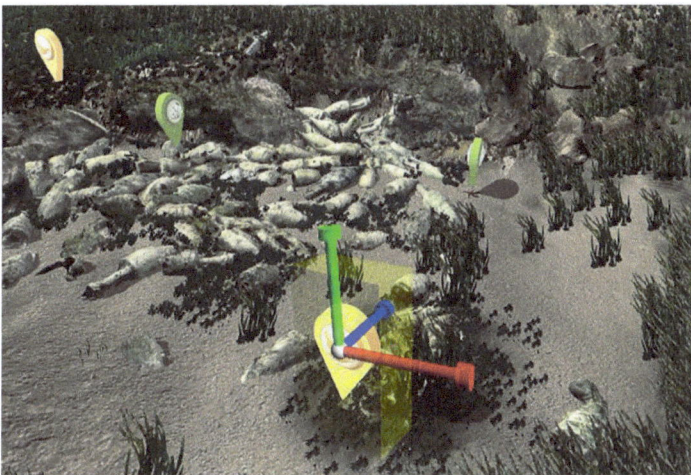

Figure 2. Adding a three-dimensional (3D) object of a Points of Interests (POI) in the virtual scene.

In particular, the virtual scene is populated with 3D models of the flora and fauna that are typical of the specific marine ecosystem, such as fishes, sponges, seagrass, and seaweed plants. The POIs are added to the virtual scene in form of 3D large head map tips, whose colour depends on the category that they belong to, e.g. yellow for the historical and archaeological information and green for biological ones. As depicted in Figure 3, 3D models of fishes and schools of fish, which are typical of that site, are settled into the underwater environment and are animated by means of artificial intelligence techniques. The vegetation is placed exactly as if it was captured during the optical survey, and it is reproduced by means of texture effects that mimic the movements of the real plants.

Figure 3. Reconstructed archaeological site with 3D POIs, flora and fauna.

3.3. Semi-Immersive Virtual Reality Environment

The semi-immersive setup of the VR system has been designed on the basis of a user-centered design (UCD) (Barbieri et al., 2017) by evaluating different embodiments, such as monitors and projectors for the visualization and touch-screen consoles and trackballs for the interaction. After an accurate evaluation of the pros and cons of the different solutions, a HD monitor and a multi-touch tablet have been adopted for the user interaction and exploitation of the virtual diving. About the touch-screen remote control, it could be a handheld device, i.e., tablet, or fixed in a specific position (Figure 4). The first solution can usually be adopted when an operator stands over the system. Instead, the second solution can be employed when the system is intended for unattended operation, and, since the console cannot be moved, it is possible to increase the screen size of the touch-screen to enhance its legibility.

The 3D HD monitor is based on passive technology. It has been preferred to the active one because passive 3D glasses are lighter and more comfortable than active glasses, which, on the contrary, are expensive and need batteries to work.

The user interface displayed on the touch-screen console provides all of the input functionalities that are needed to explore the 3D environment and to get access to the multimedia data. In particular, the User Interface (UI) features two large command buttons, respectively, to go back and forth and to rotate the camera's point of view. On the top left side of the UI, a slider controls the depth of the camera view from the water surface. While, on the top right side, two circle buttons allow for the user to switch between two different exploration modalities: guided tour modality and free navigation mode.

Figure 4. Semi-immersive VR system with a handheld device (left) or a fixed remote control (right).

When the user approaches a POI element, the UI displays a visual warning, and, at the same time, provides two command buttons that allow the user to question or skip it. In the first case, a new window appears in the centre of the UI containing textual, graphical and audio information related to the specific POI (Figure 5).

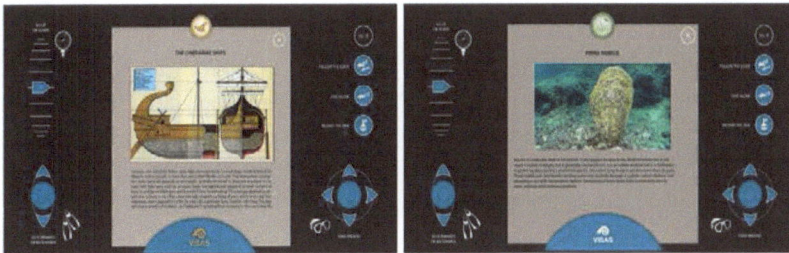

Figure 5. Multimedia data appear on the touch-screen console when questioning POIs.

3.4. Game Logic Design

The game logic design of the semi-immersive VR environment has been developed according to the best practices related to the gamification design process (Wei and Li, 2010; Kapp, 2012; Morschheuser et al., 2017) to maximize enjoyment and engagement through capturing the interest of learners (Huang and Soman, 2013).

The exploration of the underwater archaeological site starts above the water surface in the diving spot (Figure 6). In order to make a more attractive and engaging experience, the terrestrial environment has been added and constructed in the most realistic way possible. The buoy and the inflatable boat have been added to the virtual scene, as well as the stretch of coastline that overlooks the diving site.

Figure 6. Diving spot of the virtual session.

Once the user dives in the submerged virtual environment, he/she is guided by a directional 3D arrow toward the archaeological site. The exploration can be performed in two different modes, which can be selected by the user on the UI: free or guided tour. In both cases, the player is engaged in an active state of learning where he/she is motivated to create his/her own knowledge rather than to receive information passively. In particular, in the first mode, the user can dive freely in the archaeological area and he/she is free to pick the desired POI or simply take an overview of the submerged area. In the other case, the guided tour mode features a virtual diver who guides the user during the exploration of the underwater archaeological site (Figure 7). In particular, the virtual diver implements a logical follow-on of the POIs that, according to a storyline approach, allows for users to follow one or more itineraries and 'themed' routes. In the guided tour mode, the 3D POIs are hidden and they become visible, one at a time, when the scuba guide moves closer to them. At that point, the user can decide to skip the POI and pass to the next one, or to activate it and receive specific information by means of multimedia content (audio, video, and text).

Moreover, the guided tour modality allows for users to simulate a real diving session, taking into account the shortest path to safely visit the POIs of the underwater sites. The scope is to raise awareness of the time constraints and human body's limitations, arising from scuba diving related problems, which typical of the underwater environment. This modality has been implemented developing a pathfinding algorithm that: recognizes obstacles within the 3D environment; generates the shortest paths through the POIs avoiding obstacles and minimizing the pressure at which the diver is subjected; verifies the generated paths' admissibility according to constraints (the air in the diving cylinder and the decompression stops) imposed by the underwater environment.

Figure 7. Scuba guide for the exploration of the underwater archaeological site.

3.5. Immersive Virtual Reality Environment

The immersive diving environment is built by means of HMD technology in order to provide to its users a totally immersive experience of the underwater archaeological site. This technology, in fact, allows for achieving the best results in terms of immersivity because it absorbs completely the users in the virtual environment. The light-weight helmet isolates the user from the distractions of the actual physical environment and encompasses the entire field of view, including the peripheral space. It contains a high-resolution stereoscopic display; adjustable optics (usually based on Fresnel lenses); an optical tracking system capable of tracking both the position and the orientation of user's head; and, usually, a stereo audio output. Most of the products currently available on the market still require a cable connection to the computer because of the very high demand on data throughput, image latency, and power consumption.

The HMD is usually coupled with one or two wireless handheld controllers for a better experience in the VR environment. Each controller is equipped with several buttons, joystick, or touchpad as a means of human-computer interaction. They could also be tracked in terms of position and orientation in 3D space. The use of this configuration (HMD and a pair of tracked controllers) provides a very immersive and natural interaction with virtual worlds; the user can freely walk and look around, reach out their hands, and interact with virtual objects.

The immersive VR system has been developed in Unity for the HTC Vive VR headset (Figure 8). In particular, the HTC Vive features a resolution of 1080x1200 per eye, 90 Hz refresh rate, and a field of view of 110°). Furthermore, it comes bundled with two motion tracked controllers and laser based tracking system called Lighthouse, providing six degrees of freedom (6-DOFs) tracking in an up to 4.5m x 4.5m area with two beacons.

Figure 8. Immersive VR system.

When the user wears the HMD, he/she experiences the immersive virtual environment from the scuba diving viewpoint simulating a real diving session. The scenario that appears at the beginning of the virtual experience is above the water surface in the diving spot. Once he/she dives in the submerged virtual environment, he/she is guided by a directional 3D arrow to the archaeological underwater site. When he/she arrives at the site, the 3D arrow disappears and lets him/her free to interact with the 3D POIs to discover historical and archaeological information.

The user interaction is focused on navigating the virtual environment and receiving information about the archaeological and biological assets. He/she navigates in the virtual environment by moving his/her head and interacting with the virtual scene using the wireless handheld controllers. Through the input devices, the user instructs the software about the desired orientation and direction to follow for exploring the submerged virtual area. During the navigation, the controllers are mainly used for directional inputs for exploring the virtual underwater environment. Nevertheless, when the user reaches a POI, this needs to be enabled to get access to its multimedia contents. In the immersive environment, the POIs are enabled directly by the user, by pointing and selecting them by squeezing the hair trigger of the handheld controller. In this case, the multimedia contents are displayed within the virtual scenario into a 3D frame (Figure 9).

Figure 9. Archaeological asset description.

4. User Study

The semi-immersive and immersive environments have been evaluated by means of user studies (Barbieri et al., 2017) that were carried out at the Department of Mechanical, Energy and Management Engineering (DIMEG) of the University of Calabria (Italy). In particular, a comparative user study has been performed for evaluating the usability and enjoyment provided by the two different environments.

The minimum number of participants has been defined on the basis of the most influential articles on the topic of sample size in user studies (Lewis, 1994; Dumas et al., 1995) adopting a problem discovery rate of 95%. This value has been adopted as a threshold to deploy participants among the groups. Nevertheless, as many users as possible have been involved in the test in order to collect a greater number of feedbacks and personal opinions.

Four representative user groups (Figure 10), divided by age, have been involved in the comparative study:

- G1: 28 preteens from 10 to 13 years old (mean=12 - standard deviation=0,82);
- G2: 24 teenagers from 14 to 16 years old (mean=14,83 - standard deviation=0,76);
- G3: 18 male young adults from 17 to 22 years old (mean=18,11 - standard deviation=1,23);
- G4: 31 female young adults from 17 to 24 years old (mean=18,06 - standard deviation=1,21).

The comparative procedure consisted of three main steps. In the first step, after a short presentation of the virtual diving system, a quick demo of its main features and of the different user interactions in the semi-immersive and immersive environments have been shown to the participants. In the second step, each participant performed a free exploration of both the environments without any limitation in time. By the end of the second step, users have been invited to fulfill a satisfaction questionnaire and to perform a one-on-one personal interview aimed to comprehend their enjoyment and catch all their possible personal judgments. The satisfaction questionnaire has been developed on the basis of standard questionnaires based on psychometric methods (Lewis, 2006) whose items' evaluation can be expressed by users by means a seven-point graphic scales (Likert scale), anchored at the end points with the term "Strongly disagree" for 1 and "Strongly agree" for 7.

Figure 10. Participants of the user study while interacting with the semi-immersive and immersive environments.

The user satisfaction questionnaire results revealed that both the semi-immersive and immersive environments of the VR system gained similar levels of learnability, efficacy, and enjoyment. These levels reached high values in all of the groups, with a minimum score of six out of seven. In particular, little differences in subjective opinions between the semi-immersive and the immersive environment have been observed in each group and among the groups. But, statistical analysis, based on t-test and ANOVA techniques, revealed insufficient evidence to confirm these small gaps.

Differently from the user satisfaction questionnaire results, when participants were asked to express their preference, the vast majority of them clearly expressed their choice, as depicted in the following figure, for the immersive environment, and consequently, for the HMD device.

This result has been unexpected because even if all of the users were confident in the use of 3DTV and multi-touch devices, only a few of them had previous experience in the use of HMDs. In the one by one interviews, the subjects justified their preference asserting that, even if they felt comfortable in the adoption of 3DTV, the HMD technology provided a much more immersive virtual experience and allowed for them to behave in a more natural way: in fact, they were able to turn their heads or their whole body naturally.

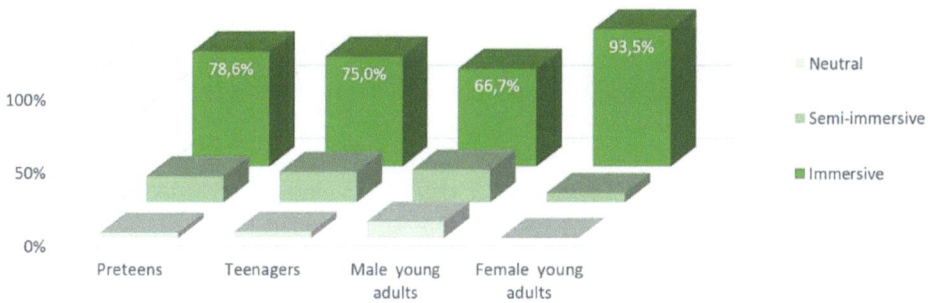

Figure 11. Subjective preference results.

In conclusion, on the basis of this user study, the semi-immersive and the immersive environments provide to the users high equivalent levels of usability and enjoyment. Then, the level of the immersion of the VR system can be chosen according to factors that are related to the specific application context or to subjective preferences, rather than usability metrics. In fact, as emerged from the participants' personal interviews, the HMD technology is an appropriate choice, especially when users need to make frequent turns to look around and enjoy the virtual environment. On the other side, the semi-immersive environment could be more adequate for a museum and school contexts in which there is the need to make the virtual experience available to a large number of visitors, or to allow a shared exploitation of the virtual scenario to more than one person at a time.

5. Conclusions

The paper has presented a virtual diving system for the exploitation of the Underwater Cultural Heritage. This VR system allows for users to live a virtual experience inside the reconstructed 3D model of underwater archaeological sites and explore the various and different POIs that populate the archaeological area by means of a free or guided tour.

The VR system has been designed to entertain users, but its added pedagogical value is explicitly emphasized too. In fact, the ludic activity, consisting in the simulation of a real diving session from the point of view of a scuba diver, is combined with the educational purpose by means of a storyline approach that is based on a virtual scuba diver that guides the user among the various POIs that populate the underwater site. When activated, POIs offer specific information about historical and archaeological peculiarities, flora and fauna are also described with a particular attention on their interaction with the submerged artifacts. The user study has demonstrated that both the semi-immersive and immersive environments present high levels of usability and enjoyment for all of the categories of participants, which differ for age and sex. Since the user studies have been carried out among teenagers and young adults, further usability tests will be performed recruiting adult and elder participants too. Furthermore, the transmission of knowledge to the user, after having experienced the virtual diving environment, will be explored to assess the learning benefit of the system.

Acknowledgments: The VISAS Project (Ref. Start-Up PAC02L2-00040), has been financed by the MIUR under the PAC Programme. The authors would like to thank the Marine Protected Area of Capo Rizzuto, the Soprintendenza per i Beni culturali e ambientali del Mare della Sicilia for the permission to conduct the experimentation in the sites of Cala Minnola and the Soprintendenza per i Beni Archeologici della Calabria for the site of Punta Scifo.

References

1. Barbieri, L.; Bruno, F.; Mollo, F.; Muzzupappa, M. User-centered design of a Virtual Museum system: a case study. In *International Joint Conference on Mechanics, Design Engineering & Advanced Manufacturing*; Springer International Publishing: New York, NY, USA, 2017; pp. 155–165.
2. Barcelo, M.; Forte D.H. Virtual reality in archaeology. In *Archeopress BAR International Series 843*; Sanders, Ed.; Archeopress: Oxford, UK, 2000.
3. Bacpoles. Preserving cultural heritage by preventing bacterial decay of wood in foundation piles and archaeological sites. In *Final Report BACPOLES-project (EVK4-CT-2001-00043)*; Klaassen, R., Ed.; SHR: Wageningen, The Netherlands, 2005.
4. Bianco, G.; Gallo, A.; Bruno, F.; Muzzupappa, M. A comparative analysis between active and passive techniques for underwater 3D reconstruction of close-range objects. *Sensors* **2013**, *13*, 11007–11031.
5. Bruno, F.; Gallo, A.; Barbieri, L.; Muzzupappa, M.; Ritacco, G.; Lagudi, A.; La Russa, M.F.; Ruffolo, S.A.; Crisci, G.M.; Ricca, M; et al. The CoMAS project: new materials and tools for improving the in-situ documentation, restoration and conservation of underwater archaeological remains. *Marine Technol. Soc. J.* **2016**, *50*, 108–118.
6. Bruno F.; Lagudi A.; Muzzupappa M.; Lupia M.; Cario G.; Barbieri L.; Passaro S.; Saggiomo R. Project VISAS: Virtual and Augmented exploitation of submerged archaeological sites—Overview and first results. *Marine Technol. Soc. J.* **2016**, *50*, 119–129.
7. Bruno, F.; Bruno, S.; De Sensi, G.; Luchi, M.L.; Mancuso, S.; Muzzupappa, M. From 3D reconstruction to virtual reality: A complete methodology for digital archaeological exhibition. *J. Cult. Herit.* **2010**, *11*, 42–49, doi:10.1016/j.culher.2009.02.006.
8. Bruno, F.; Bianco G.; Muzzupappa M.; Barone, S.; Razionale, A.V. Experimentation of structured light and stereo vision for underwater 3D reconstruction. *ISPRS J. Photogramm. Remote Sens.* **2011**, *66*, 508–518.

9. Chapman, P.; Roussel, D.; Drap, P.; Haydar, M. Virtual Exploration of Underwater Archaeological Sites: Visualization and Interaction in Mixed Reality Environments. In Proceedings of The 9th International Symposium on Virtual Reality, Archaeology and Intelligent Cultural Heritage, Braga, Portugal, 2–5 December 2008; pp. 141–148.

10. Chapman, P.; Conte, G.; Drap, P.; Gambogi, P.; Gauch, F.; Hanke, K.; Richards, J. Venus, virtual exploration of underwater sites. In Proceedings of the Joint Event CIPA/VAST/EG/Euro-Med, Nicosia, Cyprus, 28 October–4 November 2006.

11. De Alteriis, G.; Passaro, S.; Tonielli, R. New, high resolution swath bathymetry of Gettysburg and Ormonde Seamounts (Gorringe Bank, eastern Atlantic) and first geological results. *Marine Geophys. Res.* **2003**, *24*, 223–244.

12. Dumas, J.; Sorce, J.; Virzi, R. Expert reviews: how many experts is enough? In Proceedings of the Human Factors and Ergonomics Society 39th Annual Meeting, HFES, San Diego, CA, USA, 9–23 October 1995, pp. 309–312.

13. EU Commission Directorate General for Maritime Affairs and Fisheries. A European Strategy for more Growth and Jobs in Coastal and Maritime Tourism. *Communication from the Commission to the European Parliament, the Council, the European and Economic and Social Committee and the Committee of the Regions*; 2014. Available online: http://ec.europa.eu/maritimeaffairs/policy/coastal_tourism/documents/com_2014_86_ en.pdf.

14. Gregory, D.J. Development of tools and techniques to survey, assess, stabilise, Monitor and preserve underwater archaeological sites: Sasmap. *Int. J. Herit. Dig. Era* **2012**, *1*, 367–371.

15. Gündoğdu, H.T.; Dede, M.İC.; Taner, B.; Ridolfi, A.; Costanzi R.; Allotta, B. An innovative cleaning tool for underwater soft cleaning operations. In Proceedings of the OCEANS 2015, Genova, Italy, 2–3 October 2015; pp. 1–8.

16. Haydar, M.; Roussel, D.; Maïdi, M.; Otmane, S.; Mallem, M. Virtual and augmented reality for cultural computing and heritage: a case study of virtual exploration of underwater archaeological site. *Vir. Real.* **2011**, *15*, 311–327.

17. Huang, W.H.Y.; Soman, D. *Gamification of Education*; Research Report Series: Behavioural Economics in Action; Rotman School of Management, University of Toronto: Toronto, ON, Canada, 2013.

18. Hunter, J.W. Underwater archaeology: The NAS guide to principles and practice. *Australasian Hist. Archaeol.* **2009**, *27*, 127.

19. Katsouri, I.; Tzanavari, A.; Herakleous, K.; Poullis, C. Visualizing and assessing hypotheses for marine archaeology in a VR CAVE environment. *J. Comput. Cult. Herit. (JOCCH)* **2015**, *8*, 10, doi:10.1145/2665072.

20. Kapp, K.M. *The Gamification of Learning and Instruction: Game-Based Methods and Strategies for Training and Education*; John Wiley & Sons: New York, NY, USA, 2012.

21. Lepouras, G.; Vassilakis, C. Virtual museums for all: employing game technology for edutainment. *Vir. Real.* **2005**, *8*, 96–106, doi:10.1007/s10055-004-0141-1.

22. Lewis, J.R. Sample sizes for usability studies: Additional considerations. *Hum. Factors* **1994**, *36*, 368–378.

23. Lewis, J.R. Usability testing. In *Handbook of Human Factors and Ergonomics*; Salvendy, G., Ed.; John Wiley: New York, NY, USA, 2006; pp. 1275–1316.

24. Liarokapis, F.; Kouřil, P.; Agrafiotis, P.; Demesticha, S.; Chmelík, J.; Skarlatos, D. 3D modelling and mapping for virtual exploitation of underwater archaeology assets. *Int. Arch. Photogramm. Remote Sens. Spat. Inf. Sci.* **2017**, *XLII-2/W3*.

25. Morschheuser, B.; Werder, K.; Hamari, J.; Abe, J. How to gamify? Development of a method for gamification. In Proceedings of the 50th Annual Hawaii International Conference on System Sciences (HICSS), Waikoloa, HI, USA, 4–7 January 2017; pp. 4–7.

26. Passaro, S.; Barra, M.; Saggiomo, R.; Di Giacomo, S.; Leotta, A.; Uhlen, H.; Mazzola, S. Multi-resolution morphobathymetric survey results at the Pozzuoli-Baia underwater archaeological site (Naples, Italy). *J. Archaeol. Sci.* **2013**, *40*, 1268–1278.

27. Pavlidis, G.; Koutsoudis, A.; Arnaoutoglou, F.; Tsioukas, V.; Chamzas, C. Methods for 3D digitization of cultural heritage. *J. Cult. Herit.* **2007**, *8*, 93–98, doi:10.1016/j.culher.2006.10.007.

28. Roussou, M. Virtual heritage: from the research lab to the broad public. *BAR Int. Series* **2002**, *1075*, 93–100.

29. Stone, R.; White, D.; Guest, R.; Francis, B. The Virtual Scylla: an exploration of "serious games", artificial life and simulation complexity. *Vir. Real.* **2009**, 13, 13–25.

30. Unesco. Convention on the Protection of the Underwater Cultural Heritage, 2 November 2001. Available online: http://www.unesco.org.

31. Varinlioğlu, G. Data Collection for a Virtual Museum on the Underwater Survey at Kaş, Turkey. *Int. J. Naut. Archaeol.* **2011**, *40*, 182–188.

32. Vote, E.; Feliz, D.A.; Laidlaw, D.H.; Joukowsky, M.S. Discovering petra: archaeological analysis in VR. *IEEE Comput. Graph. Appl.* **2002**, *22*, 38–50. doi:10.1109/MCG.2002.1028725.

33. Wei, T.; Li, Y. Design of educational game: a literature review. *In Transactions on Edutainment IV*; Springer: Berlin/Heidelberg, Germany, 2010; pp. 266–276.

Chapter 4

Underwater 3D Surveying and Modelling

Underwater Image Enhancement before Three-Dimensional (3D) Reconstruction and Orthoimage Production Steps: Is It Worth?

Panagiotis Agrafiotis [a,b], **Georgios I. Drakonakis** [a], **Dimitrios Skarlatos** [b] and **Andreas Georgopoulos** [a]

[a] National Technical University of Athens, School of Rural and Surveying Engineering, Lab. of Photogrammetry, Zografou Campus, 9 Heroon Polytechniou str., 15780, Zografou, Athens, Greece; pagraf@central.ntua.gr, georgios.i.drakonakis@gmail.com, drag@central.ntua.gr

[b] Cyprus University of Technology, Civil Engineering and Geomatics Dept., Lab of Photogrammetric Vision, 2-8 Saripolou str., 3036, Limassol, Cyprus; dimitrios.skarlatos@cut.ac.cy

Abstract: The advancement of contemporary digital techniques has greatly facilitated the implementation of digital cameras in many scientific applications, including the documentation of Cultural Heritage. Digital imaging has also gone underwater, as many cultural heritage assets lie in the bottom of water bodies. Consequently, a lot of imaging problems have arisen from this very fact. Some of them are purely geometrical, but most of them concern the quality of the imagery, especially in deep waters. In this paper, the problem of enhancing the radiometric quality of underwater images is addressed, especially for cases where this imagery is going to be used for automated photogrammetric and computer vision algorithms later. In detail, it is investigated whether it is worth correcting the radiometry of the imagery before the implementation of the various automations or not, the alternative being to radiometrically correct the final orthoimage. Two different test sites were used to capture imagery ensuring different environmental conditions, depth, and complexity. The algorithms investigated to correct the radiometry are a very simple automated method, using Adobe Photoshop®, a specially developed colour correction algorithm using the CLAHE (Zuiderveld, 1994) method, and an implementation of the algorithm, as described in Bianco et al. (2015). The corrected imagery is afterwards used to produce point clouds, which in turn are compared and evaluated.

Keywords: underwater 3D reconstruction; underwater image enhancement; SfM-MVS (Structure from Motion-Multi View Stereo)

1. Introduction

A great percentage of mankind's Cultural Heritage lies underwater. Ancient ports, seaside fortifications, and shipwrecks are only but a few examples of underwater Cultural Heritage. These assets wait to be documented according to article 16 of the Venice Charter (ICOMOS 1964). However, the underwater environment is very hostile for humans and for digital equipment alike. Nowadays, technological advances have enabled the use of automated algorithms for mapping and three-dimensional (3D) modelling using digital imagery, while the highly sensitive equipment needs special protection. In addition, digital RGB imagery is adversely affected by the underwater conditions, and its radiometry is rapidly deteriorating with depth. Because, nowadays, most of the documentation procedures are based on digital image processing (Drap 2012; Henderson et al., 2013; Johnson-Roberson et al., 2016), a thorough investigation into the resulting radiometry of the original images and the final imagery products seems necessary.

Within this context, this paper investigates the effect of several algorithms that correct the underwater imagery radiometry and assesses their suitable implementation moment, i.e. before or after processing and the production of the finale orthoimage. As illumination and colour loss is rapid underwater and this largely depends on depth, this research focuses on the investigation of the behaviour of the RGB channels and the algorithms that are available to restore their natural values.

For this purpose, two different approaches for underwater image processing are implemented according to their description in literature. The first one is image restoration. It is a strict method that is attempting to restore true colours and correct the image using suitable models, which parameterize adverse effects, such as contrast degradation and backscattering, using image formation process and environmental factors, with respect to depth (Hou et al., 2007, Treibitz and Schechner, 2009). The second one uses image enhancement techniques that are based on qualitative criteria, such as contrast and histogram matching (Ghani and Isa, 2014, Iqbal et al., 2007 and Hitam et al., 2013). Image enhancement techniques do not consider the image formation process and do not require environmental factors to be known a priori (Agrafiotis et. al., 2017).

Visual computing in underwater settings is particularly affected by the optical properties of the surrounding medium (von Lukas 2016). The goal of the presented work is to investigate the effect of the underwater imagery preprocessing on the 3D reconstruction of the scene using automated Stuctrure from Motion (SfM) and Multi View Stereo (MVS) software. An additional aim is to present measurable results regarding the effect of this preproccessing on the produced orthoimages. Since the processing of images, either separately or in batch, is a time-consuming procedure that has high compotutational cost requirements, it is critical to determine the necessity of implementing colour correction and enhancement before the SfM - MVS procedure or directly to the final orthoimage when this is the deliverable.

2. Materials and Methods

In order to address the above research issues, two different test sites were selected to capture underwater imagery ensuring different environmental conditions (i.e. turbidity, waves etc.), depth, and complexity. In addition, three different image correction methods are applied to these datasets: A very simple automated method using Adobe Photoshop, a colour correction algorithm, which was developed using the CLAHE (Zuiderveld, 1994) method, and an implementation of the algorithm described in Bianco et al. (2015). Subsequently, dense 3D point clouds (3Dpc), 3D meshes, and orthoimages were generated for each dataset using a commercial SfM - MVS software. The produced 3D point clouds were then compared using Cloud Compare (CloudCompare 2.8.1, 2016) open-source software, while the resulting orthoimages were compared using both their visual appearance and their histograms.

2.1. Test Datasets

2.1.1. Shallow Waters

The dataset created for shallow waters (Figure 1a) is a near-shore underwater site situated at depths varying from 2 to 3 meters, and presents smooth depth changes (Figure 2a). Image acquisition took place with a Nikon DSLR D5200, with 24 MP sensor and pixel size of 3.92 μm and by using an 18-55 mm lens set at 18 mm, and an Ikelite dome housing. No artificial light sources were used due to the small depth. Because of the wind, the wavy surface of the water creates dynamic sun flicker (caustics) on the seabed. Waves also resulted into water turbidity, and thus very poor visibility conditions. The camera positions of this dataset follow a typical phtogrammetric distribution (Figure 3a), with obvious parallel dive paths.

2.1.2. Deep Waters

The amphorae dataset is an artificial reef constructed using 1m long amphorae, replicas from Mazotos shipwreck (Demesticha, 2010) and presents abrupt changes on the imaged object depth (Figure 2c). They are positioned on the sea bottom at 23 meters depth, stacked together in two layers, with 25 and 16 amphorae, respectively (Figure 1b). Their placement is assumed to be similar to their original position in the cargo area of the ship. Photography took place with an action camera, Garmin Virb XE camera, with 12MP sensor with 1.5μm physical pixel size. Four LED video lights of 300LX eatc, were mounted next to the camera, to enhance the recorded colour information. In contrast with the shallow waters dataset, camera positions do not follow a specific pattern (Figure 3b). As it is expected, due to its depth for the shallow waters dataset, green and blue are the dominant colours, while a percentage of red colour is also present (Figure 4a). In contrast, in the deep waters dataset, blue is the dominant colour, while green follows. Although red channel absorption is strong in such depths, red is still present in the histogram (Figure 5b) of the images, depending on the artificial lighting and on the camera to object distance.

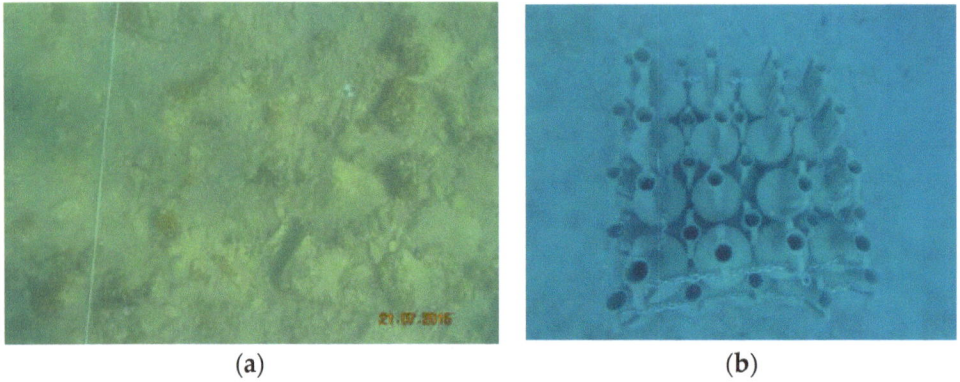

(a)   (b)

Figure 1. Sample images from shallow waters dataset (**a**) and deep waters dataset (**b**).

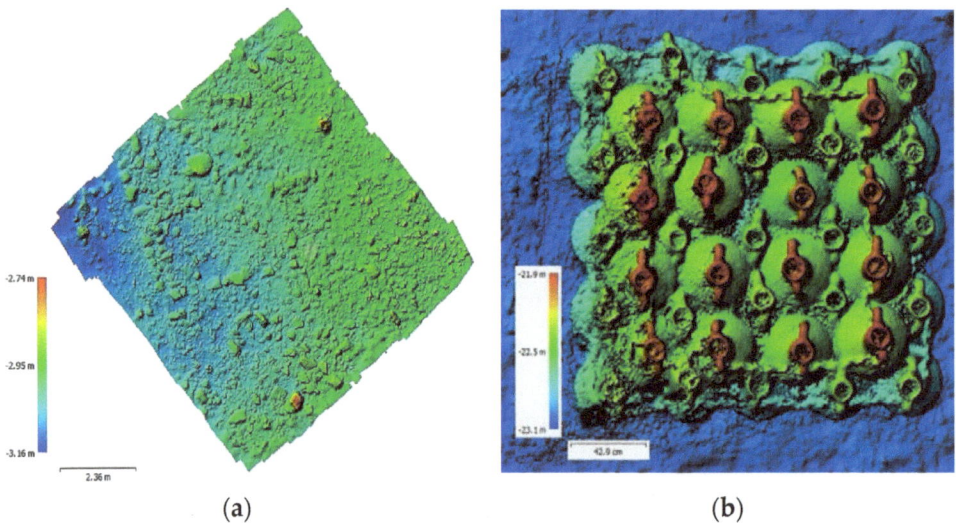

(a)   (b)

Figure 2. Resulted Digital Surface Model (DSM) for shallow waters dataset (**a**) and deep waters dataset (**b**). The difference in depth range between the two datasets is profound.

242

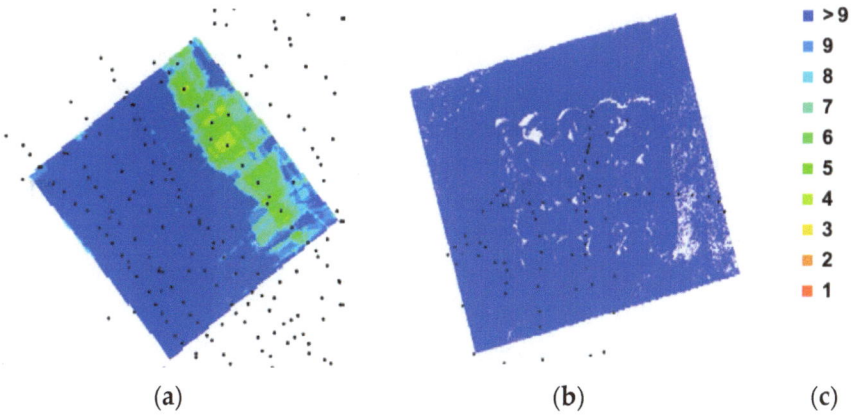

Figure 3. Camera positions and number of overlapping images used for shallow waters dataset (**a**) and deep waters dataset (**b**). In (**c**) the image overlap colour scale.

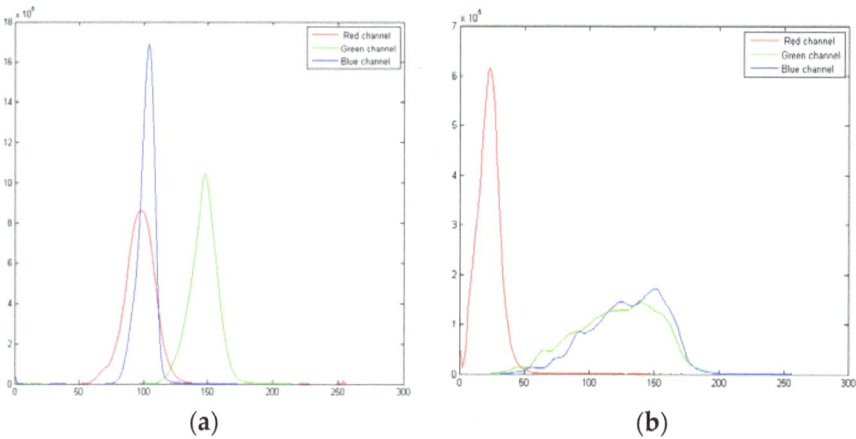

Figure 4. The respective histograms of the sample images from shallow waters dataset (**a**) and deep waters dataset (**b**).

2.2. The Applied Image Correction Algorithms

2.2.1. Clahe Based Algorithm

Adaptive histogram equalization (AHE) is a computer image processing technique that is used to improve contrast in images. Contrast Limited AHE (CLAHE) differs from ordinary adaptive histogram equalization in its contrast limiting. The CLAHE (Zuiderveld, 1994) algorithm partitions the image into contextual regions and applies the histogram equalization to each one, while it limits the contrast. This evens out the distribution of used grey values, and thus makes hidden features of the image more visible. The CLAHE algorithm has been used extensively in underwater image correction in the literature (Kumar Rai, et al. 2012; Yussof, et al., 2013; Singh, et al., 2011 and Hitam et al., 2013). Recently, CLAHE was used for the dehazing of underwater video for an augmented reality application (Bruno et al., 2017) presenting interesting results in real time scenarios.

The implemented image enhancement algorithm separates image channels of RGB colour space. Then a histogram equalization process is applied to each channel. A Rayleigh Cumulative Distribution Function (CDF) was created for the equalization. The CDF range is from 0 to 255, in order to match the pixel intensity values and its maximum value was appointed to one-third of the total range (76.5). Afterwards, the CLAHE (Zuiderveld, 1994) algorithm was applied to each channel. Partition size and contrast clipping were determined experimentally. Rayleigh distribution was used again as the transformation function parameter. Finally, the algorithm composes all three channels and the output is the colour corrected image.

In the shallow waters dataset, applying histogram equalization using a Rayleigh CDF on each colour channel results in histograms of practically identical shape. This dramatically improves image sharpness and its colours seem to be restored. In the deep waters dataset the algorithm was modified in order to cope with the large depth conditions. The red channel was not equalized before CLAHE algorithm correction, while green and blue channels were equalized with a different CDF in order to restrict their intensity into lower values.

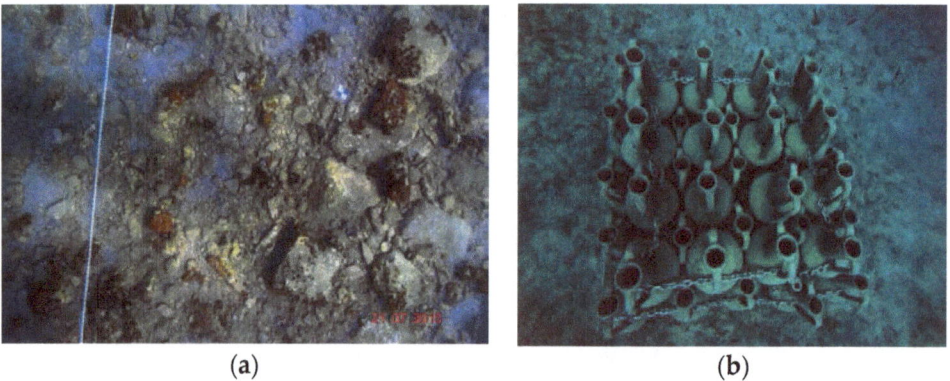

(a)   (b)

Figure 5. Corrected imagery with Contrast Limited adaptive histogram equalization (CLAHE) based algorithm for shallow (a) and deep (b) water.

2.2.2. Algorithm of Bianco et al. (2015)

The imagery was also processed using the algorithm presented by Bianco et al., (2015), where colour correction of underwater images is performed by using lab colour space. In more detail, the chromatic components are changed moving their distributions around the white point (white balancing), and histogram cut-off and stretching of the luminance component is performed to improve the image contrast. Main constrains of this method are the grey-world assumption and the uniform illumination of the scene (Bianco et al., 2015).

In the shallow water dataset, the corrected imagery looks very realistic as all of the colours are correctly enhanced. However, the sharpness of the imagery is not well improved. In the deep waters dataset, the corrected image presents an enhanced contrast despite the fact that the image looks similar to greyscale due to the absence of the red colour at such depths.

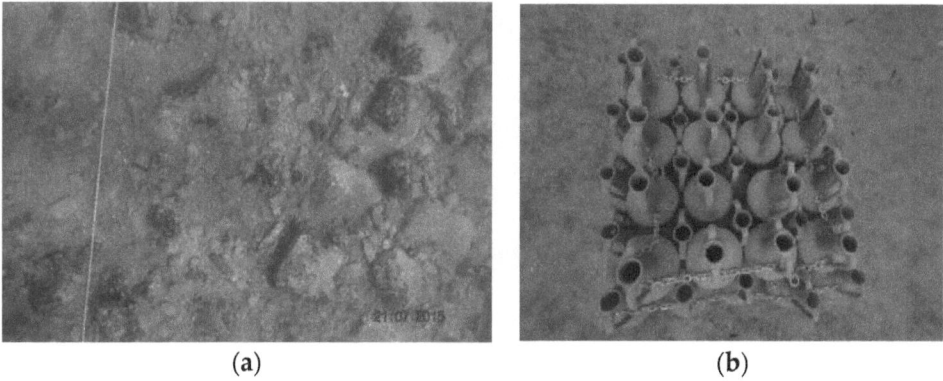

(a)   (b)

Figure 6. Corrected imagery with Lab for shallow (**a**) and deep (**b**) water.

2.2.3. Adobe Photoshop® (Adobe Photoshop CS5, (2010))

Additionally, images were processed with Adobe Photoshop in order to enhance the contrast and sharpness. Automated algorithm *"Find Dark and Light Colours"* was used for this correction. According to the software, this algorithm analyzes the image in order to find dark and light colours, and uses them as the shadow and highlight colours. The option *"Snap Neutral Midtones"* was also checked. This adjusts the midtones so that colours that are close to neutral are mapped to the target neutral colour.

In the shallow waters dataset, image blurriness is reduced and colours become more realistic. In contrast, the correction has little effect on the imagery that is acquired in deep waters. Contrast is slightly increased, while colours remain unaffected. For the deep waters dataset, this method was proven ineffective and a different algorithm was used. It clips colour channels identically in order to increase contrast while it preserves the original colours. The final image is improved partially. Red prevails on the image corners because of the chromatic aberration effect provoked by the fish-eye lens of the action camera.

To sum up, for the shallow waters dataset, all of the correction algorithms improve the colours of the imagery. The CLAHE based algorithm improves more the image sharpness, however, the more realistic colours are obtained by using the algorithm of Bianco et al., (2015). For the deep waters dataset, the results of Adobe Photoshop and CLAHE based algorithms present more percentage of red values,

245

and thus the imagery looks more appealing to the human eye. However, again, the results of the algorithm of Bianco et al., (2015) resemble more to the underwater environment reality, even if the red colours are undervalued. Important to note is the overincreased red values on the pixels of the corners of the deep water dataset (Figure 7b). This phenomenon is usually observed when robust colour correction methods are applied in imagery that is produced by action cameras using fish-eye lenses.

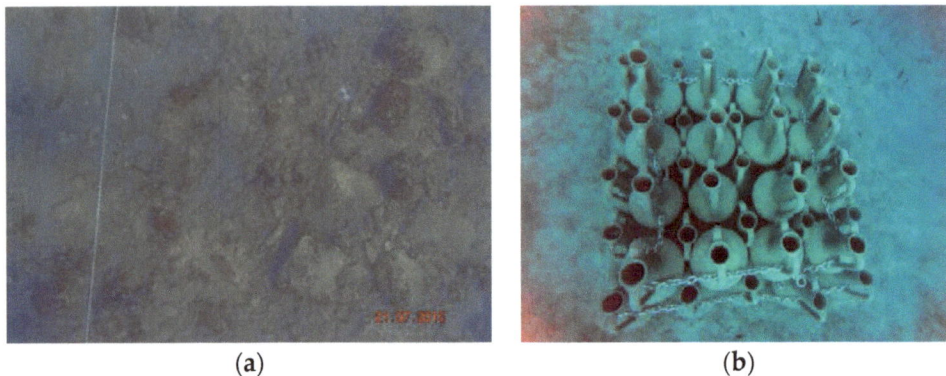

(a)   (b)

Figure 7. Corrected imagery with Adobe Photoshop® for shallow (**a**) and deep (**b**) waters.

2.3. SfM-MVS Processing

These image enhancement methods were evaluated by visual inspection and histogram comparison. Subsequently, corrected image were processed using SfM-MVS with Agisoft's Photoscan commercial software (AgiSoft PhotoScan Professional 1.2.6, 2016).). To this end, four different three-dimensional (3D) projects were created for each test site: (i) One with the original uncorrected imagery, which is considered the initial solution, (ii) a second one using the developed correction algorithm applying CLAHE (Zuiderveld, 1994), (iii) a third one using the imagery that resulted implementing the colour correction algorithm presented in Bianco et al., (2015), and (iv) a fourth one using Adobe Photoshop enhanced imagery. All three channels of the images were used for these processes.

For the created projects of each test site, the alignment parameters of the original (uncorrected) dataset were used as the corrected images were replacing the uncorrected ones by changing the image path. This ensured that the alignment parameters were the same in order to test only the number of points that were extracted for the dense cloud. In order to scale the 3D dense point clouds, predefined Ground Control Points (GCPs) were used for the shallow water projects. These GCPs were mesured using a Total Station from the shore. Scalebars were used for scaling of the deep water dataset projects since it was impossible to use traditional surveying methods due to the depth. Subsequently, 3D dense point

246

clouds of medium quality and density were created for each data set. No filtering during this process was performed in order to get the total number of dense point clouds, also the noise. It should be noted that medium quality dense point cloud means that the initial images' resolution was reduced by factor of 4 (2 times by each side), in order to be processed by the SfM-MVS software (Agisoft, 2017).

Table 1 sums up the results of the aforementioned processing for medium quality dense point cloud generation. For the shallow waters dataset, an area of 21.3 m² was covered by 155 images, having an average camera to object distance 1.57 m, and thus resulting to a ground resolution of 0.304 mm/pixel (Figure 8). For the deep water dataset, an area of 8.01 m² was covered by 89 images, having an average camera to object distance 1.68 m, and thus resulting to a ground resolution of 0.743 mm/pixel (Figure 11). It must be noted that the percentages of the differences of dense point cloud number resulting from the processed images in comparison to the dense point cloud resulted from the original ones are not significant. These differences are magnified when it comes to the deep waters dataset and this is probably due to the complexity of the object in respect with image correction methodology.

Table 1. Results of the Stuctrure from Motion (SfM)- Multi View Stereo (MVS) procedure for medium quality dense cloud generation.

Datase t used	Colour Correction Method	Focal Lenght (mm)	Pixel Size (μm)	Average Camera to Object Distance (m)	Ground Resolution (mm/pixel)	Area Covered (m²)	Reprojection Error (pixel)	Point Number in Dense Point Cloud	Differences of Dense Point Cloud Point Numbers from the Original Ones
Shallo w	Original	18	3.92	1.57	0.304	21.3	1.22	20.891.576	-
	CLAHE							20.924.857	0.16%
	Bianco et al. (2015)							19.885.863	-4.81%
	Photoshop							20.841.409	-0.24%
Deep	Original	3	1.5	1.68	0.743	8.01 (top projectio n)	2.04	6.892.271	-
	CLAHE							7.083.982	2.78%
	Bianco et al. (2015)							6.496.123	-5.75%
	Photoshop							7.222.016	4.78%

2.4. Orthoimage Generation

Orthoimages were created for both test sites. The main aim of this procedure is to determine the necessity of implementing colour correction and enhancement before the SfM-MVS procedure or directly to the final orthoimage when the orthoimage is the deliverable. To this end, orthoimages were generated for every different project using the original imagery (Figures 9a and 12a), the imagery that was corrected with Adobe Photoshop (Figures 9b and 12b), the imagery that was corrected by the CLAHE based algorithm (Figures 9c and 12c), and the imagery that was corrected by the algorithm of Bianco et al., (2015) (Figures 9d and 12d). For the two datasets, the orthoimage Ground Sampling Distance (GSD) was selected to be 0.005 m. Colour correction mode was disabled for all of the datasets since the datasets were not characterized by extreme brightness variations

(Agisoft, 2017), and enabling it would lead to false comparisons and misleading results.

Additionally to this process, the two final orthoimages resulted from the projects of the two test sites using the original imagery (Figures 10a and 13a), were processed with the three colour correction methods (Figures 7b–d and 10b–d) and the results were compared and evaluated. These results are illustrated in Section 3.

(a)   (b)   (c)   (d)

Figure 8. Medium quality point cloud for shallow waters dataset from the original imagery (**a**), the imagery corrected by the CLAHE based algorithm (**b**), the imagery corrected by the algorithm of Bianco et al., (2015) (**c**) and the imagery corrected with Adobe Photoshop (**d**).

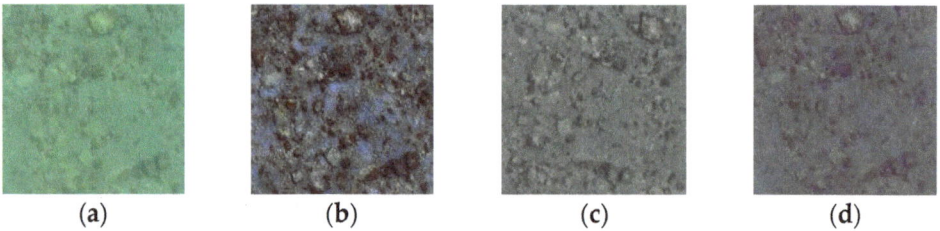

(a)   (b)   (c)   (d)

Figure 9. The orthoimage of the shallow dataset created by using the original images (**a**), by using the imagery corrected by the CLAHE based algorithm (**b**), by using the imagery corrected by the algorithm of Bianco et al., (2015) (**c**) and by using the imagery corrected with Adobe Photoshop (**d**) (the dimensions of the imaged area are 4.50 m × 4.50 m).

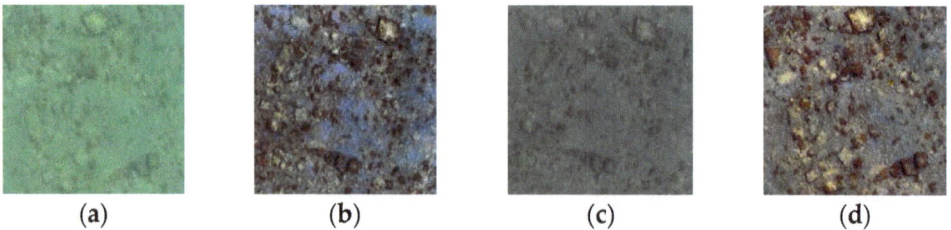

(a)   (b)   (c)   (d)

Figure 10. The orthoimage of the shallow dataset created by using the original images (**a**), by correcting the (**a**) using the CLAHE based algorithm (**b**), by correcting the (**a**) using the algorithm of Bianco et al., (**c**), (2015) and by correcting the (**a**) using the Adobe Photoshop (**d**) (the dimensions of the imaged area are 4.50 m × 4.50 m).

248

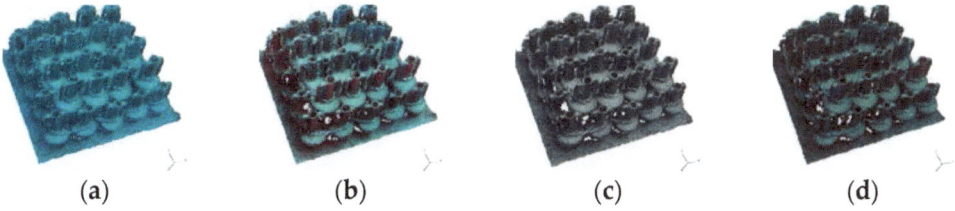

Figure 11. Medium quality point cloud for deep waters dataset from the original imagery (**a**), the imagery corrected with Adobe Photoshop (**b**), the imagery corrected by the algorithm of Bianco et al., (2015) (**c**) and the imagery corrected by the CLAHE based algorithm (**d**).

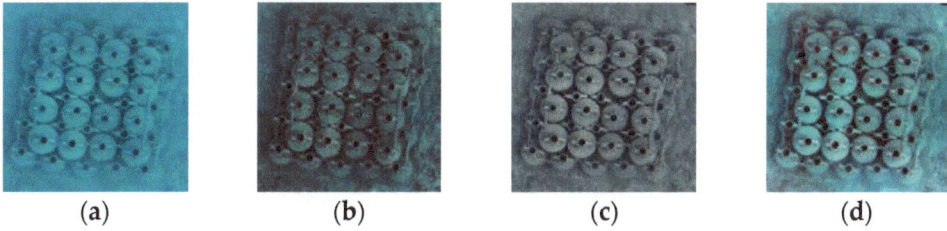

Figure 12. The orthoimage of the deep waters dataset created by using the original images (**a**), by using the imagery corrected by the CLAHE based algorithm (**b**), by using the imagery corrected by the algorithm of Bianco et al., (2015) (**c**), and by using the imagery corrected with Adobe Photoshop (**d**) (the dimensions of the imaged area are 2.30 m × 2.30 m).

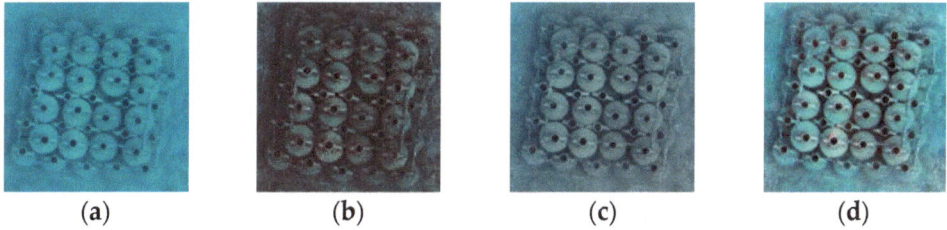

Figure 13. The orthoimage of the deep dataset created by using the original images (**a**), by correcting the (**a**) using the CLAHE based algorithm (**b**), by correcting the (**a**) using the algorithm of Bianco et al., (2015) (**c**) and by correcting the (**a**) using the Adobe Photoshop (**d**)) (the dimensions of the imaged area are 2.30 m × 2.30 m).

3. Evaluation and Results

3.1. Evaluation of the Resulting Orthoimages

When it comes to underwater archaeological projects and excavations, in most of the cases, the main aim of the photogrammetric applications is the generation of accurate and colour consistent 3D model and orthoimages. In this section, orthoimages using the correcting imagery were generated for each dataset. This was done in order to investigate the necessity of implementing colour correction and enhancement before the SfM-MVS procedure or directly to the final orthoimage, when this is the final deliverable. Two methodologies of orthoimage colour enhancement were used; by applying colour correction on the individual photos of the original data set or directly to the final orthoimage.

By visually inpecting and comparing both the visual appearance and the histograms of the results that are illustrated in Figures 9 and 10 and Figures 12 and 13, one may easily deduce that the implementation of the specific image enhancement techniques does not significantly affect the produced orthoimages. There are only some minor differences between the orthoimages resulting from the corrected imagery and the respective corrected orthoimage.

However, it must be noted that the orthoimages resulting from the corrected imagery of the CLAHE implementation and the Bianco et al., (2015) algorithm are sharper and higher contrasted than the respective directly corrected orthoimages. The opposite happens for the Adobe Photoshop orthoimages for the shallow waters dataset only. These results are confirmed both from tests that were performed by using the shallow waters dataset and the deep waters dataset. This is possibly explained by the fact that when it comes to direct orthoimage colour correction, the necessary computed histograms for colour correction algorithm takes into account a wider variety of colour values at the extent instead of the limited area of a single image.

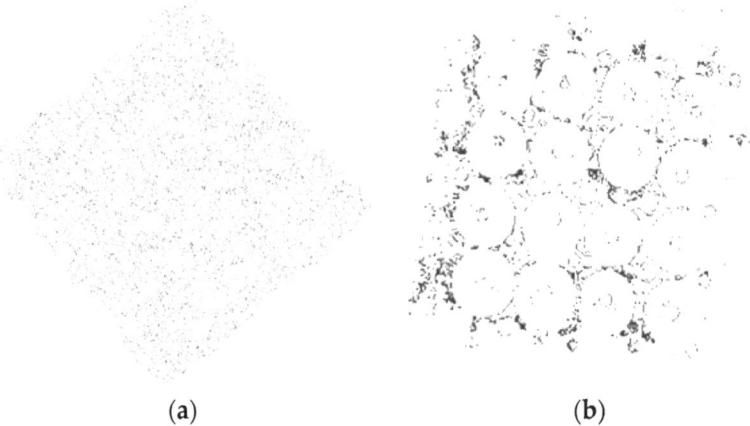

(a) (b)

Figure 14. (a) A result of the subtraction of the shallow waters dataset orthoimages and (b) a result of the subtraction of the deep waters dataset orthoimages. Results from all the subtraction are almost the same.

Additionally, orthoimages produced were compared through image subtraction in QGIS software. Orthoimages were inserted to QGIS software and subtracted in respect to the colour correction methods. Results also suggest that there is not any notable geometric difference between the orthoimages (Figure 14a,b). In more detail, regarding the shallow waters dataset, differences suggest minor geometric defferences that are mainly observed in dark and underexposed areas. The same happens for the deep water dataset. However, in this dataset, changes are more intense due to more dark and underexposed areas existing inbetween the amphorae.

3.2. Test and Evaluation of 3D Point Clouds

The generated dense point clouds were compared within Cloud Compare. Differences between 3D point clouds were illustrated with a colour scale bar (Figure 15a,b and Table 2). In order to ignore outliers, the maximum distance between the respective compared points of each 3D point cloud was set at 0.02m, and the colour scale was divided into eight levels of 0.0025m.

Regarding the tests performed for the shallow waters dataset, in Figures 15a,b and 14c, along with the top view of the result of the comparison, a side view is presented (Figure 15a,b and c bottom) to demonstrate the produced noise in the 3D point cloud. As it is observed in Figures 15a,b and 15c (bottom), these points are the main reason of the existence of points deviating more that 0.003m from the original dense point cloud.

Figures 15d,e,f ,j,k and l present the approximate number of points of the 3D dense point clouds that were created using the corrected images that deviate from the 3D dense point cloud resulting from the original images in relation to the value of that deviation. In Figure 15d,e,f it is observed that over 91% of the points resulted from the corrected images of all the algorithms deviates less than 0.0025 m from the point cloud resulting from the original images. However, all of the points deviated less than 0.005 m. In Figure 15j,k,l it is observed that for the 3D point clouds of the CLAHE corrected imagery, over 34% of the points resulted from the corrected images of all the algorithms deviates less than 0.0025 m from the point cloud resulting from the original images. The respective percentage for Bianco et al., (2015) is 26% and for Adobe Photoshop corrected images 23%. The rest 49% of the 3D point clouds of the CLAHE images represents deviations from 0.0025 m to 0.01 m, while the same percentages for Bianco et al., (2015) is 72.5%, and for Adobe Photoshop corrected images, 69.8%. It is noteworthy that althought the CLAHE has better figurs in under 0.0025 m, the sum of below 0.01m is worse that the other two. The rest 17%, 1.5%, and 7.2% of points, respectively, represent errors from 0.01 to 0.02 m.

Shallow
waters

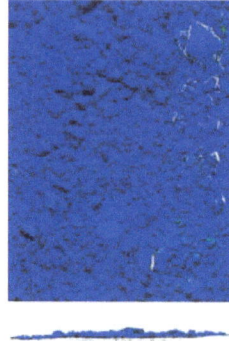

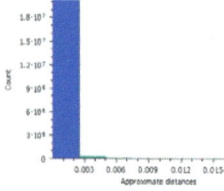

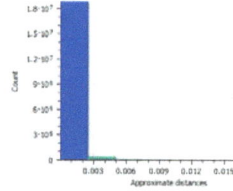

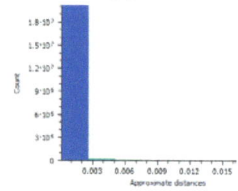

(a)

(b)

(c)

(d)

(e)

(f)

Deep
waters

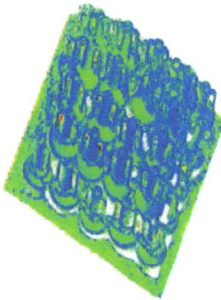

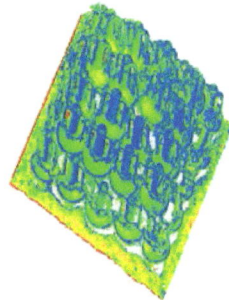

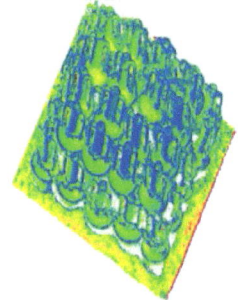

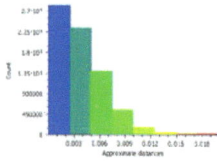

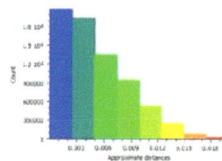

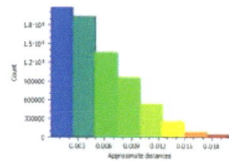

(g)

(h)

(i)

(j)

(k)

(l)

Figure 15. Three-dimensional point cloud (3Dpc) comparisons in Cloud Compare: The results of the comparison of the 3Dpc of the shallow waters test site (**a–c**) and histograms of deviations from the 3Dpc of the original images for shallow waters: CLAHE-shallow (**d**), Bianco et al. (2015) (**e**), Photoshop (**f**) and for deep waters the results of the comparisons (**g–i**) and histograms of deviations from the 3Dpc of the original images CLAHE-deep (**j**), Bianco et al. (2015) (**k**), Photoshop (**l**).

Table 2. Differences between 3Dpcs in Cloud Compare. 3Dpc from the original photos is considered as reference.

Compared 3D Point Clouds		Point Cloud Differences (m)			Percentage of Points That Deviate		
		Max	Mean	Sigma	0–0.0025m	0.0025–0.01m	0.01–0.02m
Shallow waters	Original- CLAHE based Algorithm	0.02	1.01-05	0.00005	91%		9%
	Original- Lab based algorithm	0.02	0.65-05	0.00040			
	Original.-Photoshop	0.02	0.40-05	0.00030			
Deep waters	Original- CLAHE based Algorithm	0.02	0.001	0.00400	34%	49%	17%
	Original- Lab based algorithm	0.02	0.002	0.00500	26%	72.5%	1.5%
	Original.-Photoshop	0.02	0.002	0.00600	23%	69.8%	7.2%

As it is observed, the results of the comparison of the 3D point clouds of the deep waters dataset have larger percentages of errors in relation with the 3D point clouds of the shallow water dataset. This is because of the different complexity of the objects in the dataset, as well as the impact of the increased depth to the original colours. Additionally, it is noted that for the shallow depth dataset, the 3D point cloud comparisons resulted in almost the same percentages, and this is due to the small depth and mainly due to the non-complex captured environment. These are also confirmed by Table 2, showing that the mean and the sigma of the differences between 3D point clouds in Cloud Compare are smaller for the shallow test site. Regarding the differences in the GSD of the two datasets, they are considered of minor importnace since both of the GSDs are sub-millimetered (Table 1). Moreover, even if the 3D point cloud of the CLAHE based corrected images presents the largest percentage of points that differ less than 0.0025 m from the point cloud resulting from the original images, the 3D point cloud of the images resulting from the correction of Bianco et al., (2015) has most of its points between 0.0025 and 0.01 m while it presents the smallest percentage of deviations larger than 0.01 m for deep waters dataset. However, the 3D point clouds resulting from all the correction algorithms do not present any important deviation from the 3D point cloud of the original imagery. Most of these insignificant differences are resulting mainly due to the noise that is introduced by the alteration of the image radiometry, a fact that may mislead the image matching algorithm. Following the figures of Table 2, one may assume that the colour correction technique with more effect to the 3Dpc with respect to the original, is the CLAHE. If total percentage of difference below 0.01m is taken into consideration, the remaining two techniques, have less effect to the 3Dpc. Nevertheless, this effect may be either a diminishment or an improvement.

As already mentioned, the underwater environment affects the underwater image radiometry, even in shallow waters. Results show that the number of automatically generated 3Dpc points is increased for each image enhancment method, without providing useful extra information, where, in some cases, adding noise to the object. The specific image enhancement techniques on shallow and deep depth underwater imagery do not seem to affect the 3D reconstruction. Colour correction before automated photogrammetric procedure does not seem to have any important impact on the final orthoimage, and, as such, the stage of using the specific image enhancement processes is subjective. However, this result it seems is strongly dependeny on the colour correction method used. Results showed that this is valid for the 5/6 of the tests and comparisons perfomed and presented in this article, while for the rest 1/6; the comparison between Figures 9d and 10d (Adobe photoshop results for shallow water dataset) does not apply since the direct colour correction of the orthoimage using Adobe Photoshop has results.

4. Concluding Remarks

When it comes to underwater orthoimage production applications, the underwater image pre-processing seems unnecessary, since simply the colour enhancement and correction of the produced orthoimages is sufficient and time efficient. In more detatil, the tested enhancement and correction methods do not seem to significantly improve 3D reconstruction effectiveness in Agisoft's Photoscan software and for the selected types of test sites and depths using the specific cameras. Point cloud comparisons in Cloud Compare software showed minor differences in the relevant accuracy, when the noise is ignored. In addition, orthoimages subtraction did not suggest any important differences. The image enhancement methods that are mentioned above improve the image visual quality and make them more appealing for the human eye. However, they do not improve feature detection and matching on the SfM process and overall 3D reconstruction in the specific SfM - MVS software in non turbid water, while in high turbidity water seems to be effective enough (Mahiddine et al., 2012). This result can be explained by the fact that the colour correction and enhancement methods used exploit the already stored information of the image. This also points out the robustness of the keypoint detection algorithm that is used. Additonally, these results might suggest that Agisoft Photoscan software, while using three-channel RGB images, might create and use a different channel (i.e. luminance channel) for applying SfM-MVS processing, and, as such, the colour correction and enhancment methods do not affect directly the SfM-MVS results.

Future work should include more tests by using more test sites of different depths and complexity, and by exploiting more colour correction and enhancment algorithms and SfM-MVS software, as well as all-purpose photogrammetric tools, like GRAPHOS (González-Aguilera et al, 2016), which integrates different algorithms and methodologies for automated image orientation and dense 3D reconstruction. Finally, recent research results indicate that the colour correction of the textures of a 3D model is a fast and reliable way to improve the visual quality of an underwater 3D model without enchanching the source image dataset.

Acknowledgments: The authors would like to acknowledge the Department of Fisheries and Marine Research of Cyprus, for the creation and permission to use the artificial amphorae reef, in this paper. Additionally, the authors would like to acknowledge the 3D Research s.r.l. member Antonio Lagudi for providing the implementation of the algorithm Bianco et al. 2015. The contribution of Dimitrios Skarlatos and Panagiotis Agrafiotis was partially supported by iMARECULTURE project (Advanced VR, iMmersive Serious Games and Augmented REality as Tools to Raise Awareness and Access to European Underwater CULTURal heritagE, Digital Heritage) that has received funding from the European Union's Horizon 2020 research and innovation programme under grant agreement No 727153.

References

1.   Adobe Photoshop CS5 (Software), 2010. Available online: https://www.adobe.com/products/photoshop.html.
2.   Agisoft LLC. Agisoft PhotoScan User Manual: Professional Edition; 2017.
3.   Agrafiotis, P.; Drakonakis, G.I.; Georgopoulos, A.; Skarlatos, D. The effect of underwater imagery radiometry on 3d reconstruction and orthoimagery. *Int. Arch. Photogramm. Remote Sens. Spat. Inf. Sci.* **2017**, *XLII-2/W3*, 25–31, doi:10.5194/isprs-archives-XLII-2-W3-25-2017.
4.   Bianco, G.; Muzzupappa, M.; Bruno, F.; Garcia, R.; Neumann, L. A new color correction method for underwater imaging. *Int. Arch. Photogramm. Remote Sens. Spat. Inf. Sci.* **2015**, *XL-5/W5*, 25–32, doi:10.5194/isprsarchives-XL-5-W5-25-2015.
5.   Bruno, F.; Lagudi, A.; Ritacco, G.; Agrafiotis, P.; Skarlatos, D.; Čejka, J.; Kouřil, P.; Liarokapis, F.; Philpin-Briscoe, O.; Poullis, C.; et al. Development and integration of digital technologies addressed to raise awareness and access to European underwater cultural heritage. An overview of the H2020 i-MARECULTURE project. In Proceedings of OCEANS 2017, Aberdeen, UK, 19–22 June 2017; pp. 1–10, doi:10.1109/OCEANSE.2017.8084984.
6.   CloudCompare (Version 2.8.1) [GPL Software], 2016. Available online: http://www.cloudcompare.org/.
7.   Demesticha, S. The 4th-Century-BC Mazotos Shipwreck, Cyprus: A preliminary report. *Int. J. Naut. Archaeol.* **2011**, *40*, 9–59.
8.   Drap, P. Underwater photogrammetry for archaeology. In *Special Applications of Photogrammetry*; InTech Open: Rijeka, Croatia, 2012.
9.   Ghani, A.S.A.; Isa, N.A.M. Underwater image quality enhancement through composition of dual-intensity images and rayleigh-stretching. *SpringerPlus* **2014**, *3*, 757.
10. González-Aguilera, D.; López-Fernández, L.; Rodriguez-Gonzalvez, P.; Guerrero, D.; Hernandez-Lopez, D.; Remondino, F.; Menna, F.; Nocerino, E.; Toschi, I.; Ballabeni, A.; et al. Development of an all-purpose free photogrammetric tool. *Int. Arch. Photogramm. Remote Sens. Spat. Inf. Sci.* **2016**, *XLI-B6*, 31–38, doi:10.5194/isprs-archives-XLI-B6-31-2016.
11. Henderson, J.; Pizarro, O.; Johnson-Roberson, M.; Mahon, I. Mapping submerged archaeological sites using stereo-vision photogrammetry. *Int. J. Naut. Archaeol.* **2013**, *42*, 243–256.

12. Hitam, M.S.; Awalludin, E.A.; Yussof, W.N.J.H.W.; Bachok, Z. Mixture contrast limited adaptive histogram equalization for underwater image enhancement. In Proceedings of the 2013 International Conference on Computer Applications Technology (ICCAT), Sousse, Tunisia, 20–22 January 2013; pp. 1–5.

13. Hou, W.; Weidemann, A.D.; Gray, D.J.; Fournier, G.R. Imagery-derived modulation transfer function and its applications for underwater imaging. *Proc. SPIE* **2007**, *6696*, doi: 10.1117/12.734953.

14. Iqbal, K.; Abdul Salam, R.; Osman, M.; Talib, A.Z. Underwater image enhancement using an integrated colour model. *IAENG Int. J. Comput. Sci.* **2007**, *32*, 239–244.

15. Johnson-Roberson, M.; Bryson, M.; Friedman, A.; Pizarro, O.; Troni, G.; Ozog, P.; Henderson, J.C. High-resolution underwater robotic vision-based mapping and three-dimensional reconstruction for archaeology. *J. Field Robot.* **2017**, *34*, 625–643.

16. Kumar Rai, R.; Gour, P.; Singh, B. Underwater image segmentation using clahe enhancement and thresholding. *Int. J. Emerg. Technol. Adv. Eng.* **2012**, *2*, 118–123.

17. Mahiddine, A.; Seinturier, J.; Boi, D.P.J.-M.; Drap, P.; Merad, D.; Long, L. Underwater image preprocessing for automated photogrammetry in high turbidity water: An application on the arles-rhone xiii roman wreck in the Rhodano river, France. In Proceedings of the 2012 18th International Conference on Virtual Systems and Multimedia (VSMM), Milan, Italy, 2–5 September 2012; pp. 189–194.

18. Singh, B.; Mishra, R.S.; Gour, P. Analysis of contrast enhancement techniques for underwater image. *Int. J. Comput. Technol. Electron. Eng.* **2011**, *1*, 190–194.

19. Treibitz, T.; Schechner, Y.Y. Active polarization descattering. *IEEE Trans. Pattern Anal. Mach. Intell.* **2009**, *31*, 385–399.

20. Von Lukas, U.F. Underwater visual computing: The grand challenge just around the corner. *IEEE Comput. Graph. Appl.* **2016**, *36*, 10–15.

21. Yussof, W.; Hitam, M.S.; Awalludin, E.A.; Bachok, Z. Performing contrast limited adaptive histogram equalization technique on combined color models for underwater image enhancement. *Int. J. Interact. Digit. Media* **2013**, *1*, 1–6.

22. Zuiderveld, K. Contrast Limited Adaptive Histogram Equalization. In *Graphics GEMS IV*; Academic Press Professional, Inc.: San Diego, CA, USA, 1994; pp. 474–485.

Agrafiotis, P.; Drakonakis, G.I.; Skarlatos, D. Underwater Image Enhancement before Three-Dimensional (3D) Reconstruction and Orthoimage Production Steps: Is It Worth? In *Latest Developments in Reality-Based 3D Surveying and Modelling*; Remondino, F., Georgopoulos, A., González-Aguilera, D., Agrafiotis, P., Eds.; MDPI: Basel, Switzerland, 2018; pp. 239–256.

Underwater Photogrammetry and Visual Odometry

Mohamad Motasem Nawaf, Jean-Philip Royer, Jerôme Pasquet, Djamal Merad and Pierre Drap

Aix-Marseille Université, CNRS, ENSAM, Université De Toulon, LSIS UMR 7296, Domaine Universitaire de Saint-Jérôme, Bâtiment Polytech, Avenue Escadrille Normandie-Niemen, 13397 Marseille, France

Abstract: We propose an improved visual odometry approach that is adapted to low computational resources systems in an underwater environment. The aim is to guide underwater photogrammetry surveys in real time. The visual odometry relies on stereo image stream that is captured by an embedded system. An improved pose estimation procedure underlying fast stereo matching approach is followed by a semi-global bundle adjustment. Computed trajectory is maintained stochastically and a divergence measure is used for more realistic optimization zone selection. In particular, we propose a new approach to find an approximation of the uncertainty for each estimated relative pose based on machine learning manifesting on simulated data. This allows the user to find potential overlaps in the estimated trajectory for better drifts handling and loop closure. The evaluation of the proposed method demonstrates the gain in terms of computation time w.r.t. other approaches. The built system opens promising areas for further development and integration of embedded vision techniques.

Keywords: photogrammetry; visual odometry; underwater imaging; embedded systems; probabilistic modelling

---

1. Introduction

Mobile systems nowadays undergo a growing need for self-localization to accurately determine its absolute/relative position over time. Despite the existence of very efficient technologies that can be used on-ground (indoor/outdoor) and in-air, such as Global Positioning System (GPS), optical, radio beacons, etc. However, in the underwater context most of these signals are jammed so that the corresponding techniques cannot be used. On the other side, solutions based on active acoustics, such as imaging sonars and Doppler Velocity Logs (DVL) devices remain expensive and require high technical skills for their deployment and operation. Moreover, their size specifications prevent their integration within small mobile systems or even being hand held. The research for an alternative is

257

ongoing, notably, the recent advances in embedded systems outcome relatively small, powerful, and cheap devices. This opens interesting perspectives to adapt a light visual odometry approach that provides relative path in real-time, this describes our main research direction. The developed solution is integrated within underwater archaeological site survey where it plays an important role to facilitate image acquisition. An example of targeted sites is shown in Figure 1.

In underwater survey tasks, mobile underwater vehicles (or divers) navigate over the target site to capture images. The obtained images are treated in a later phase to obtain various information and to also form a realistic three-dimensional (3D) model using photogrammetry techniques Drap (2012). In such a situation, the main problem is to totally cover the underwater site before ending the mission. Otherwise, we may obtain incomplete 3D models and the mission cost will raise significantly as further exploitation is needed. However, the absence of an overall view of the site especially under bad lighting conditions makes the scanning operation blind. In practice, this yields to over-scanning the site, which is a waste of time and cost. Moreover, the quality of the taken images may go below an acceptable limit. This mainly happens in terms of lightness and sharpness, which is often hard to quantify visually on the fly. In this work, we propose solutions for the aforementioned problems. Most importantly, we propose to guide the survey based on a visual odometry approach that runs on a distributed embedded system in real-time. The output ego-motion helps to guide the site scanning task by showing approximate scanned areas. Moreover, an overall subjective lightness and sharpness indicators are computed for each image to help the operator to control the image quality. Overall, we provide a complete hardware and software solution for the problem.

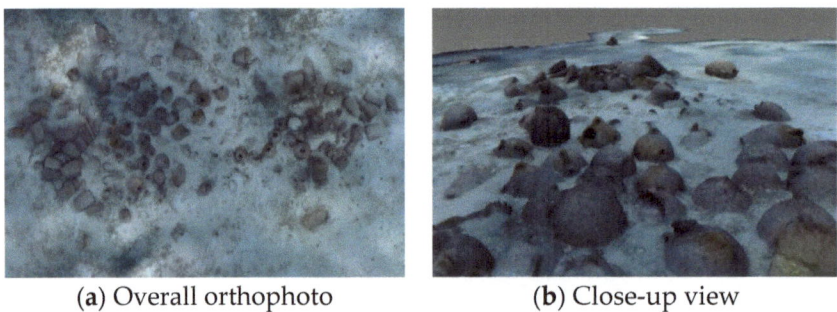

(a) Overall orthophoto　　　　　　　　(b) Close-up view

Figure 1. Example of a three-dimensional (3D) model of an underwater site; a Phoenician shipwreck located near Malta.

In common approaches of visual odometry, a significant part of the overall processing time is spent on feature points detection, description, and matching. In the tested baseline algorithm, the aforementioned operations represent ~65% of processing time in case of local/relative bundle adjustment (BA) approach, which occupies in return the majority of the time left. In our proposed method, we rely on low level Harris based detection and template matching procedure, which significantly speeds up the feature matching speed. Further, whereas in traditional stereo matching the search for correspondence is done along the epipolar line within certain fixed range, in our method we proceed first by computing *a priori* rough depth belief based on image lightness and following the law of light divergence over distance. This is only valid for a configuration where the light source is fixed to the system, which is the case here. Hence, our first contribution is that we benefit from the rough depth estimation to limit points correspondence search zone to reduce processing time.

From another side, traditional visual odometry methods based on local BA suffers from rotation and translation drifts that grows with time Mouragnon et al. (2009). In contrary, the solutions based on using features from the entire image set, such as global BA Triggs et al. (2000), require more computational resources that are very limited in our case. Similarly, the simultaneous localization and mapping (SLAM) approaches, Thrun et al. (2005), which are known to perform good loop closure, are computationally intensive, especially when complex particle filters are used Montemerlo and Thrun (2007), and they can only operate in moderate size environments if real-time processing is needed. In our method, we adopt a semi-global approach Nawaf et al. (2016), which proceed in the same way as local method in optimizing a subset of image frames. However, it differs in the way of selecting the frames subset, as local methods use Euclidean distance and deterministic pose representation to select frames, ours represents the poses in a probabilistic manner, and uses a divergence measure to select such sub set.

The rest of the paper is organized as follows: We survey related works in Section 2. In Section 3 we describe the designed hardware platform that we used to implement our solution. Our proposed visual odometry method is explained in Section 4. The analytical results are verified through simulation experiments presented in Section 5. Finally, we present a summary and conclusions. We note that parts of this work have been presented in Nawaf et al. (2016) and Nawaf et al. (2017).

2. Related Works

2.1. Ego-Motion Estimation

Estimating the ego-motion of a mobile system is an old problem in computer vision. Two main categories of methods are developed in parallel, namely; simultaneous localization and mapping (SLAM) Davison (2003), and visual odometry Nistér et al. (2004). In the following, we highlight the main characteristics for both of the approaches.

SLAM family of methods uses probabilistic model to handle vehicle pose, although this kind of methods is developed to handle motion sensors and map landmarks, they work efficiently with visual information solely. In this case, a map of the environment is built, and at the same time it is used to deduce the relative pose, which is represented using probabilistic models. Several solutions to SLAM involve finding an appropriate representation for the observation model and motion model, while preserving efficient and consistent computation time. Most methods use additive Gaussian noise to handle the uncertainty which imposes using extended Kalman Filter (EKF) to solve the SLAM problem Davison (2003). In case of using visual features, computation time and used resources grows significantly for large environments. A remarkable improvement of SLAM is the FastSLAM approach Montemerlo and Thrun (2007), which improves largely the scalability, it uses recursive Monte Carlo sampling to directly represent the non-linear process model, although the state-space dimensions are reduce using Rao-Blackwellisation approach Blanco et al. (2008), the method remains not scalable to long autonomy. In the context of long trajectories, several solutions are proposed to handle relative map representations, such as Eade and Drummond (2008), Davison et al. (2007), Piniés and Tardós (2007). In particular, by breaking the estimation into smaller mapping regions, called sub-maps, then computing individual solutions for each sub-map. The issues with this kind of approaches arise in sub-mapping creation, overlapping, fusion of sub-maps, and map size selection, especially in our context where the S-shape scanning causes very frequent sub-maps switches, which is time consuming.

In all of the reviewed SLAM methods, the measurement noise is modeled by diagonal covariance matrix with equal values that are set empirically for the case of using pure visual information. This modeling leads to produce spherical measurement uncertainty (though estimated pose has an associated full degrees of freedom (DOF) uncertainty) in 3D when using only visual features. This does not approve with practical cases where uncertainty is not spherical. Although there exist several works in literature that studied the uncertainty of 3D reconstructed points based on their distance from the camera and the baseline distance between frames, such as in Eade and Drummond (2006) and Montiel et al. (2006), the effect of the relative motion parameters on the uncertainty of the pose estimation have not been taken into account. For a complete review for SLAM methods, we refer the reader to Bailey and Durrant-Whyte (2006).

From another side, visual odometry methods uses structure from motion methodology to estimate the relative motion Nistér et al. (2004). Based on multiple view geometry fundamentals, Hartley and Zisserman (2004), approximate relative pose can be estimated, this is followed by a BA procedure to minimize re-projection errors, which yields in improving the estimated structure. Fast and efficient BA approaches are proposed simultaneously to handle larger number of images Lourakis and Argyros (2009). However, in case of long time navigation, the number of images increases dramatically and prevents applying global BA if real time performance is needed. Hence, several local BA approaches have been proposed to handle this problem. In local BA, a sliding window copes with motion and select a fixed number of frames to be considered for BA Mouragnon et al. (2009). This approach does not suit S-Type motion since the last n frames to the current frame are not necessarily the closest. Another local approach is the relative BA proposed in Sibley et al. (2009). Here, the map is represented as Riemannian manifold based graph with edges representing the potential connections between frames. The method selects the part of the graph where the BA will be applied by forming two regions, an active region that contains the frames with an average re-projection error changes by more than a threshold, and a static region that contains the frames that have common measurements with frames in active region. When performing BA, the static region frames are fixed, whereas active region frames are optimized. The main problem with this method is that distances between frames are metric, whereas the uncertainty is not considered when computing inter-frames distances.

Recently, a novel relative BA method is proposed by Nawaf et al. (2016). Particularly, an approximation of the uncertainty for each estimated relative pose is estimated using a machine learning approach manifesting on simulated data. Neighboring observations that are used for the semi-global optimization are established based on a probabilistic distance in the estimated trajectory map. This helps to find the frames with potential overlaps with the current frame, while being robust to estimation drifts. We found this method most adapted to our context.

2.2. Feature Points Matching

Common ego-motion estimation methods rely on feature points that are matching between several poses Nistér et al. (2004). The choice of the used approach for matching feature points depends on the context. For instance, features matching between freely taken images (six degrees of freedom), must be invariant to scale and rotation changes. Scale invariant feature descriptors (SIFT) Lowe (2004) and the Speeded Up Robust Features (SURF) Bay et al. (2006) are well used in this context Nawaf and Tremeau (2014). In this case, the search for a point's correspondence is done w.r.t. all of the points in the destination image.

In certain situations, some constraints can be imposed to facilitate the matching procedure. In particular, limiting the correspondence search zone. For instance, in case of pure forward motion, the focus of expansion (FOE), being a single point in the image, the search for the correspondence for a given point is limited to the epipolar line Yamaguchi et al. (2013). Similarly, in case of sparse stereo matching, the correspondence point lies on the same horizontal line in the case of rectified stereo or on the epipolar line otherwise. This speeds up the matching procedure first by having less comparisons to perform, and second low-level features can be used Geiger et al. (2011). According to our knowledge there is no method that proposes an adaptive search range following a rough depth estimation from lightness in underwater imaging.

3. Hardware Platform

As mentioned earlier, we use an embedded system platform for our implementation. Being increasingly available and cheap, we choose the popular Raspberry Pi © (RPi) [1] as the main processing unit of our platform. This allows to run smoothly most of image processing and computer vision techniques. A description of the built system is shown in Figure 2, which is composed of two RPi's computers, where each is connected to one camera module to form a stereo pair. The cameras are synchronized using a hardware trigger. Both computers are connected to one more powerful computer that can be either within the same enclosure or on-board in our case. Using this configuration, the embedded computers are responsible for image acquisition. The captured stereo images are first partially treated on the fly to provide image quality information, as will be details in Section 4.1. Images are then transferred to the main computer, which handles the ego-motion computation that the system undergoes. For visualization purposes, we use two monitors that are connected to the embedded computers to show live navigation and image quality information (See Figure 2).

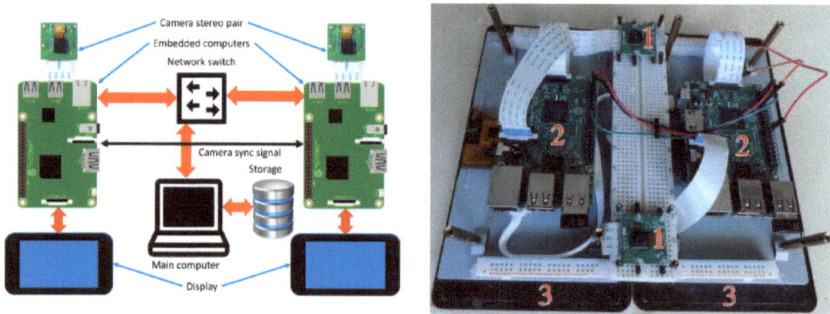

Figure 2. The hardware platform used for image acquisition and real-time navigation; it is composed mainly of (1) stereo camera pair, (2) Raspberry Pi © computers, and (3) monitors.

---

[1] A credit-card size ARM architecture based computer with 1.2 GHz 64-bit quad-core CPU and 1GB of memory, running Rasbian ©, a Linux based operating system.

4. Visual Odometry

Starting by computing and displaying image quality measures, the images are transferred over the network to a third computer as shown in Figure 2. This computer is responsible for hosting the visual odometry process, which will be explained in this section. We start first by introducing the used feature matching approach, and then we present the ego-motion estimation, finally, we explain the semi-global BA approach.

4.1. Image Quality Estimation

Real-time image quality estimation provides two benefits, first, it can alert the visual odometry process of having bad image quality, two reactions can be taken in this case, either pausing the process until taken image quality is recovered, or predicting position estimation based on previous poses and speed. We go for the first case while leaving the second for further development in future. Second, image quality indicators provide direct information to the operator to avoid going too fast in case of blur, or changing the distance to the captured scene when going under or over-exposed.

The first indicator is the image sharpness, we rely on image gradient measure that detects high frequencies that are often associated with sharp images, hence, we use a Sobel kernel based filtering, which computes the gradient with smoothing effect. This removes the effect of dust that is commonly present in underwater imaging. We consider the sharpness measure to be the mean value of the computed gradient magnitude image. The threshold can be easily learned from images by fixing a minimum number of matched feature points that are needed to correctly estimate the ego-motion. Similarly, an image lightness indicator is estimate as the average of L channel in CIE-LAB color space.

4.2. Sparse Stereo Matching

Matching feature points between stereo images is essential to estimate the ego-motion. As the two cameras alignment is not perfect, we start by calibrating the camera pair. Hence, for a given point on the right image, we can compute the epipolar line containing the corresponding point in the left image. However, based on the known fixed geometry, the corresponding point position is constrained by a positive disparity. Moreover, given that at deep water, the only light source is the one used in our system, the furthest distance that feature points that can be detected is limited, see Figure 3 for illustration.

This means that there is a minimum disparity value that is greater than zero. Furthermore, when going too close to the scene, parts of the image will become overexposed, similar to the previous case, this imposes a limited maximum disparity. Figure 4 illustrates the aforementioned constraints by dividing the epipolar line into four zones, in which only one is an acceptable disparity range. This range can be straightforward identified by learning from a set of captured images (oriented at 30 degrees for better coverage).

Figure 3. An example of underwater image showing minimum disparity (red dots, ~140 pixels) and maximum disparity (blue dot, ~430 pixels).

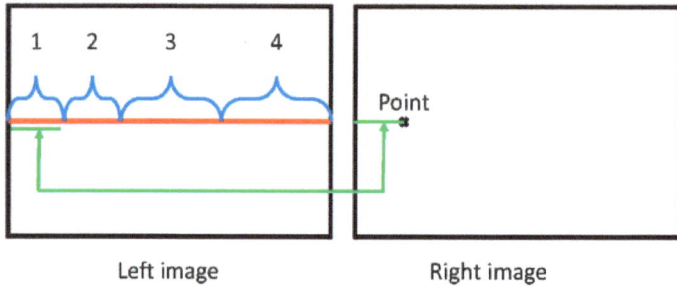

Figure 4. Illustration of stereo matching search ranges. (1) Impossible (2) Impossible in deep underwater imaging due to light's fading at far distances (3) Possible disparity (4) The point is very close so it becomes overexposed and undetectable.

In our approach, we propose to constraint the so-called acceptable disparity range further, which corresponds to the third range in Figure 4. Given the used lighting system, we can assume a light diffuse reflection model where the light reflects equally in all directions. Based on inverse-square law that relates light intensity over distance, image pixels' intensities are roughly proportional to their squared disparities. Based on such an assumption, we could use pixels' intensity to constraint the disparity and hence limiting the range of searching for a correspondence. To do so, we are based on a dataset of stereo images. For each pair, we perform feature points matches. Each point match (x_i, y_i) and (x_i', y_i'), x being the coordinate in the horizontal axis, we compute the squared disparity $d_i^2 = (x_i - x_i')^2$. Next, we associate each d_i^2 to the mean lightness value of a window centered at the given point computed from L channel in CIE-LAB color space.

We assign a large window size (≈ 12) to compensate for using Harris operator that promotes local minimum intensity pixels as salient feature points. The computed $(\bar{l}_{x_i,y_i}, d_i^2)$ pair shows the linear relationship between the squared disparity and the average lightness. A subset of such pairs is plotted in Figure 5.

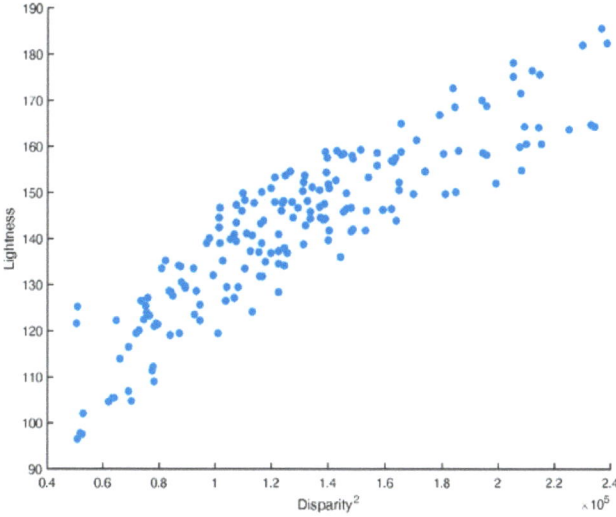

Figure 5. A subset of matched points squared disparity plotted against average pixel lightness.

In addition to finding the linear relationship between both variables, it is also necessary to capture the covariance that represents how rough is our approximation. More specifically, given the diagram shown in Figure 6, we aim at defining a tolerance t that is associated to each disparity as a function of lightness l. In our method, we rely on the Principal Component Analysis (PCA) technique to obtain this information. In details, for a given lightness l_i, we first compute the corresponding squared disparity d_i^2 using a linear regression approach as follows:

$$d_i^2 = -\alpha l_i - \beta \tag{1}$$

$$\alpha = \frac{Cov(L, D^2)}{Var(L)} \tag{2}$$

$$\beta = \bar{l} - \alpha \bar{d}^2 \tag{3}$$

where D and L are the disparity and lightness training set, d and l are their respective means.

Second, let $V_2 = (v_{2,x}, v_{2,y})$ be the computed eigenvector that correspondences to the smallest eigenvalue λ_2. Based on the illustration shown in Figure 6, the tolerance t associated to d_i^2 can be written as:

$$t = \sqrt{\lambda_2^2 \left(\frac{v_{2,x}^2}{v_{2,y}^2} + 1 \right)} \tag{4}$$

By considering a normal error distribution of the estimated rough depth, and based on the fact that t is equal to one variance of D^2, we define the effective disparity range as:

$$d_i \pm \gamma \sqrt[4]{t} \tag{5}$$

where γ represents the number of standard deviations. It is trivial that γ is a trade-off between computation time and the probability of having points correspondences within the chosen tolerance range. We set $\gamma = 2$, which means that there is 95% probability to cover the data.

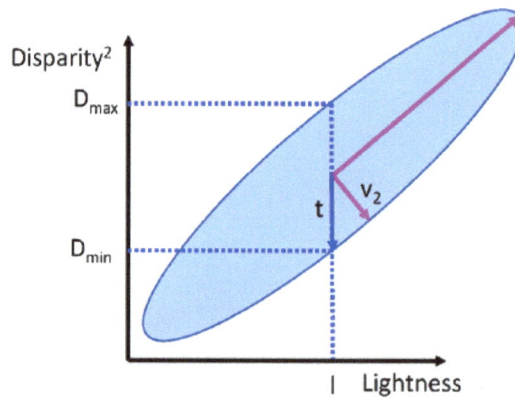

Figure 6. Illustration of disparity tolerance t given a lightness value l.

4.3. Initial Ego-Motion Estimation

Given left and right frames at time t (we call them previous frames), our visual odometry pipeline consists of four stages (an illustration is shown in Figure 7):

- Feature points matching for every new stereo pair $t + 1$. As described in Subsection 4.2.
- 3D reconstruction of the matched feature points using triangulation as described in Hartley and Zisserman (2004). Two displaced point clouds are obtained at this step.

266

- Relative motion computation using adaptation between the point clouds for the frames at t and $t+1$. Semi-Global BA procedure Nawaf et al. (2016) is applied to minimize re-projection errors; to be explained in the following subsections.

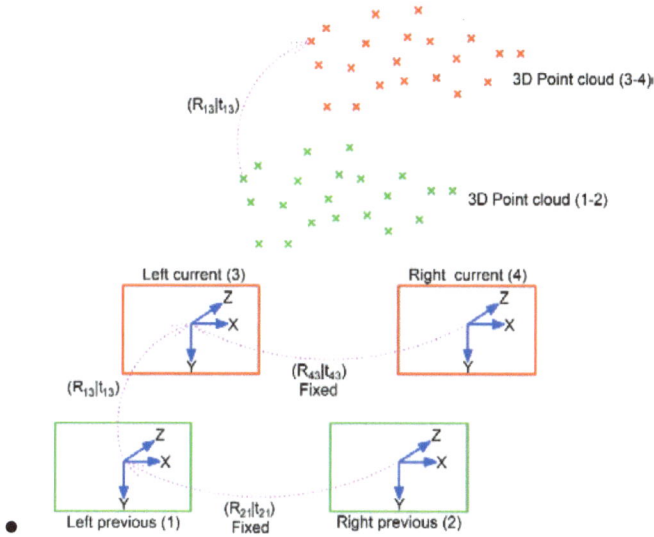

Figure 7. Image quadruplet, current (left and right) and previous (left and right) frames are used to compute two 3D point clouds. The transformation between the two points clouds is equal to the relative motion between the two camera positions.

In details, let (f_1, f_2, f_3, f_4) denote the previous left, previous right, current left and current right frames, respectively. For each new captured image pair, we compute a 3D point cloud using triangulation, as described in Hartley and Zisserman (2004) for the matched feature points that are obtained using the method proposed in the previous subsection.

The rigid transformation $[R|T]$ that is required for expressing the frames at time $t+1$ in the reference frame at time t is the rigid transformation that is required to move the 3D point cloud at time t to the one obtained at time $t+1$. Hence, the problem of calculating the orientation of the cameras at time $t+1$ in relation to time t leads back to the calculation of the transformation used to move from one point cloud to the other. This is possible under our configuration, with small rotation. We note here that there is no scale problem between both point clouds, which is specific to stereo systems. We consider here the left previous to left current frames $f_1 \rightarrow f_3$ positions to represent the system relative motion, and their relative transformation denoted $[R_{13}|T_{13}]$.

267

Below, we present the method to compute the transformation for passing from the point cloud calculated at time $t + 1$, denoted P, to the one calculated at time t, denoted P'. So, we have two sets of n homologous points $P = P_i$ and $P' = P_i'$ where $1 \leq i \leq n$. We have:

$$P_i' = R_{13}P_i + T_{13} \tag{6}$$

The best transformation the minimizes the error r, the sum of the squares of the residuals:

$$r = \sum_{i=1}^{n} \|R_{13}P_i + T_{13}P_i'\|^2 \tag{7}$$

To solve this problem, we use the singular value decomposition (SVD) of the covariance matrix C:

$$C = \sum_{i=1}^{n} (P_i - \bar{P})(P_i' - \bar{P}') \tag{8}$$

where $\bar{P}$ and $\bar{P}'$ are the centers of mass of the 3D points sets P and P', respectively. Given the SVD of C as: $[U, S, V] = SVD(C)$, the final transformation is computed as:

$$R_{13} = VU^T \tag{9}$$

$$T_{13} = -R_{13}\bar{P} + \bar{P}' \tag{10}$$

Once the image pair $t + 1$ is expressed in the reference system of the image pair t, the 3D points can be recalculated using the four observations that we have for each point. A set of verifications are then performed to minimize the pairing errors (verification of the epipolar line, the consistency of the y-parallax, and re-projection residues). Once validated, the approximated camera position at time $t + 1$ are used as input values for the BA, as described earlier.

4.4. Uncertainty in Visual Odometery

Like any visual odometry estimation, the estimated trajectory using the method mentioned in the previous section is exposed to a computational error, which translates to some uncertainty that grows in time. A global BA may handle this error accumulation, however it is time consuming. From another side, a local BA is a trade-off for precision and computational time. The selection of n closest frames is done using standard Euclidean distance. Loop closure may occur when overlapping with already visited areas, which in turn enhances the precision. This approach remains valid as soon as the uncertainty is equal in all directions. However, as uncertainty varies across dimensions, the selection of the closest frames based on Euclidean distance is not suitable. In the following, we are going to prove that it is the case in any visual odometry method. Also, we will provide a more formal definition of the uncertainty.

Most visual odometry and 3D reconstruction methods rely on matched feature points to estimate relative motion between two frames. The error of matched features is resulting from several accumulated errors. These errors are due, non-exclusively, to the following reasons; the discretization of 3D points projection to image pixels, image distortion, the camera internal noise, salient points detection, and matching. By performing image un-distortion, and constraining the points that are matching with the fundamental matrix, the aforementioned errors are considered to follow a Gaussian distribution, so as their accumulation. This is actually implicitly considered in most computer vision fundamentals. Based on this assumption, we can prove that the error distribution of the estimated relative pose is unequal among dimensions. Indeed, it can be fitted to a multivariate Gaussian whose covariance matrix has non-equal Eigen values as we will see later. Formally, given a pair of matched points between two frames $m \leftrightarrow m'$. Based on our assumption, each matched point can be represented by a multivariate Gaussian distribution:

$$\mathcal{N}(m, \Sigma) \leftrightarrow \mathcal{N}(m', \Sigma)$$ (11)

$$\Sigma = \begin{bmatrix} \sigma^2 & 0 \\ 0 & \sigma^2 \end{bmatrix}$$ (12)

The pose estimation procedure relies on the fundamental matrix that satisfies $m'Fm = 0$. Writing $m = [x \; y \; 1]^T$ and $m' = [x' \; y' \; 1]^T$. The fundamental matrix constraint for one matching pair of points can be written as:

$$x'xf_{11} + x'yf_{12} + x'f_{13} + y'xf_{21} + y'yf_{22} + y'f_{23} + xf_{31} + yf_{32} + f_{33} = 0$$ (13)

To show the variance of error distribution of estimated pose, without the loss of generality, we consider one example of configuration; identity camera intrinsic matrix $K = diag(1 \; 1 \; 1)$. Let us now take the case of pure translational motion between the two camera frames, $T = [T_x \; T_y \; T_z]^T$, and $\theta = [\theta_x \; \theta_y \; \theta_z]^T = [0 \; 0 \; 0]$, the fundamental matrix in this case is given as:

$$F = K^{-1^T}EK^{-1} = K^{-1^T}[T]_x R \; K^{-1} = \begin{bmatrix} 0 & -T_Z & T_Y \\ T_Z & 0 & -T_X \\ -T_Y & T_X & 0 \end{bmatrix}$$ (14)

where $[T]_x$ is the skew-symmetric cross-product matrix of T, and R is the rotation matrix, which is the identity in this case. Hence, equation 14 simplifies to:

$$-x'yT_Z + x'T_Y + y'xT_Z - y'T_X - xT_Y + yT_X = 0$$ (15)

By using enough matched points (seven points in this case), we can recover the translation vector T by solving a linear system. However, the Gaussian noise whose covariance matrix is expressed by equation 12 will propagate to the variables T_X and T_Y, whereas for T_Z the error distribution is different due to the product of two variables, where each is a Gaussian distribution. So the covariance is equal to $\Sigma/2$. Moreover, the recovered translation variables are correlated even though the observations are un-correlated. This is due to the usage of least square approach through SVD Strutz (2010). This leads to have the estimated pose follow a Gaussian distribution (proved experimentally in the following) with a full DOF covariance matrix (within the positive semi-definite constraint).

4.5. Pose Uncertainty Modeling

Pose uncertainty is difficult to estimate straightforward. This is due to the complexity of the pose estimation procedure and the number of variables. In particular, noise propagation through two consecutive SVDs (used for Fundamental matrix computation and Essential matrix decomposition). Instead, inspired by the unscented Kalman filter approach as proposed in Wan and Van Der Merwe (2000), we proceed similarly by simulating noisy input and trying to characterize the output error distribution in this case. This process is illustrated in Figure 8. In our work, we propose to learn the error distribution based on finite pose samples. This is done using a Neural Network approach which fits well to our problem as it produces soft output.

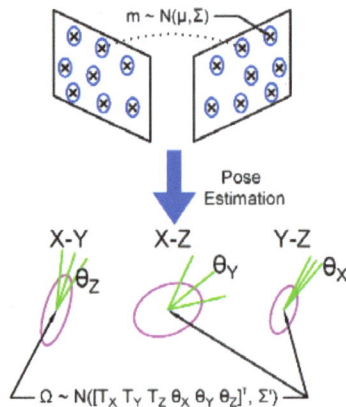

Figure 8. Illustration of error propagation through the pose estimation procedure. Estimated pose uncertainty is shown for each of the six degrees of freedom DOF. Full covariance matrix can result from diagonal error distribution of matched two-dimensional (2D) feature points.

There are two factors that plays role in the estimated pose uncertainty. First, the motion $\Omega = [T\ \theta]^{\mathsf{T}}$ between the two frames expressed by a translation T and a rotation θ, which is explained in the previous section. Second, the 3D location of the matched feature points. Although their location is not computed explicitly in our method, their distance from the camera affects the computation precision. In particular, the further the points are from the camera, the less precise is the estimated pose. This is due to the fact that close points yield larger 2D projection disparity which is more accurate to estimate after the discretization. For instance, in pure translation motion, if all of the matched points are within the blind zone of the vision system (yield zero-pixels disparity after discretization), the estimated motion would be equal to zero. In the contrary, it will be more accurate when points are closer. Both mentioned factors are correlated to some point. For instance, given some points in 3D ($n > 7$), the estimated pose precision is a function of their depth, but also to the baseline distance. Hence, considering one factor is sufficient. In our work, we consider the motion as a base to predict the uncertainty.

Formally, given a motion vector $\Omega_j = [T_j\ \theta_j]^\mathsf{T}$, ideally, we want to find the covariance matrix that expresses the associated error distribution. Being a positive semi-definitive (PSD), such $n \times n$ covariance matrix has unique $(n^2 + n)/2$ entries, where $n = 6$ in our case, this yields 21 DOF, in which six are the variances. However, learning this number of parameters freely violates the PSD constraint. Whereas finding the nearest PSD in this case distorts largely the diagonal elements (being much fewer). At the same time, we found experimentally that the covariance between T and θ variables is relatively small when compared to such of inter T and inter θ. Thus, we propose to consider two covariance matrices Σ_T and Σ_θ. So, in total, we have 12 parameters to learn, in which six are the variances.

For the aim of learning Σ_T and Σ_θ, we have created a simulation of the pose estimation procedure. For a fixed well distributed 3D points $\{X_i \in R^3 : i = 1..8\}$, we simulate two cameras with known relative rotation and translation. The points are projected according to both cameras to 2D image points, let us say $\{x_i \in R^2\}$ and $\{x_i' \in R^2\}$. These points are disturbed with random Gaussian noise as given by the equations 11 and 12. Next, the 3D relative pose is estimated based on the disturbed points. Let $\tilde{\Omega}_j = [\tilde{T}_j\ \tilde{\theta}_j]^\mathsf{T}$ be the estimated relative motion. Repeating the same procedure (with the same motion Ω_j) produce a motion cloud around the real one. Now, we compute the covariance matrices [2] Σ_T and Σ_θ of the resulting motion cloud in order to obtain the uncertainty associated to the given motion Ω_j. Further, we repeat this procedure for a wide range of motion values[3]. Now, having the output covariance matrices (two for each motion vector Ω_j), we proceed to build a system that learns the established correspondences (motion $\leftrightarrow$ uncertainty). So, that in case of new motion, we will be able to estimate the uncertainty. This soft output is offered by Neural Networks by nature, which is the reason that we adopt

---

[2]   We increase the number of simulation runs until the output mean is close enough to the input real motion Ω_j, in our case we run the simulation 10000 times for each pose.

[3]   In the performed simulation, we use the range [0-1] with 0.25 step size for each of the 6 dimensions, these values are in radians in case of rotation. This raises up to 15625 test case.

this learning method. In our experiments, we found that a simple Neural with single hidden layer Bishop (1995) was sufficient to fit well the data. The input layer has six nodes that correspond to motion vector. The output layer has 12 nodes, which corresponds to the unique entries in Σ_T and Σ_θ, hence, we form our output vector as:

$$O = [\Sigma_T^{11}\ \Sigma_T^{22}\ \Sigma_T^{33}\ \Sigma_T^{12}\ \Sigma_T^{13}\ \Sigma_T^{23}\ \Sigma_\theta^{11}\ \Sigma_\theta^{22}\ \Sigma_\theta^{33}\ \Sigma_\theta^{12}\ \Sigma_\theta^{13}\ \Sigma_\theta^{23}]^\mathsf{T} \tag{16}$$

where Σ^{ij} is the element of row i and column j of the covariance matrix Σ .

In the learning phase, we use a gradient-descent based approach Levenberg-Marquardt backpropagation, which is described in Hagan et al. (1996). Further, by using the mean-squared error as a cost function we could achieve around 3% error rate. The obtained parameters are rearranged in symmetric matrices. In practice, the obtained matrix is not necessarily PSD, although this is rare to happen in the case of small variances. We proceed to find the closest PSD as $Q\Lambda_+ Q^{-1}$, where Q is the eigenvector matrix of the estimated covariance, and Λ_+ the diagonal matrix of Eigen values in which negative values are set to zero.

4.6. Semi-Global Bundle Adjustment

After initiating the visual odometry, the relative pose estimation at each frame is maintained within a table that contains all pose related information (18 parameters per pose, in which six for the position, and 12 for two covariance matrices). At any time, it is possible to get the observations in the neighborhood of the current pose being estimated in order find potential overlaps to consider while performing BA. Since we are dealing with statistical representations of the observations, a divergence measure has to be considered. Here, we choose Bhattacharyya distance (Modified metric version can also be used Comaniciu et al. (2003)) for being reliable and relevant to our problem. In our case, the distance between two observations $\{\Omega^1, \Sigma_T^1, \Sigma_\theta^1\}$ and $\{\Omega^2, \Sigma_T^2, \Sigma_\theta^2\}$ is given as:

$$D = \frac{1}{8}(\Omega^1 - \Omega^2)^\mathsf{T}\Sigma^{-1}(\Omega^1 - \Omega^2) + \frac{1}{2}\ln\left(\frac{det\ \Sigma}{\sqrt{det\ \Sigma_1 + det\ \Sigma_2}}\right) \tag{17}$$

where

$$\Sigma = \begin{pmatrix} \Sigma_T & 0 \\ 0 & \Sigma_\theta \end{pmatrix}, \Sigma = \frac{\Sigma_1 + \Sigma_2}{2} \tag{18}$$

Having selected the set of frames F in the neighborhood of the current pose statistically, we perform BA as follows; First, we divide F into two subsets, similar to Sibley et al. (2009), the first subset F_d contains the current and previous frames in time, whereas the other sub-set F_s contains the remaining frames, mostly resulting from overlapping with an already scanned area. Second, BA is performed on both subsets, however, although F_s parameters are included in the optimization, they are masked as static so that they are not optimized in contrary to F_d . This strategy is necessary in order to keep past trajectories consistent.

After determining the error distribution arising with a new pose, it has to be compounded with propagated error from the previous pose. Similar to the SLAM approach, we propose to use a "Kalman filter" like gain, which allows controllable error fusion and propagation. Given an accumulated previous pose estimation defined by $\{\Omega^p , \Sigma_T^p, \Sigma_\theta^p\}$ and a current one $\{\Omega^c , \Sigma_T^c, \Sigma_\theta^c\}$, the updated current pose is calculated as:

$$\Omega^u = \Omega^c \tag{19}$$

$$\Sigma_T^u = \left(I - \Sigma_T^p (\Sigma_T^p + \Sigma_T^c)^{-1}\right) \Sigma_T^p \tag{20}$$

$$\Sigma_\theta^u = \left(I - \Sigma_\theta^p (\Sigma_\theta^p + \Sigma_\theta^c)^{-1}\right) \Sigma_\theta^p \tag{21}$$

5. Evaluation

The proposed method is desired to represent a trade-off between precision and computation time, the maximum precision being the case of global BA, whereas the fastest computation time is pure visual odometry. Moreover, a performance improvement is expected w.r.t local method due for better selection of neighboring observations. Therefore, we analyze the performance of our method from two points of view; computation time and precision.

5.1. Computation Time

We tested and compared the computation speed of our method as compared to using high level feature descriptors, specifically SIFT and SURF. At the same time, we monitor the precision for each test. The evaluation is done using the same set of images.

We run our experiments using the speed optimized BA toolbox as proposed in Lourakis and Argyros (2009). In the obtained results, the computation time when using the reduced matching search range, as proposed in this work is ~72% when compared[4] to the method using the whole search range (range 3 in Figure 4). Concerning SIFT and SURF, the computation time is 342% and 221%, respectively, as compared to the proposed method. The precision of the obtained odometry is reasonable which is within the limit of 3% for the average translational error and 0.02 [deg/m] for the average rotational error.

---

[4]   The time evaluation is shown in percentage because the evaluation is carried out on three platforms with different computational power, in which one is an embedded unit. The minimum computation time being 220 ms.

5.2. Simulation Using Orthophoto

Our work falls within a preliminary preparation for a real mission. All of the experiments are tested within a simulated environment which uses images from previously reconstructed orthophoto in Drap et al. (2015) which is illustrated in Figure 9. The area covered is approximately $60m^2$ with very high resolution ~330 megapixels. The advantage of using simulated environment is that we can define precisely the trajectory, and then, after running the visual odometry method we can evaluate the performance and tune different components. Especially, with the lack of real sequences provided with odometry ground truth. Hence, we created a dataset of images based on simulating stereo camera motion which is shown superimposed on the orthophoto in Figure 9. The motion has an S-shape type scaled in one direction. The reason is to test the visual odometry method in two cases; when there is an overlap with previously scanned area and another case when there is not. Our method is more adapted to the first case scenario.

We evaluate the proposed semi-global BA as compared to three cases, using global BA, local BA, and without using BA. As expected, the method that uses global BA performs the best in this context. The translation error is 1.2%, while the rotation error 0.009 [deg/m]. Followed by our method, with 2.44% of translation and 0.011 [deg/m] of rotation errors. This is fairly ahead of the local BA method that achieved 3.68% of translation and 0.012 [deg/m] of rotation errors. The optimization free visual odometry showed the largest divergence with a translation error of 6.8% and rotation error of 0.08 [deg/m]. Figures 10 and 11 show the obtained trajectories for our method and the mentioned methods, respectively.

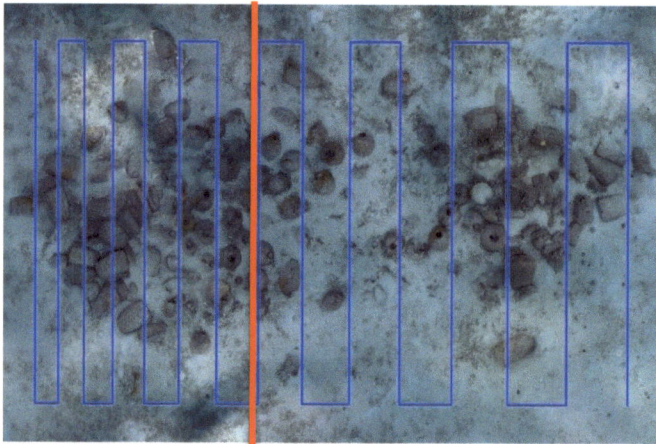

Figure 9. Simulation scenario with modified S-shape scanning profile which covers two situations; neighboring observations. Red border divides the map in overlapping/non overlapping path.

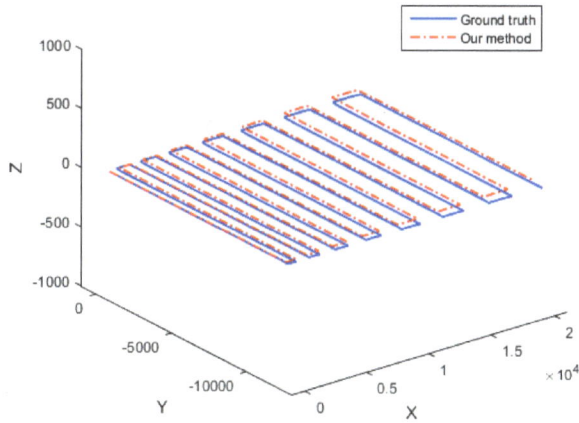

Figure 10. Estimated 3D trajectory using our method compared to ground truth.

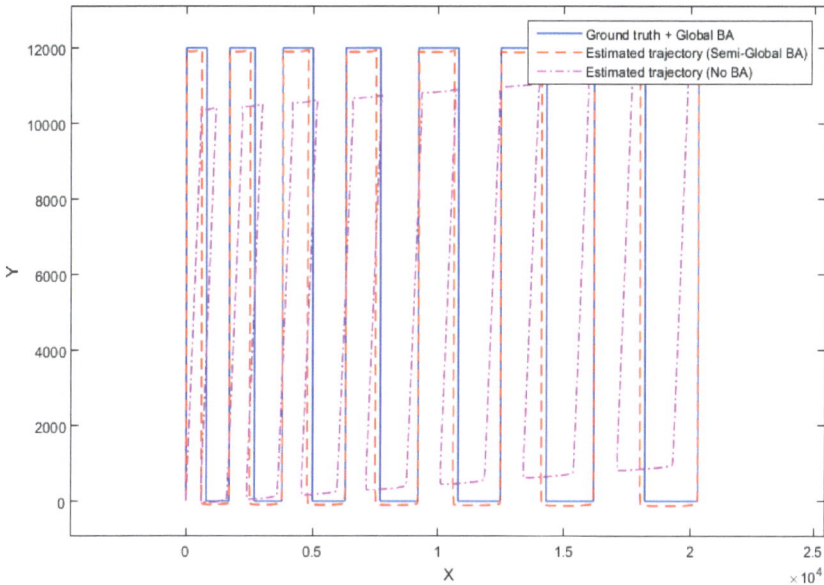

Figure 11. Comparison between several cases of visual odometry in terms of using BA. Note that the trajectory produced by the method without BA is scaled by ~1.8 for visualization purpose.

6. Conclusions and Perspectives

In this work, we introduced several improvements to the current traditional visual odometry approach in order to serve in the context of underwater surveys. The goal is to be adapted to embedded systems that are known for their lower resources. The sparse feature points matching guided with a rough depth estimation using lightness information is the main factor beyond most of the gain in computation time when compared to sophisticated feature descriptors combined with brute-force matching. Also, using stochastic representation and selection of frames in the semi-global BA improved the precision as compared to local BA methods, while remaining within real-time limits.

Our future perspectives are mainly centered on reducing the overall system size, for instance, replacing the main computer in our architecture with a third embedded unit, which in turn does not keep evolving. This also allows to the user to reduce the power consumption, which increases the navigation time. On the other hand, dealing with visual odometry failure is an important challenge specially in the context of underwater imaging, which is mainly due to bad image quality. The ideas of failing scenarios discussed in this paper can be extended to deal with the problem of interruptions in the obtained trajectory.

Acknowledgments: This work has been partially supported by both a public grant overseen by the French National Research Agency (ANR) as part of the program *Contenus numériques et interactions* (CONTINT) 2013 (reference: ANR-13-CORD-0014), GROPLAN project (*Ontology and Photogrammetry; Generalizing Surveys in Underwater and Nautical Archaeology*)[5], and by the French Armaments Procurement Agency (DGA), DGA RAPID LORI project (*LOcalisation et Reconnaissance d'objets Immerges*). Logistic support for underwater missions is provided by COMEX[6].

References

1. Bailey, T.; Durrant-Whyte, H. Simultaneous localization and mapping (SLAM): Part II. *IEEE Robot. Autom. Mag.* **2006**, *13*, 108–117.
2. Bay, H.;Tuytelaars, T.; Van Gool, L. Surf: Speeded up robust features. In Proceedings of the 9th European Conference on Computer Vision, Graz, Austria, 7–13 May 2006; pp. 404–417.
3. Bishop, C.M. *Neural Networks for Pattern Recognition*; Oxford University Press: Oxford, UK, 1995.
4. Blanco, J.-L.; Fernandez-Madrigal, J.-A.; González, J. A novel measure of uncertainty for mobile robot slam with rao—Blackwellized particle filters. *Int. J. Robot. Res.* **2008**, *27*, 73–89.
5. Comaniciu, D.; Ramesh, V.; Meer, P. Kernel-Based Object Tracking. *IEEE Trans. Pattern Anal. Mach. Intell.* **2003**, *25*, 564–575.
6. Davison, A.J. Real-Time Simultaneous Localisation and Mapping with a Single Camera. In Proceedings of the Ninth IEEE International Conference on Computer Vision—Volume 2, Washington, DC, USA, 13–16 October 2003.

---

[5] http://www.groplan.eu
[6] http://www.comex.fr/

7. Davison, A.J.; Reid, I.D.; Molton, N.D.; Stasse, O. MonoSLAM: Real-time single camera SLAM. *IEEE Trans. Pattern Anal. Mach. Intell.* **2007**, *29*, 1052–1067.

8. Drap, P. *Underwater Photogrammetry for Archaeology*; InTech Open: Rijeka, Croatia, 2012.

9. Drap, P.; Merad, D.; Hijazi, B.; Gaoua, L.; Nawaf, M.; Saccone, M.; Chemisky, B.; Seinturier, J.; Sourisseau, J.-C.; Gambin, T.; et al. Underwater Photogrammetry and Object Modeling: A Case Study of Xlendi Wreck in Malta. *Sensors* **2015**, *15*, 29802.

10. Eade, E.; Drummond, T. Scalable Monocular SLAM. In Proceedings of the IEEE Computer Society Conference on Computer Vision and Pattern Recognition, New York, NY, USA, 17–22 June 2006; pp. 469–476.

11. Eade, E.; Drummond, T. Unified Loop Closing and Recovery for Real Time Monocular SLAM. In Proceedings of the British Machine Vision Conference 2008, Leeds, UK, 1–4 September 2008.

12. Geiger, A.; Ziegler, J.; Stiller, C. Stereoscan: Dense 3d reconstruction in real-time. In Proceedings of the IEEE Intelligent Vehicles Symposium, Baden-Baden, Germany, 5–9 June 2011; pp. 963–968.

13. Hagan, M.T.; Demuth, H.B.; Beale, M.H.; De Jesús, O. *Neural Network Design*; PWS publishing company Boston: 1996.

14. Hartley, R.I.; Zisserman, A. *Multiple View Geometry in Computer Vision*; Cambridge University Press: Cambridge, UK, 2004.

15. Lourakis, M.I.; Argyros, A.A. SBA: A software package for generic sparse bundle adjustment. *ACM Trans. Math. Softw.* **2009**, *36*, 2.

16. Lowe, D.G. Distinctive image features from scale-invariant keypoints. *Int. J. Comput. Vis.* **2004**, *60*, 91–110.

17. Montemerlo, M.; Thrun, S. FastSLAM 2.0. In *FastSLAM: A Scalable Method for the Simultaneous Localization and Mapping Problem in Robotics*; Springer: Berlin, Heidelberg, Germany, 2007; pp. 63–90.

18. Montiel, J.; Civera, J.; Davison, A.J. Unified inverse depth parametrization for monocular SLAM. *Analysis* **2006**, *9*, 1.

19. Mouragnon, E.; Lhuillier, M.; Dhome, M.; Dekeyser, F.; Sayd, P. Generic and real-time structure from motion using local bundle adjustment. *Image Vis. Comput.* **2009**, *27*, 1178–1193.

20. Nawaf, M.M.; Drap, P.; Royer, J.P.; Merad, D.; Saccone, M. Towards Guided Underwater Survey Using Light Visual Odometry. *Int. Arch. Photogramm. Remote Sens. Spat. Inf. Sci.* **2017**, *XLII-2/W3*, 527–533.

21. Nawaf, M.M.; Hijazi, B.; Merad, D.; Drap, P. Guided Underwater Survey Using Semi-Global Visual Odometry. In Proceedings of the 15th International Conference on Computer Applications and Information Technology in the Maritime Industries, Lecce, Italy, 9–11 May 2016; pp. 288–301.

22. Nawaf, M.M.; Tremeau, A. Monocular 3D Structure Estimation for Urban Scenes. In Proceedings of the IEEE International Conference on Image Processing (ICIP), Paris, France, 27–30 October 2014; pp. 526–535.

23. Nistér, D.; Naroditsky, O.; Bergen, J. Visual odometry. In Proceedings of the 2004 IEEE Computer Society Conference on Computer Vision and Pattern Recognition, Washington, DC, USA, 27 June–2 July 2004; Volume 1, pp. I-652–I-659.

24. Piniés, P.; Tardós, J.D. Scalable SLAM building conditionally independent local maps. In Proceedings of the IROS 2007. IEEE/RSJ International Conference on Intelligent Robots and Systems, San Diego, CA, USA, 29 October–2 November 2007; pp. 3466–3471.

25. Sibley, D.; Mei, C.; Reid, I.; Newman, P. Adaptive relative bundle adjustment. In Proceedings of Robotics: Science and Systems V, Seattle, WA, USA, 28 June–1 July 2009.

26. Strutz, T. *Data Fitting and Uncertainty: A Practical Introduction to Weighted Least Squares and Beyond*; Vieweg + Teubner: Wiesbaden, Germany, 2010.

27. Thrun, S.; Burgard, W.; Fox, D. *Probabilistic Robotics*; MIT Press: Cambridge, MA, USA, 2005.

28. Triggs, B.; Mclauchlan, P.; Hartley, R.; Fitzgibbon, A. Bundle adjustment: A modern synthesis. In *Vision Algorithms: Theory and Practice*; 2000; pp. 153–177.

29. Wan, E.A.; Van Der Merwe, R. The unscented Kalman filter for nonlinear estimation. In Proceedings of the Adaptive Systems for Signal. Processing, Communications, and Control. Symposium 2000, Lake Louise, AB, Canada, 4 October 2000; pp. 153–158.

30. Yamaguchi, K.; Mcallester, D.; Urtasun, R. Robust Monocular Epipolar Flow Estimation. In Proceedings of the IEEE Conference on Computer Vision and Pattern Recognition, Portland, OR, USA, 23–28 June 2013; pp. 1862–1869.

Nawaf, M.M.; Royer, J.-P.; Pasquet, J.; *et al.* Underwater Photogrammetry and Visual Odometry. In *Latest Developments in Reality-Based 3D Surveying and Modelling*; Remondino, F., Georgopoulos, A., González-Aguilera, D., Agrafiotis, P., Eds.; MDPI: Basel, Switzerland, 2018; pp. 257–278.

Photogrammetric Modelling of Submerged Structures: Influence of Underwater Environment and Lens Ports on Three-Dimensional (3D) Measurements

Fabio Menna, Erica Nocerino and Fabio Remondino

3D Optical Metrology (3DOM) unit, Bruno Kessler Foundation (FBK), Trento, Italy;
(fmenna, nocerino, remondino) @fbk.eu; http://3dom.fbk.eu

Abstract: Underwater three-dimensional (3D) measurements and modelling is gaining interest thanks to technological developments of diving equipment and underwater technologies. Also, in this environment, photogrammetry is regarded as competitive and flexible method, which may lead to impressive and valuable results at different depths and in a broad range of application fields. To guarantee fit-to-purpose digitization results, a deep understanding of the physical peculiarities of the underwater environment is crucial, equipment and tools, which rule the image formation and affect the quality of the results. This paper aims to address these topics, providing a brief overview of the main properties of water and deepening the optical behaviour of photographic apparatus underwater. The difference between dome and flat ports is investigated, both in terms of image quality, geometric calibration, and accuracy potential. The two ports are employed for the 3D modelling of a semi-submerged structure; the experimental results show lower metric performances of the flat port, with significant differences per colour channel.

Keywords: underwater photogrammetry; calibration; image aberrations; dome port; flat port

1. Introduction

Since remote times, mankind has been bound to water bodies, either in their natural forms, such as oceans, lakes, rivers, wetlands, or in their man-made counterparts, such as structures, flumes, channels, basins, dams, etc. Evidence of human life from the very beginning hides under the water level, off the coasts, under shallow seas or deep oceans, but also inland water bodies of countries all around the world. Consequently, underwater cultural heritage (UCH) is very vast, probably much more than one can imagine, encompassing wrecks, ruins, submerged landscapes, caves, and all traces of exploitation of marine resources. At

279

the same time, exploring, documenting, and measuring the underwater environment and submerged objects is gaining importance in many application fields, ranging from biological studies to monitor climate changes and man-induced environmental alterations, to the investigation, assessment, and exploitation of seabed mineral and oil deposits.

Accordingly, there is a clear need to develop efficient techniques and methods to face the challenges and problems of underwater prospection, documentation, and monitoring.

With these aims, underwater photogrammetry has gained a remarkable success over the last few years even using consumer grade photographic equipment (Balletti et al., 2015; Demesticha et al., 2014; Diamanti and Vlachaki, 2015; Menna et al., 2013). Indeed, diver operated underwater housings are available for a wide range of digital cameras, sometimes designed and sold by the manufacturer of the cameras themselves, sometimes from third-party companies. For applications where cameras are used at very great depth, specialised housings are manufactured ad hoc to resist the high pressures and are not operated by a diver but installed to AUV, ROV, or a submarine (Drap et al, 2015; Kwasnitschka et al, 2016; Roman and Mather 2010).

This contribution reports the current state of a research that the authors have been carrying out for several years and part of a wider project called OptiMMA (optical metrology for maritime applications, http://3dom.fbk.eu/projects/underwater-photogrammetry-maritime-applications) with the twofold aim of (i) investigating the accuracy potential of different optical sensors when immersed in water, and (ii) providing operational guidelines to meet predefined quality requirements.

Testing underwater photographic equipment and assessing the accuracy potential of cameras enclosed in a waterproof housing is even more crucial than in photogrammetry applications above the water, with additional difficulties given by the underwater environment. Although underwater accuracy requirements are generally less demanding, assessing and evaluating the quality of photogrammetric measurements underwater is yet necessary, albeit far from being an easy task, to guarantee a fit-to-purpose underwater three-dimensional (3D) acquisition.

Stating the peculiarities of the hostile underwater environment, probably even more than in air, a basic understanding of its physical characteristics, together with a deep knowledge of optical behaviour of underwater photographic equipment is fundamental for a successful approach to underwater photogrammetry.

The planning of an underwater photographic acquisition for photogrammetric purposes needs to be carried out considering water physical properties and conditions, directly (i.e., water turbidity, lighting conditions) or indirectly (i.e., currents, depth, temperature) influencing the quality of acquired images.

The knowledge about photographic equipment and its performances in different conditions is a further fundamental topic to be investigated. Photographic cameras normally used on land, above the water, need a special housing with a flat or dome port to be used also in water. Looking to the underwater scene through a flat or dome port has many optical consequences, of which the most known is that the field of view of the lens mounted on the camera is preserved in case of a dome port and reduced by a factor almost equal to the refractive index of water for the flat port. In general, this common rule is satisfied, but there are many other factors that intervene in the optical formation of the image, some with very important practical implications that may make the choice of a type of port with respect to another one not as trivial, as described above.

For instance, lenses that are used in photography are designed to minimise optical aberrations throughout the entire image format. Residual aberrations are always present and their amount is depending on the optical design, quality of glasses, and therefore cost of the lens.

Nowadays, even the cheap zoom kit, bundled with consumer cameras, provide an acceptable image quality for less demanding photogrammetric purposes. Nevertheless, when used underwater, behind flat and dome ports, image quality, even for the most expensive cameras, undergoes quite a visible degradation due to the modification of the entire optical design. Depending on the combination of lens and port used (spherical dome or flat), the consequences on the overall image quality may be disappointing.

Moving from these considerations, in the first part of this contribution, the main physical properties of water are explained in relation to how they affect underwater photography (section 2). Then, a brief, but necessary comparative analysis about how dome and flat ports influence an underwater photogrammetry project is provided (Section 3). The analysis is the result of literature review as well as computer simulations and tests carried out by the authors above the water in laboratory as well as underwater (in swimming pool and at sea). Underwater camera calibrations, as well as image quality analyses, are carried out with flat and dome ports using a specific test object designed by the authors (Section 4). The paper continues with experimental tests that aim to analyse the accuracy potential of underwater photogrammetry when working with minimum number of reference control measurements (i.e., ground control points and reference lengths): the tests are reported for an industrial archaeology 3D modelling project of a maritime heritage building (Section 5). A comparative analysis is carried out using a same camera and lens encased in a waterproof housing using both dome and flat ports. Then, the archaeological structure is surveyed twice using the same camera-lens-housing but mounting the dome port first and flat port after. Internal quality figures from the bundle adjustment and from comparison against measured reference lengths are reported and discussed.

2. The Underwater Environment

Seawater is about 800 times denser than air, and is characterised by an optical index of refraction that varies as function of temperature, salinity, pressure, and light wavelength (Austin and Halikas, 1976). All these quantities are correlated: density increases as temperature decreases, salinity increases as pressure increases, pressure increases linearly with depth (every 10 m the pressure increases of 1 atmosphere, equal to 1.033 N/cm²).

The optical properties of seawater (Mobley, 1995) are highly variable and rule the propagation of light, consequently affecting all of the disciplines that entail optical measurements. Optical properties are usually divided into two classes: (i) the inherent properties, such as the absorption coefficient and volume scattering, which depend on the medium; and, (ii) the apparent optical properties (irradiance reflectance, attenuation coefficients, etc.) that are also functions of the light field, i.e., the way in which light particles (photons) travel in all directions through space.

According to the National Snow and Ice Data Centre (NSIDC), almost 94% of sunlight radiation that hits the sea surface penetrates and is then absorbed by water. The height of sun, depending on the location on Earth, time of day, and season, and the sea conditions influence the actual amount of light that is absorbed or reflected upward.

Figure 1. An underwater image of a rock in shallow water taken at noon (**left**) and later in the afternoon (**right**). Light caustics are visible in the left image.

Light is more absorbed in rough seas, whereas is highly reflected in calm waters that act as a mirror surface. Light rays are refracted by wavy surface, producing bright patterns, known as caustics, which are unfavourable for photogrammetric applications because they could affect the automatic extraction of two-dimensional (2D) features in images, as well as produce poor quality object texture. Figure 1 shows the same scene with (Figure 1-left) and without (Figure 1-

right) caustics. The intensity of light caustics depends on sun elevation, water turbidity and depth, with the effect that is significantly reduced already after few meter depth (Floor, 2005).

A major effect of water on light propagation is that it acts as a selective filter: the great amount of light entering the sea is absorbed or attenuated, i.e., it is converted in heat, within the first meters; only 1% of light entering the sea reaches 100 m. Not all of the components of light are absorbed at the same rate: both long (orange, red, and infrared) and short (ultra-violet) wavelengths are rapidly attenuated; green, and, especially, blue components penetrate more. Water composition affects the maximum depth at which light penetrates: in turbid coastal waters light hardly goes deeper than 20 m; while in open ocean blue light may even arrive at 200 m, depth limit after which there is almost no light.

The use of artificial light sources, such as strobes, is essential in underwater environment, especially in deep water, to restore the full range of colours, and compensate for light attenuation limiting the visibility range.

The presence in water of suspended particles (phytoplankton, organic matter, pollution, etc.) causes turbidity in water, and, that in turn, produces light scattering. Turbidity of water is generally quantified using the Secchi distance (Wernand, 2010; Cialdi and Secchi, 1865), an empirical method introduced in 1865 that makes use of a white circular disk. One Secchi distance is defined as the maximum depth or distance at which the disk is still visible.

Scattering or diffuse reflection is an optical phenomenon that arises when the light rays are randomly deviated from their straight paths. Scattering limits image quality, reducing the contrast and producing blurred images.

When strobes are used, also backscattering may occur; similar to scattering, the difference is that the light from the artificial source is reflected from the particles mainly back to the camera. To reduce backscattering, strobes should be carefully positioned, trying to keep the light cones of the strobes outside of the field of view of the camera as much as possible.

This translates in lightening the subject with the strobes far from the camera lens. The closer the flash to the camera, the more backscatter is produced. The closer to the subject the picture, the less water and particles between the camera and subject, and consequently, the less backscatter is produced.

From what has been highlighted above, the environment peculiarities heavily affect both the quality and how the images are to be acquired underwater. Turbidity is a remarkable limiting constraint, and scatter and backscatter reduce the contrast of the scene and the final quality of the image. Even in very clear water, the image acquisition can be difficult. Indeed, when a Secchi distance corresponds to several meters, a strong illumination would be required to light the object, and a wide baseline would be necessary between two lateral light sources and the camera lens to avoid backscatter. In extreme cases, a single diver might not be able to operate alone the system of cameras and strobe lights. An example of this phenomenon is visible in Figure 2. Two underwater photographs are taken, respectively, with (Figure 2-left) and without the strobe (Figure 2-right). The prominent back scatter effect is here caused by the short strobe-to-lens distance (less than 1m) between the strobe and the camera lens while the object is 7-10 m away from the camera lens.

Figure 2. Back scatter effect (**left**) caused by a short strobe-to-lens distance. The same image taken in natural light, without strobes, results in a higher contrast and sharpness image (**right**).

3. Lens Ports for Underwater Photography

As well known among underwater photographers and photogrammetrists, lens ports, coupled with waterproof camera housings, are designed to be either flat or (hemi)spherical. Their different shape rules and highly changes the geometry of image formation underwater, consequently affecting the survey planning and image acquisition. Table 1 reports the main characteristics and differences between the two ports. The most remarkable differences are due to the planar shape of flat ports that acts as a separation surface between two media, i.e., water outside and air inside the waterproof housing, causing a deviation (refraction) of optical rays from the ideal path, according to Snell's law. Instead, the hemispherical ports are made of two spherical surfaces, having ideally the same centre of curvature that coincides with the centre of the lens (the so-called entrance pupil, EP, or perspective centre). The closest the EP to the centre of curvature of the port, the less the light rays are refracted, and thus the closer is the image to the hypothesis of central perspective geometry.

Table 1. Characteristics of flat and dome ports.

	Hemispherical Dome Port	Flat Port
Description	Concentric lens acting as additional optical element (negative or diverging lens)	Flat plane of optically transparent glass or plastic
Field of view (FOV) WRT the camera-lens system	Equal	Reduced
Focal length WRT the camera-lens system	Equal	Increased (by a factor equal to approximately the ratio between the refraction indices of water and air)
Magnification WRT the camera-lens system	Equal	Increased (by a factor equal to approximately the ratio between the refraction indices of water and air)
Effect on the observed object	An upright, smaller virtual image of the object is formed at a distance from the dome surface equal to 3 times the curvature radius of the dome. The camera-lens system focuses on this virtual image.	The object appears closer to the camera by a factor of about 25%, i.e., approximately the reciprocal of the refraction index of water.
Maximum FOV	Not limited	Limited to 96°
Lens distortion	No significant distortion	Pincushion distortion
Other effects	• Increase of Depth of field (DOF) by a factor equal to approximately the ratio between the refraction indices of water and air • Spherical aberration • Field curvature	• Chromatic aberration. • Astigmatism
Costs	More expensive	Cheaper
Typical use	For DSLR cameras	For compact digital cameras

4. Underwater Equipment Calibration

The first step in the experimental work consists of the calibration of the photographic system made of a digital camera, plus lens in a waterproof housing with both dome and flat ports (Section 4.1) to assess its optical quality and accuracy potential. The tests are performed on a portable test object (Section 4.2), which is specifically designed by the authors, made of three Dibond® aluminium composite sheets with a thickness of 3 mm. The image quality of two systems, one equipped with the dome port and the other with the flat port, is analysed (Section 4.3.1). Finally, their accuracy potential is assessed comparing the results of self-calibration with respect to measurements that are derived from the calibration of the camera-lens system in air (Section 4.3.2).

4.1. Photographic System

A Nikon D750 24 Mpx full frame camera (pixel size 5.97 μm) mounting a Nikkor AF 24mm f2.8/D wide angle lens was put in a NiMAR NI3D750ZM pressure housing. The system is shown in Figure 3. In order to guarantee the highest accuracy, each image acquisition was carried out with a fixed focus set for the first image of the sequence. The distance to the object was kept constant through both of the visual references and using ropes with marks. Between the two underwater surveys just the port was changed and then the adjustment of the focus done. A Nikon SB700 strobe mounted in a dedicated NiMAR housing was used for the underwater calibrations.

Figure 3. Nikon D750 camera, Nikkor AF 24 mm f/2.8D, NiMAR NI3D750ZM pressure housing and the two different ports used (from left to right).

4.2. Underwater Test Object

The test object (Figure 4) is made of three Dibond® aluminium composite sheets with a thickness of 3 mm. The Dibond® material consists of 2 layers of 0.3mm thick aluminium sandwiching a core containing UV stabilized virgin low density polyethylene (http://alucobond.com.au/specify-now/dibond/#sthash.uKarrJ0o.dpuf). The three panels, sided together, form a 1500x1000mm² board, fixed on the back to a structural frame made with Rexroth aluminium profiles, which add rigidity and mechanical stability to the structure. To provide the structure with depth variation that is suitable for photogrammetric camera calibration, six square plates with additional targets can be mounted using optical breadboard support rods made of stainless steel, currently up to 200 mm long. The linear thermal expansion coefficient of Dibond® panels is 0.024 mm/(m·K). A total of 160 circular coded targets, designed with a black square background that allows for MTF measurements, are regularly distributed over the test object. Other resolution wedges and colour checkboards are also present.

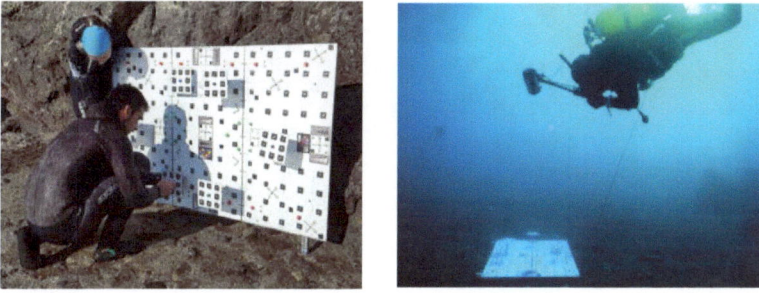

Figure 4. The test object used for underwater calibrations during the assembly (**left**) and image acquisition (**right**).

4.3. Underwater Camera Calibrations

The portable test object is laid down at a depth of about five meters and photographed from an average distance of about 1.2 m for the dome port and 1.6m for the flat port, providing a ground sampling distance (GSD) of about 0.3mm for both of the calibrations. An aperture value of f/11 is used for both the flat and dome ports. About 30 images per each port are collected, using a standard self-calibration protocol with multi-view convergent images and roll diversity (Fraser, 1997). The image acquisitions are carried out in sequence, the dome port first and the flat port after.

4.3.1. Analysis of Image Quality

As expected, from a visual analysis of the acquired images, it is easily recognisable that, while the dome port preserves the barrel distortion of the lens almost unchanged, (Figure 5-centre), the flat port introduces a heavy pincushion distortion (Figure 5-bottom). Furthermore, the image quality for the flat port is severely different between the centre (Figure 6a) and the corners, showing some heavy chromatic aberrations (Figure 6b) and blur astigmatism, different per red, green, and blue channels, with the blue channel behaving the worst (Figure 6c–e).

Above Water No Port

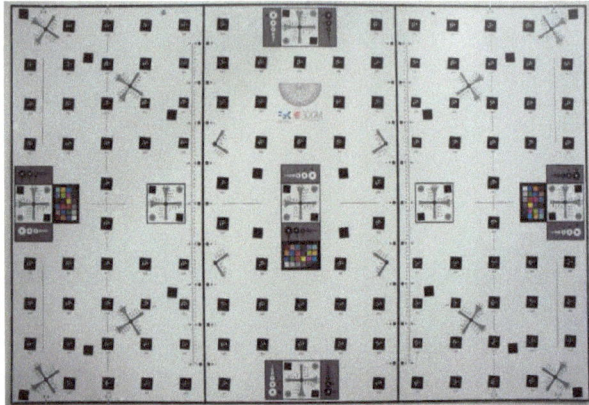

UW Dome Port

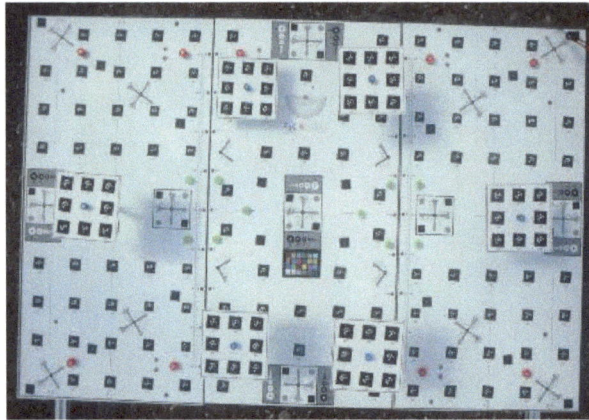

UW Flat Port

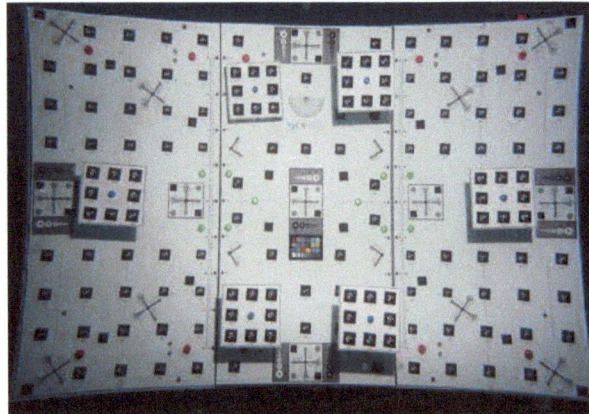

Figure 5. The portable test object as imaged above the water without the pressure housing (up), underwater with the dome port (centre) and flat port (bottom).

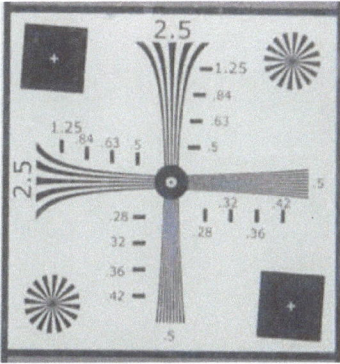

(a) RGB_centre (b) RGB_corner

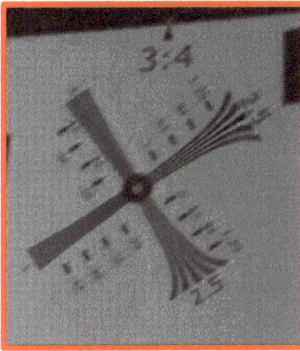

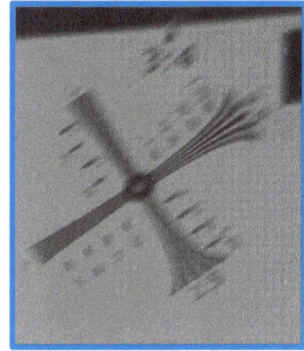

(c) RED_corner (d) GREEN_corner (e) BLUE_corner

Figure 6. Visual image analysis for the flat port. Resolution patches respectively at the centre (**a**) and at the corner (**b**) of the field of view as seen in the colour image. Single red (**c**), green (**d**), blue (**e**) channels at the corner of the image displaying a colour dependant astigmatism.

Figure 7 shows three resolution patches that are placed along a diagonal (first quadrant) of the test object, respectively, in the centre, at 2/3 of maximum radius and at corner both for the dome and flat ports. All of the underwater images display a reduction in contrast; however, while the centre patch is well resolved, going towards the upper right corner results in a significant worsening of image quality for both dome and flat ports. Especially at the corner, the dome port shows some blur due to field curvature, while the flat port shows severe chromatic aberrations and blur due to astigmatism already at half of the maximum radial distance. Note how for the flat port the astigmatism is colour dependent. To visually highlight this behaviour for the flat port, Figure 6 also shows for the upper right corner patch, the red, green, and blue channels. For the red channel, a slight astigmatism makes tangential limiting resolution (stripes with tangential edges) worse than the radial one. A cross section extracted at subject resolution of 1.25 mm still shows a good reproduction of the signal that is instead flattened for the blue channel.

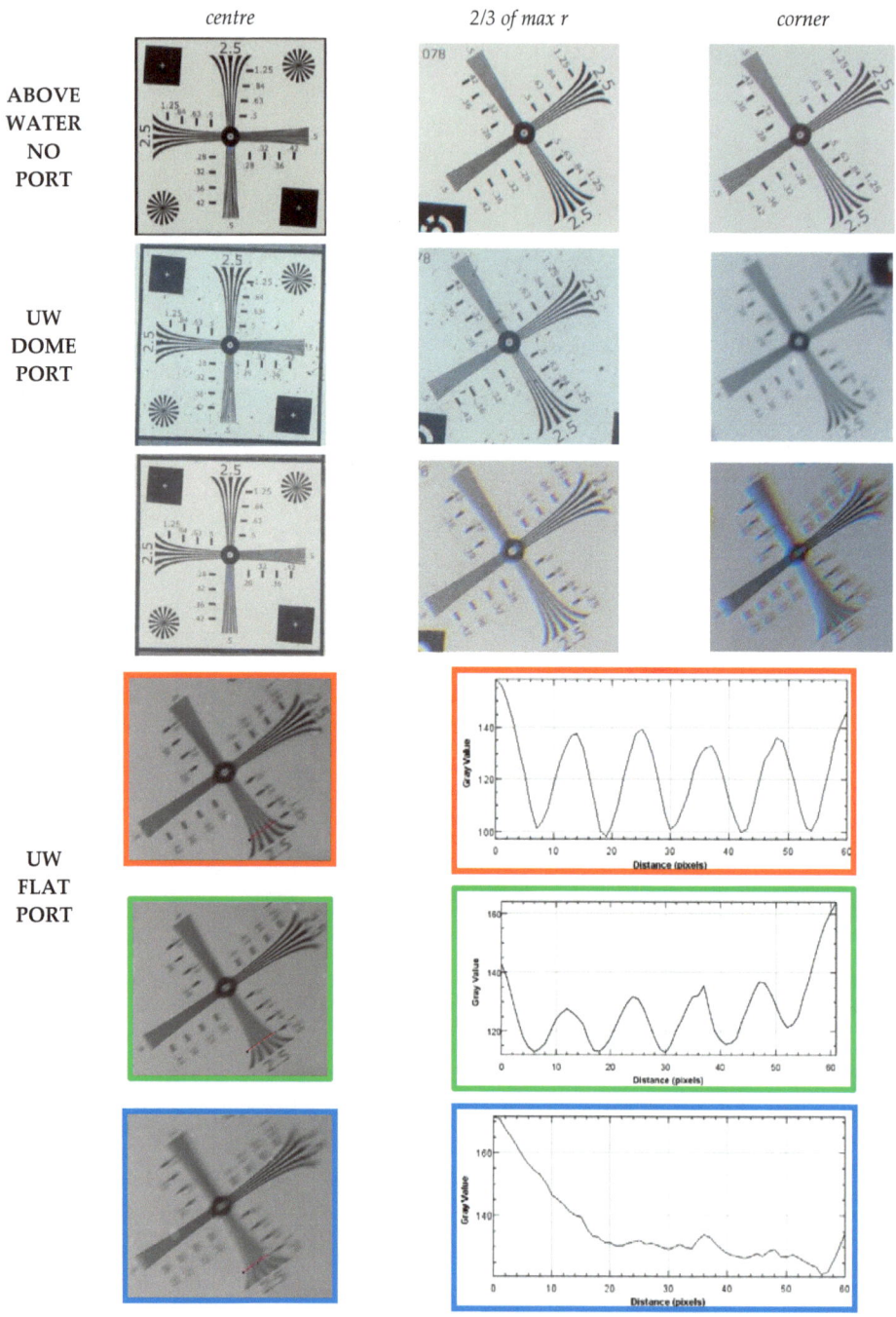

Figure 7. Resolution patches as imaged above the water without the pressure housing, with the dome port and with the flat port. For the flat port, a section along the radius is reported for the red and blue channels showing a colour-dependent astigmatism.

4.3.2. Geometric Calibration

The two underwater datasets and the reference in air (or above water) dataset are processed using the open source damped bundle adjustment toolbox (DBAT) v0.6.2.0 (Börlin and Gussenmeyer, 2013) for MATLAB environment. The least squares method implemented in PhotoModeler for the automatic marking of circular coded targets is employed. The method extracts the target centroids in the green channel as default. Table 2 synthetically summarises the results of the calibration processing. As it can be noted, the flat port performs quite significantly worse than the dome port, which displays a higher potential accuracy (image observation from green channel for both ports).

Table 2. Results of self-calibrating bundle adjustment: interior orientation and additional parameters are reported along with internal assessment in image and object space.

	D750-24mm *No Port above Water* Calibration	D750-24mm *Dome Port* UW Calibration	D750-24mm *Flat Port* UW Calibration
Principal distance c [mm]	24.624 ± 4.7e-004	26.001 ± 0.001	33.110 ± 0.004
Principal point ppx [mm]	0.0391 ± 5.5e-004	-0.0437 ± 0.002	-0.0831 ± 0.005
Principal point ppy [mm]	0.0158 ± 5.0e-004	-0.1486 ± 0.002	-0.1341 ± 0.005
$k1$ [mm^{-2}]	1.649e-004 ± 2.2e-007	1.679e-004 ± 7.3e-007	-1.965e-004 ± 9.5e-007
$k2$ [mm^{-4}]	-2.461e-007 ± 1.2e-009	-2.873e-007 ± 5.7e-009	-1.917e-007 ± 5.8e-009
$k3$ [mm^{-6}]	2.593e-011 ± 2.0e-012	2.200e-010 ± 1.3e-011	1.943e-010 ± 1.0e-011
$P1$ [mm^{-1}]	-3.506e-006 ± 3.4e-007	-8.022e-006 ± 7.9e-007	4.445e-005 ± 2.9e-006
$P2$ [mm^{-1}]	-1.233e-006 ± 2.9e-007	3.168e-005 ± 7.4e-007	8.984e-005 ± 2.6e-006
Re-projection error RMS [pixel]	0.21	0.32	0.91
Point error vector length RMS [mm]	0.03	0.06	0.12
Point error vector length maximum [mm]	0.10	0.22	0.30
Relative precision (wrt a maximum dimension of 1800 mm)	≈1:62000	≈1:29000	≈1:15000

From the output of bundle adjustments with self-calibration (Brown model formulation with radial and decentring distortions), we can highlight the following considerations:

- the obtained precisions of interior orientation parameters for dome and flat ports are considerably poorer than laboratory calibration. The flat port performed significantly worse than the dome port with a standard deviation of the principal distance 4 times greater than the dome port and nine times

greater than the laboratory calibration;
- a higher potential accuracy for the dome port with respect to the flat port (image observation from green channel for both ports) is notable;
- the principal point for both dome and flat ports is significantly different with respect to the one computed above water without the pressure housing. In particular, the principal point variation in y is systematically greater towards the negative values, probably caused by a camera misalignment in the pressure housing;
- the RMS of image residuals for the flat port is three times greater than the dome port and more than four times with respect to the above water calibration; and,
- as expected, due to the effect of refraction, the principal distance of the imaging system with the flat port is about 34% greater than the above the water calibration.

The accuracy assessment is carried out using 23 known distances on the three panels of the test object. The influence of deformations of the panels are taken into account so that only those lengths whose potential errors (chord against arc distance) are below 15 μm are considered.

According to Luhmann et al (2014), the theoretical length measurement error (TLME) and the length measurement error (LME) are defined as follows:

$$TLME = 3\sqrt{2} \cdot s_{XYZ} \tag{1}$$

$$LME = D_m - D_r \tag{2}$$

where S_{xyz} are the standard deviations of object coordinates, and $D_{r,m}$ are the reference and measured distances, respectively. Table 3 summarises the length measurement error (LME) for the dome port and flat port; it is computed according to (2) as the difference between the measured distance D_m for the dome and flat ports, and the distance from the above water calibration, assumed as reference length D_r. For the underwater tests reported in this paper, the theoretical length measurement error (TLME) of the reference at 99% confidence level is 0.078mm. The relative length measurement accuracy (RLMA) is defined as:

$$RLMA = 1 : ROUND\big(D_r / (D_m - D_r)\big) \tag{3}$$

Table 3. Length measurement error for the underwater bundle adjustment results.

	D750-24mm *Dome Port* UW Calibration	D750-24mm *Flat Port* UW Calibration
LME [mm]	0.082	0.212
RLMA (wrt to the maximum reference length of 900 mm)	≈1:11000	≈1:4250

5. Real Case Scenario Experiment

The two underwater photographic systems are tested in a real case scenario to quantify the effect of image quality and camera calibration that is analysed in Section 4 when dealing with the 3D modelling project of underwater heritage.

5.1. The Modelled Heritage Structure

A semi submerged industrial structure of about 20x10m² located in the Bay of Rogiolo, near Livorno, Italy, today abandoned and under consideration for restoration (Figure 8), is used as test site. Originally, the structure was a harbour, serving as support for cave and cement plan activities at the beginning of the twentieth century. A combination of close range photogrammetry above and under the water according to the procedure described in Menna et al. (2015, 2013) is chosen for modelling the structure. The part of the structure underwater is photographed twice, once with the dome port and a second time with the flat port. During the survey, the sky was overcast limiting the water caustics effect on the underwater structures.

Figure 8. The surveyed port structure in the bay of Rogiolo near Livorno, Italy. At the top row, historical images of the bay and the industrial structure (Ciompi, 1991).

5.2. Targeting of the Industrial Structure

Eight plates with eight coded target are each placed across the waterline (Figure 9-a) of the rectangular basin chosen to perform the comparative photogrammetric tests (Figure 9-b).

The coded targets are placed for the twofold aim of: (i) allowing to register the underwater and above the water 3D models, and (ii) having well and uniquely defined 3D points to perform comparisons between flat and dome port underwater surveys. The relative positions of the targets on the plates are known by laboratory calibration, thus by measuring at least three non-collinear targets in the underwater or above the water photogrammetric surveys, the 3D coordinates of the remaining targets can be computed through a similarity transformation. By means of this procedure, common points between underwater and above the water surveys can be derived and the two 3D models are registered together. Some tape length measurements are used to scale the object.

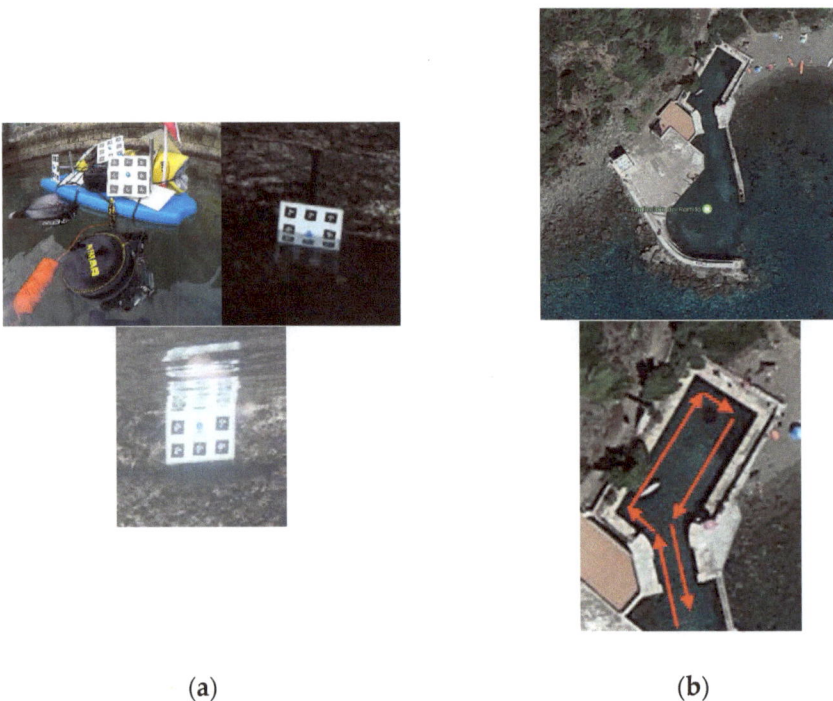

(a) (b)

Figure 9. Plates with coded targets used in the experiment (**a**). An aerial view of the port structure with an enlarged sight of the rectangular basin chosen for the tests with a schematic view of the photogrammetric strip acquired (**b**).

5.3. Planning and Acquisition of the Underwater and Above-The-Water Camera Network

Since the results of the preliminary calibrations showed a significant different behaviour of the two ports, a much evident discrepancy between the two would be expected in elongated strips, such as those needed for surveying big structures. Indeed, systematic residual errors that are not properly modelled by camera calibration parameters are expected to accumulate along the strip, thus leading to global object deformation (Nocerino et al., 2014).

The camera network planned to survey the rectangular basin consists in a singular open loop strip taken at about 2 m from the vertical walls with the dome port, and 2.6 meters with the flat port to obtain a GSD of about 0.5 mm in both cases. 80% overlap was considered along the strip and some convergent and rolled images were acquired to improve the self-calibration (especially considering the geometric characteristics of the object that results flat within the field of view of the single images). The image acquisitions (Figure 10a) were carried out in sequence, the dome port first and the flat port soon after. The maximum depth was 1.5m, water temperature about 15 degrees, and the underwater image acquisition required about three hours in total. The part above the water was surveyed with the same camera without the pressure housing. The available side walking path was used to photograph the structure on the opposite side, thus leading to an average distance to the object of about 12 m (GSD about 3 mm). Same 80% overlap with rolled and convergent image acquisition protocol was used above the water.

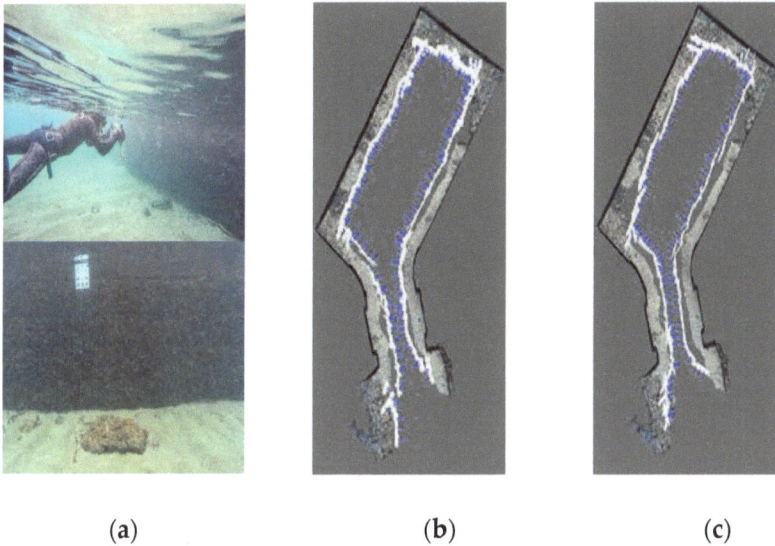

|(a)|(b)|(c)|

Figure 10. Underwater image acquisition and a sample image from the dome port (**a**). Final camera networks for the dome port (**c**) and flat port (**d**).

5.4. Self-Calibrating Bundle Adjustment

The three image datasets, two underwater and one above the water, were processed with the same procedure. The images were automatically oriented using Agisoft PhotoScan where self-calibration with radial and decentering distortion parameters were computed. The final camera network for the dome and flat ports is shown in Figure 10 (b-c).

A preliminary comparative analysis of the bundle adjustment parameters that were retrieved by PhotoScan for the two underwater datasets confirmed a less precise solution with the flat port.

A very important difference was observed on the self-calibration parameters. While for the dome port the standard deviations for the focal length is 0.4 μm, for the flat port the standard deviation resulted more than three times worse, i.e., 1.4 μm. In general, the self-calibration parameters of the flat port were an order of magnitude worse than those that were computed for the dome port. Such a worse precision is expected to be a source of systematic errors that accumulates along the photogrammetric strip and "vent" into the object space leading to a stronger global deformation of the 3D model for the flat port. Therefore, as shown in Nocerino et al. (2014), over the 70 meters linear perimeter of the underwater basin, the global deformations can reach some centimetres, even if the GSD is sub-millimetric.

For the above the water dataset, as expected, the precision of calibration parameters was much greater, providing a standard deviation for the focal length of some 0.1 μm, which is more than ten times better than that of the flat port underwater. The three datasets were scaled using a combination of length measurements that were provided by the plates and some tape measurements. A maximum scaling error of about 0.2% was estimated from the residuals on the reference known lengths.

The approach described in Nocerino et al. (2014) was employed to further analyse the three processings, namely the above the water (no port) and underwater, with both dome and flat ports, acquisitions. The tie points extracted in PhotoScan were imported in PhotoModeler Scanner where a bundle adjustment was executed and the precision vector length of the 3D object coordinates of the triangulated tie points are shown in Figure 11 for the three different acquisitions.

The precision vector length is the square root of the sum of the squared theoretical precision of the coordinates in the three directions. It is the expected variability of estimated 3D object coordinates, resulting from the adjustment process and depending on camera network and precision of image observations.

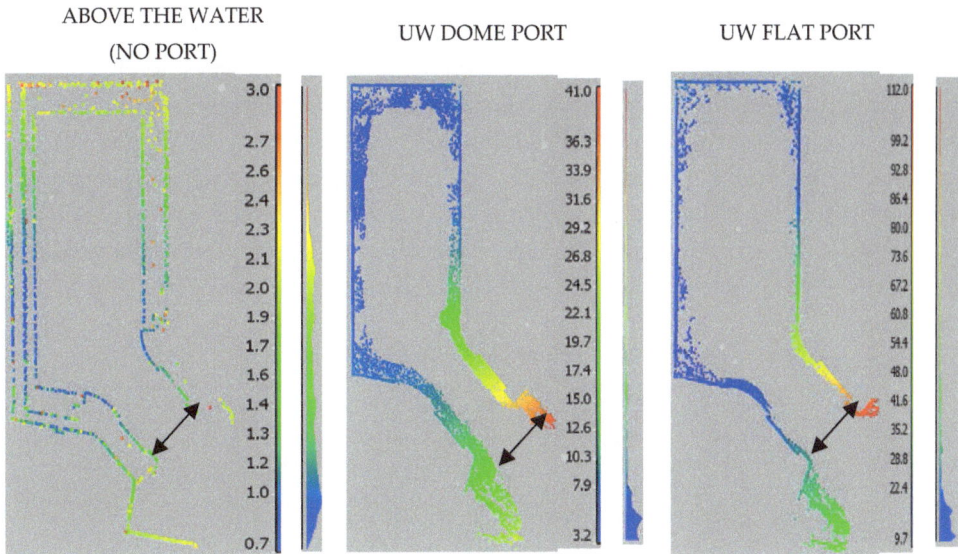

Figure 11. Three-dimensional (3D) tie point cloud coloured according to the precision vector length in mm for above the water (no port) and underwater acquisitions, with both dome and flat ports. Black arrows show where reference measurement was taken.

The precision is computed according to error propagation theory. The theoretical precision would coincide with the accuracy of object coordinates if all of the systematic errors had been properly modelled.

Three are the most remarkable outcomes. It is worth noting that the expected precision for the above the water survey is two to three orders of magnitude better than the underwater cases. The behaviour of the precision distribution is very similar for the two underwater acquisitions, where a significant degradation at the extremities of the structure is visible (the different histogram ranges of colour scales in figure 11 were chosen to highlight the weak point of each photogrammetric processing). However, the theoretical precision of the dome port is determined to three times better than that of the flat port.

Table 4. Statistics of 3D precision vector length for the different solutions.

Precision Vector Length	D750-24mm *No Port* above Water	D750-24mm *Dome Port* UW	D750-24mm *Flat Port* UW
MIN [mm]	0.7	3.2	9.7
MAX [mm]	3.0	41.0	112.0
MEAN [mm]	1.5	10.5	27.2
RMS [mm]	0.5	8.3	22.6

5.5. Accuracy and 3D Analysis in Object Space

A simple evaluation was carried out to assess the accuracy of the two underwater surveys. A reference distance by tape measurement (estimated accuracy ca. 1cm) was taken between the two plates facing each other at the entrance of the rectangular basin (black arrows in Figure 11) and compared with those obtained from the underwater survey (Table 5). Being the two plates at the beginning and end of the strip, the resulting discrepancy can be considered as a loop closure error. An error about 30 cm was observed for the flat port.

Table 5. Summary of length check.

Reference Distance	D750-24mm *Dome Port* UW	D750-24mm *Flat Port* UW
4.723 m	4.729 m	5.015 m
Error→	0.006 m (≈12xGSD)	0.292 m (≈600xGSD)

5.6. Photogrammetric Processing of the Single R, G, B Channels

As already mentioned in Section 4.3.1, the three channels for the flat port display different image quality, especially at the corners.

The three R, G, B channels for both the flat port and dome port were extracted from the RGB images and were saved as single channel images to be processed separately.

For the flat port, only the Red and Green channels succeeded the orientation stage. On the contrary, the images in the Blue channel probably appeared too blurred and were only partially oriented. The three channels for the dome port were oriented without any difficulty.

As the images in the three channels were acquired from the same position and exactly with the same camera network, the results in object space are expected not to be significantly different. An inner comparison between the three channels of each port was performed by comparing the 3D coordinates of the plates that were obtained separately from each channel. According to PhotoScan manual, the default processing considers a combination of the three R, G, B channels. Consequently, the previous results (Sections 5.4 and 5.5), obtained in default mode, were used as reference for the relative comparisons between the different colour channels.

A similarity transformation with isotropic scale factor was computed to compare the 3D coordinates. The Euclidean distances between same points are used as a measure of discrepancy. Table 6 summarizes the relative comparison for each channel and port reported as RMS and maximum discrepancy between 3D points. A maximum difference of 23 cm was observed between the red channel of the flat port and the RGB combination for the same flat port. The solutions between the three channels of the dome port proved more consistent among themselves, meaning that the 3D shape is much more invariant for the dome port when different single channels are used.

Table 6. Summary of the relative comparison between 3D coordinates obtained from the single R,G,B channels and the one obtained as RGB combination for each port.

	D750-24mm *dome port* UW		D750-24mm *flat port* UW	
	RMS [m]	Max [m]	RMS [m]	Max [m]
RED	0.005	0.013	0.096	0.229
GREEN	0.003	0.006	0.023	0.055
BLUE	0.010	0.023	n/a	n/a

The accuracy analysis shown in Section 5.5 was repeated for the different colour channels for both ports. Since the blue channel images were not successfully oriented, only the error for the red and green channels were reported in Table 7 for the flat port. The discrepancy between the reference distance and the one measured in the green channel of the flat port reduced from 29 to 22 cm, and from 29 to 7 cm for the red channel. Slight variations were seen for the dome port, with the blue channel behaving the worst again.

Table 7. Summary of the loop closure error for both dome and flat ports (reference distance: 4.723 m).

	D750-24mm *dome port* UW			
	RGB	RED	GREEN	BLUE
Measured distance	4.729 m	4.725 m	4.718 m	4.766 m
Error	0.006 mm (≈12xGSD)	0.002 m (≈4xGSD)	−0.005 m (≈10xGSD)	−0.043 m (≈86xGSD)
	D750-24mm *flat port* UW			
	RGB	RED	GREEN	BLUE
Measured distance	5.015 m	4.656	4.940	n/a
Error	0.292 m (≈580xGSD)	−0.067 m (≈134xGSD)	0.217 m (≈434xGSD)	n/a

5.7. 3D Modelling of the Structure

Dense point clouds were computed at 1/8 and ¼ linear resolution, respectively, for the dome port and the above-the-water photogrammetric surveys corresponding to a spatial resolution of 4 mm in the object space. An optimized mesh according to Rodriguez et al. (2015) was wrapped over the manually cleaned point clouds for each dataset. The joint alignment procedure presented in Menna et al. (2015, 2013) was used to register the underwater mesh with the one above the water. The RMS of the transformation was some 3 cm for the dome port and 13 cm for the flat port. Figure 12 shows some renderings of the basin after the alignment of the underwater (dome port) and above-the-water 3D models.

Figure 12. Renderings of the basin after the alignment of the underwater and above-the-water 3D models.

6. Concluding Remarks

This paper investigated the effect of the diverse image quality of flat and dome ports over the accuracy of the final 3D model obtained through photogrammetric procedures. The paper highlighted the influence of image quality over the global accuracy of the final 3D model. Image quality underwater undergoes a very evident degradation due to the sum of optical phenomena, arising from both the pressure housing and port used and the physical and environmental properties of water itself. Indeed, due to the combination of optical aberrations, such as astigmatism, heavy distortions, and chromatic aberrations, plus a non-complete modelling of unknown systematic image errors, strong global deformations were observed and assessed through both off the internal parameters that were provided by the bundle adjustment and simple length measurements for the two ports. A very high error of some 29 cm was found with the flat port data set. Preliminary calibrations on a portable test object anticipated a degradation of accuracy when using the flat port by reporting high RMS of image residuals, a less precise calibration (worse standard deviations for camera parameters), and a lower 3D point precision in object space. A significant different image quality per colour channel was observed. From the visual inspection of the images, the red channel resulted less influenced by blur effects towards the corners showing a higher contrast and sharpness across the whole sensor format. Different photogrammetric processing for each colour were carried out to investigate how the different image quality affects the final metric results. RGB combination was compared against the single red, green, and blue channels. As expected, the green channel performed

300

more similarly to the RGB combination than the other channels, as the digital sensor of the Nikon D750 uses the Bayer filter array. The red channel provided a significant improved accuracy with respect to the processing that was obtained from the combination of the R, G, B channels. The blue channel proved to be the most problematic and might probably degrade the accuracy when combined with the other channels. This test was important because software applications may combine the three channels by default, which may not be the best procedure for underwater photogrammetry.

The issues raised by this study may deserve more experimental tests for example using different housings and ports. Having observed a strong difference between image quality between the centre and corners, successive tests will take into account a different weighting for image observations according to optical quality parameters (e.g. Modulation Transfer Function-MTF).

References

1. Austin, R.W.; Halikas, G. *The Index of Refraction of Seawater*; SIO Ref. No. 76–1; National Technical Information Service: Springfield, VA, USA, January 1976.
2. Balletti, C.; Beltrame, C.; Costa, E.; Guerra, F.; Vernier, P. Underwater photogrammetry and 3D reconstruction of marble cargos shipwreck. *Int. Arch. Photogramm. Remote Sens. Spat. Inf. Sci.* **2015**, *40*, 7.
3. Börlin, N.; Grussenmeyer, P. Bundle adjustment with and without damping. *Photogramm. Rec.* **2013**, *28*, 396–415.
4. Cialdi, A.; Secchi, P.A. Sur la transparence de la mer. *Comptes Rendus Hebdomadaire de Séances de l'Academie des Sciences* **1865**, *61*, 100–104.
5. Ciompi, L. Dalla costa fiorita di Quercianella. Edizione Stella del Mare Livorno. 1991. Available online: http://www.quercianellasonnino.it/bibliografia.php (accessed on 3 May 2017).
6. Demesticha, S.; Skarlatos, D.; Neophytou, A. The 4th-century BC shipwreck at Mazotos, Cyprus: new techniques and methodologies in the 3D mapping of shipwreck excavations. *J. Field Archaeol.* **2014**, *39*, 134–150.
7. Diamanti, E.; Vlachaki, F. 3D Recording of Underwater Antiquities in the South Euboean Gulf. The International Archives of Photogrammetry. *Remote Sens. Spat. Inf. Sci.* **2015**, *40*, 93.
8. Floor, J.A. Water and Light in Underwater Photography. 2005. Available online: http://www.seafriends.org.nz/phgraph/water.htm.
9. Fraser, C.S. Digital camera self-calibration. *ISPRS J. Photogramm. Remote Sens.* **1997**, *52*, 149–159.
10. Kwasnitschka, T.; Köser, K.; Sticklus, J.; Rothenbeck, M.; Weiß, T.; Wenzlaff, E.; Schoening, T.; Triebe, L.; Steinführer, A.; Devey, C.; et al. DeepSurveyCam—A deep ocean optical mapping system. *Sensors* **2016**, *16*, 164.

11. Helmholz, P.; Long, J.; Munsie, T.; Belton, D. Accuracy assessment of GOPRO Hero 3 (Black) camera in underwater environment. *Int. Arch. Photogramm. Remote Sens. Spat. Inf. Sci.* **2016**, *41*, 477–483.

12. Luhmann, T.; Hastedt, H.; Tecklenburg, W. Modelling of chromatic aberration for high precision photogrammetry. *Int. Arch. Photogramm. Remote Sens. Spat. Inf. Sci.* **2006**, *36*, 173–178.

13. Luhmann, T.; Robson, S.; Kyle, S.; Boehm, J. *Close-Range Photogrammetry and 3D Imaging*; Walter de Gruyter: Berlin, Germany, 2014.

14. Matsuoka, R.; Asonuma, K.; Takahashi, G.; Danjo, T.; Hirana, K. Evaluation of correction methods of chromatic aberration in digital camera images. *ISPRS Ann. Photogramm. Remote Sens. Spat. Inf. Sci.* **2012**, *I-3*, 49–55.

15. McCarthy, J.; Benjamin, J. Multi-image photogrammetry for underwater archaeological site recording: an accessible, diver-based approach. *J. Marit. Archaeol.* **2014**, *9*, 95–114.

16. Menna, F.; Nocerino, E.; Fassi, F.; Remondino, F. Geometric and optic characterization of a hemispherical dome port for underwater photogrammetry. *Sensors* **2016**, *16*, doi:10.3390/s16010048.

17. Menna, F.; Nocerino, E.; Troisi, S.; Remondino, F. Joint alignment of underwater and above-the-water photogrammetric 3D models by independent models adjustment. *Int. Arch. Photogramm. Remote Sens. Spat. Inf. Sci.* **2015**, *40*, 143.

18. Menna, F.; Nocerino, E.; Troisi, S.; Remondino, F. A photogrammetric approach to survey floating and semi-submerged objects. *Proc. SPIE* **2013**; *8791*, doi:10.1117/12.2020464.

19. Nocerino, E.; Menna, F.; Fassi, F.; Remondino, F. Underwater calibration of dome port pressure housings. *Int. Arch. Photogramm. Remote Sens. Spat. Inf. Sci.* **2016**, *40*, 127–134.

20. Nocerino, E.; Menna, F.; Remondino, F. Accuracy of typical photogrammetric networks in cultural heritage 3D modeling projects. *Int. Arch. Photogramm. Remote Sens. Spat. Inf. Sci.* **2014**, *40*, 465–472.

21. Roman, C.; Mather, R. Autonomous underwater vehicles as tools for deep-submergence archaeology. Proceedings of the institution of mechanical engineers. *Part M J. Eng. Marit. Environ.* **2010**, *224*, 327–340.

22. Reznicek, J.; Luhmann, T.; Jepping, C. Influence of raw image preprocessing and other selected processes on accuracy of close-range photogrammetric systems according to VDI 2634. *Int. Arch. Photogramm. Remote Sens. Spat. Inf. Sci.* **2016**, *41*, 107–113.

23. Robson, S.; MacDonald, L.; Kyle, S.A.; Shortis, M.R. Multispectral calibration to enhance the metrology performance of C-mount camera systems. *Int. Arch. Photogramm. Remote Sens. Spat. Inf. Sci.* **2014**, *40*, 517.

24. Rodríguez-González, P.; Nocerino, E.; Menna, F.; Minto, S.; Remondino, F. 3D surveying & modeling of underground passages in WWI fortifications. *Int. Arch. Photogramm. Remote Sens. Spat. Inf. Sci.* **2015**, *40*, 17–24.

25. Shortis, M. Calibration techniques for accurate measurements by underwater camera systems. *Sensors* **2015**, *15*, 30810–30826.

26. Telem, G.; Filin, S. Photogrammetric modeling of underwater environments. *ISPRS J. Photogramm. Remote Sens.* **2010**, *65*, 433–444.

Menna, F.; Nocerino, E.; Remondino, F. Photogrammetric Modelling of Submerged Structures: Influence of Underwater Environment and Lens Ports on Three-Dimensional (3D) Measurements. In *Latest Developments in Reality-Based 3D Surveying and Modelling*; Remondino, F., Georgopoulos, A., González-Aguilera, D., Agrafiotis, P., Eds.; MDPI: Basel, Switzerland, 2018; pp. 279–303.

Chapter 5
BIM and HBIM

Automating Parametric Modelling From Reality-Based Data by Revit Api Development

Xiucheng Yang, Mathieu Koehl and Pierre Grussenmeyer

Photogrammetry and Geomatics Group, ICube Laboratory UMR 7357, INSA, Strasbourg, France; (xiucheng.yang, mathieu.koehl, pierre.grussenmeyer)@insa-strasbourg.fr

Abstract: The main objective of the project that is presented in this chapter is to explore the application of Building Information Modelling (BIM) on parametric modelling of beam frame system in the field of cultural heritage. Reality-based data has been widely accepted to realize the heritage geometric modelling, and the current challenge comes to the element segmentation and information management aiming at heritage conservation. The recently developed as-built BIM technique is increasingly utilized to parametrically describe the entity and its elements with attribute and spatial relationship information. This project exploits the typical BIM software, *Autodesk Revit*, to transfer the measured total station points and recorded terrestrial laser scanning data to the parametric beam model. Two parametric models from total station points and point cloud are manually built. A plugin tool dedicated to automating the modelling process is in preparation by *Revit Application Programming Interface (API)*.

Keywords: Cultural Heritage; Building Information Modelling; Parametric Modelling; Point Cloud; Autodesk Revit

1. Introduction

Built heritage is our unique and irreplaceable treasure from the past. The recording medium of cultural built heritage has changed from paper materials to digital materials, from two-dimensional (2D) drawings to three-dimensional (3D) models, and it is changing from geometric model to information model. ICOMOS (2003) has recommended the need of further structural analysis and management of the architectural heritage on the basis of geometric documentation. That is, a complete documentation of architectural heritage requires a unified information platform with 3D geometric model and additional semantic, materials, and relationship information. It not only records the graphic and non-graphic aspects of the structure, but it also provides the base for further analysis and management.

However, current heritage projects are still struggling with such data management and information exchange, especially for the complex structures. For example, 3D geometric information is acquired by photogrammetry and laser scanning equipment, algorithms, and software; the attributes, material, and relationship information can be added and managed in Building Information Modelling and Geographic Information System environment; and, further structural analysis is conducted in civil engineering software. The separate processing, the software incompatibility (not interoperable), and data exchange increase the difficulty about heritage conservation analysis among multi-field specialists.

1.1. Documentation of Beam Frame Structure

The above mentioned problem is particularly true for timber roof structures (Yang et al., 2017). Timber roof structure is the typical architectural style in mostly Asian historic buildings, and it is generally supported by a beam frame system. The connected and joint beams are organized as a structural system to sustain the load bearing of the roofs. However, for the wooden beam frame supporting the building roofs, the bearing mechanisms still have not received the due attention and consideration that they certainly deserve. Semplici and Tampone (2006) explored the widely existing historic timber architectures and load bearing structures in the UNESCO World Heritage List. They reported that timber beam structures were not the object of conservation and appropriate repair in many countries and were suffering from neglect and alteration.

Structural analysis is the only way to assess the structural condition and load-bearing capacity of the beam structure (Chapman et al., 2006; Sanchez-Aparicio et al., 2015; Sanchez-Aparicio et al., 2016), which is extremely important to the building roof conservation. The structural analysis depends highly on how the as-built model is close to the real situation, including various aspects of input parameters (geometry, materials, and joint relationship). That is, both reality-based geometry modelling and structural analysis of the beam frame system need to be addressed for the conservation of historic timber roof. On the one hand, the reality-based remote sensing data can model the accurate geometry model of the current condition of the heritage and monitor the subsequent changes. The typical beam frames have been geometrically modelled from reality-based data, such as the roof of historical castle (Koehl et al., 2015; Bertolini-Cestari et al., 2016) and towers (Leonov et al. 2015). On the other hand, Finite Element Method (FEM) based computational software can conduct structural analysis by introducing 3D geometry files that are obtained by reality-based modelling (Armesto et al., 2009, 2015), provided that the beam connection has been built.

Currently, it is no longer a problem to obtain an accurate 3D geometric model, either by terrestrial laser scanning or dense image matching approaches, focusing on the outline structures and measurement information for the real object. However, it is still highly anticipated to realize semantic segmentation of sub-elements and to build their connection information (Díaz-Vilariño et al., 2015). When it comes to timber beam frame structure, a detailed model of the beam frame should provide the material information and spatial relationships as joints, which is important for further analysis and heritage conservation. Therefore, the gap (shown in Figure 1) between the geometry modelling and structural analysis needs a 3D information model with segmented beam elements and their connection information.

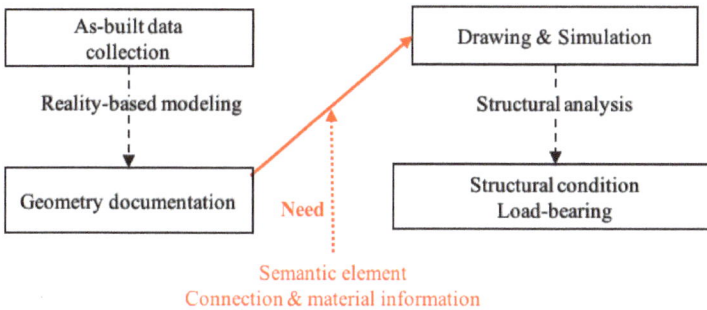

Figure 1. The separation between remote sensing reality-based modelling and civil engineering structural analysis dealing with timber beam frame.

1.2. As-Built Building Information Modelling/Management

The recently developed Building Information Modelling (BIM) technique is in accordance with the above mentioned need (Park, 2011; Chi et al., 2015) about the uniform platform for geometric and parametric modelling, connection information management, ,and structural analysis of the beam framed roof structure. BIM contributes to the creation of a digital representation having all of the physical and functional building characteristics in several dimensions, as e.g., XYZ (3D), time and non-architectural information that are necessary for construction and management of building and its elements. BIM technique can help to segment and parameterize the entity-based geometric reconstruction to element-based model enriched with measurement, semantic, attribute, relationship, and dynamic information (Yang et al. 2016).

With recent developments in "as-built" BIM techniques, the well-known traditional geometry model is increasingly developing to information model. Some BIM software have provided a platform for manual semantic modelling on reference of reality-based data. Once the elements have been parametrically created, they are simultaneously linked together with strict spatial relationships. The spatial relationships are fixed, even if the sizes or shapes of the elements change. The new model offers us a uniform platform for the whole information representing the heritage and further structural and material analysis (Barazzetti et al. 2015; Murphy et al. 2013; Saygi and Remondino, 2013).

1.3. Autodesk Revit

Autodesk Revit, a widely used software in most practices, was one of the BIM-related computer program to pay attention on the reverse engineering based as-built parametric modelling. *Revit structure*, as one of its typical workspaces, is also specialized software with the necessary theoretical model to analyse the global structure of the frame. Thus it combines the geometric modelling, element parameterization, connection information management, and structural analysis together, which avoids the transformation among diverse platforms and data formats.

Autodesk Revit has been well supportive of point cloud to aid the reality-based parametric modelling process, by direct manual family creation (Garagnani and Manferdini, 2013) or commercial plugins (Klein et al., 2015) such as *Scan-to-BIM* and *Leica CloudWorx*. However, the manual elements segmentation and parameterization process is time-consuming, especially when addressing complex structures. The plugins are expensive and address to regular buildings and the manual creation becomes time-consuming with the complex 3D space distribution of the elements.

Fortunately, *Revit* provides a rich and powerful *.Net Application Program Interface* (API), with some free and friendly tools to the development process, such as *Revit SDK*[1], *RevitLookup*[2], and *AddinManager*. *Revit SDK* is the *Revit API* document, which provides the *API* class name and function methods. *RevitLookup* interactively and visually look up the built-in information in *Revit*. The digitalized parameters link the *Revit* and developed plugins together to interactively process the elements. *AddinManager* provide the totally interactive processing between *Revit* and API program in real-time.

The *Revit API* combines the BIM parametric modelling and programming functions (Table 1). The interacting programming methods offer designers the ability to interactively design and manipulate *Revit* elements using algorithms and computational logic. *Revit* can provide the UI platform, serves as the basic view platform and database, and parametrically represents the element and builds the relationship automatically. The program can reduce the manual operation, and realize automatic and batch processing aiming at specific functions. Besides, the existing algorithms and libraries can be combined and called directly in the developed program to improve the point cloud processing. Therefore, they can

---

[1]   http://www.autodesk.com/revit-sdk
[2]   https://github.com/jeremytammik/RevitLookup

simultaneously automate the element segmentation and parametric representation procedure in BIM environment by specific functions.

Table 1. Revit Application Program Interface (API) development.

	Revit **Software**	*API* **Programming Functions**
Merits	UI platform Viewing platform Information storage and management Three-dimensional (3D) block representation Automatic relationship building	Reduced manual operation Automatic and batch processing Specific functions Calling existing algorithms External library (PCL, OpenCV, NumPy.)
Limits	Low efficiency Accurate position information Reality-based segmentation	Information storage Relationship management Parametric modification

1.4. Research Aims

Bassier et al. (2016) noted the important role of BIM from scan data to structural analysis model for heritage timber roof structures. They utilized the BIM technique to connect the geometry model (*SolidWorks ScanTO3D*) and structural analysis (*ANSYS*) (Figure 1). That is, *Revit* is utilized to store the parametric information about the beams and build their relationship, while the geometric modelling and analysis needed to be conducted in other environments.

The goal of our research is to explore the potential of such uniform platform for heritage documentation and management (Figure 2). Total station and terrestrial laser scanning data is utilized in situ to create the parametric models in *Revit* environment. The *Revit* platform serves as the visualization platform, spatial database, and the base for structural analysis. A *Revit* plug-in is in preparation to automate the semantic segmentation and parametric modelling process.

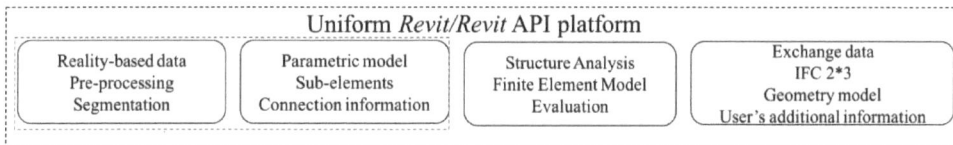

Uniform *Revit/Revit* API platform

| Reality-based data
Pre-processing
Segmentation | Parametric model
Sub-elements
Connection information | Structure Analysis
Finite Element Model
Evaluation | Exchange data
IFC 2*3
Geometry model
User's additional information |

Figure 2. Uniform Building Information Modelling (BIM) environment for beam frame structure construction and analysis.

2. Case Study and Data Source

2.1. Study Area

The case study is a historical building roof with wooden beam framed structure, the so-called "Castle of Haut-Kœnigsbourg", Alsace, France (Figure 3). It is a medieval castle and has been restored (from 1900–1908) following a close study of the remaining walls, archives, and other fortified castles that were built at the same period. It reflects the romantic nationalist ideas of the past and has been officially designated as a national historic site by the French Ministry of Culture.

The timber roof is supported by a beam frame and truss structure (Figure 3). The beams are normally leaning and oblique distributed in the 3D space. The beams are of very regular shape and are not broken, which makes a total station based approach feasible.

Figure 3. Castle of Haut-Kœnigsbourg (top left) and details of its truss structure (top right and bottom).

2.2. Total Station Data Source

In this project, conventional manual measurement surveying data was utilized. The beams were identified through a total station recording (Leica TS02), taking measurements on the edges of them. The project was georeferenced using GNSS reference points around the site in order to define a consolidated geodetic network (Figure 4). For this manual survey, we needed 18 surveying stations. The field operation lasted six days. Totally, 1,710 *XYZ* points (Figure 5) were collected in order to obtain the 3D model. At least six points were acquired on each beam, located in the three or four parallel edges of this beam.

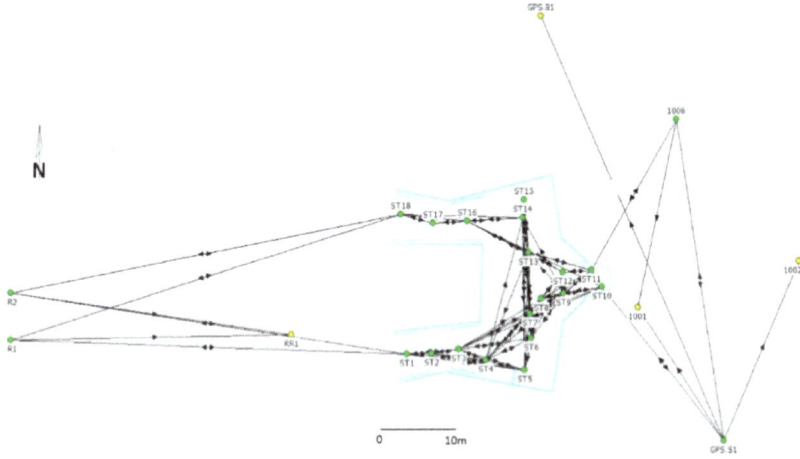

Figure 4. Geodetic network of the total station recording.

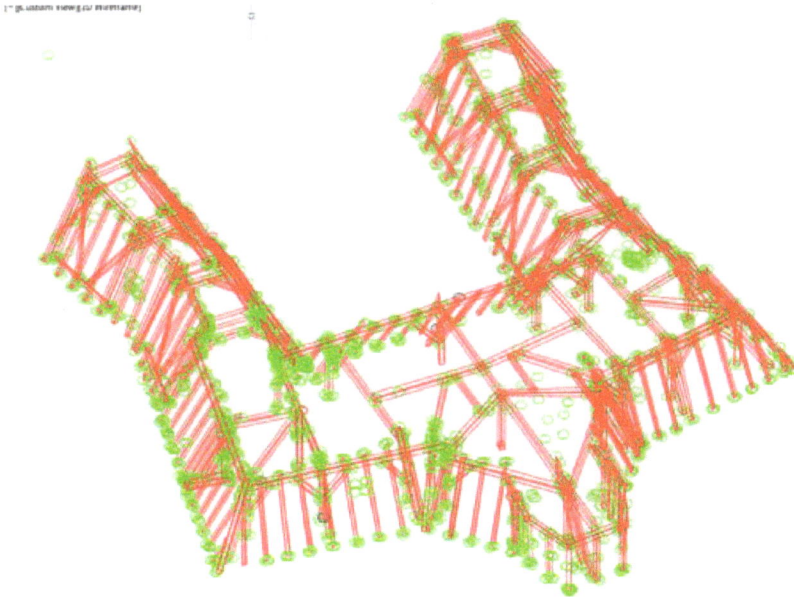

Figure 5. The total station data recording of the beams (AutoCAD). Each beam collects at least six total station points, and the green circles (left) show the distribution of the original total station points.

For the detail, the position recorded whether the collected point was the corner of the beam. In horizontal beams, the points are labelled with "n" (north), "s" (south), "e" (east) or "o" (ouest, means "west" in French) to display its position in

313

the cardinal oriented beam edge. For vertical beams, the points are labelled with "h" (haut means top) or "b" (bottom). The oblique beams are labelled with the combination of "h" or "b" and "n", "s", "e", "o". Some typical points are also marked with additional "p" to identify it as points that were to be prolonged on the edge (Figure 6). Anyway, when collecting the total station points, it is difficult for the operator to identify correctly the end of the beams, which can be a source of error, therefore a strict codification has been used during the recording. The collected data such as point IDs and XYZ coordinates were saved in ASCII text formats. The ID indicates the selected beam, the point position, and edge number.

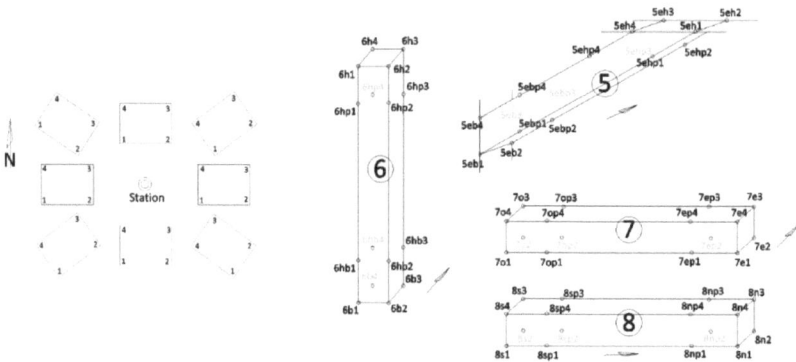

Figure 6. Point numbering and codification used by total station survey.

2.3. Terrestrial Laser Scanning

The laser scanning directly captures the 3D geometric information of the object, which provides a highly detailed and accurate representation of the shapes. Laser scanning is one of the intense developments in geomatics, and has been established as a standard tool for built heritage reconstruction. The pipeline of accurate geometric modelling derived from laser scanning generally consists of data acquisition, point cloud registration, segmentation, mesh generation, and texture mapping. In this project, point clouds are the data source for further parametric modelling in *Revit*. So, just point cloud registration and segmentation is needed for the acquired data. The obtained point cloud is shown in Figure 7.

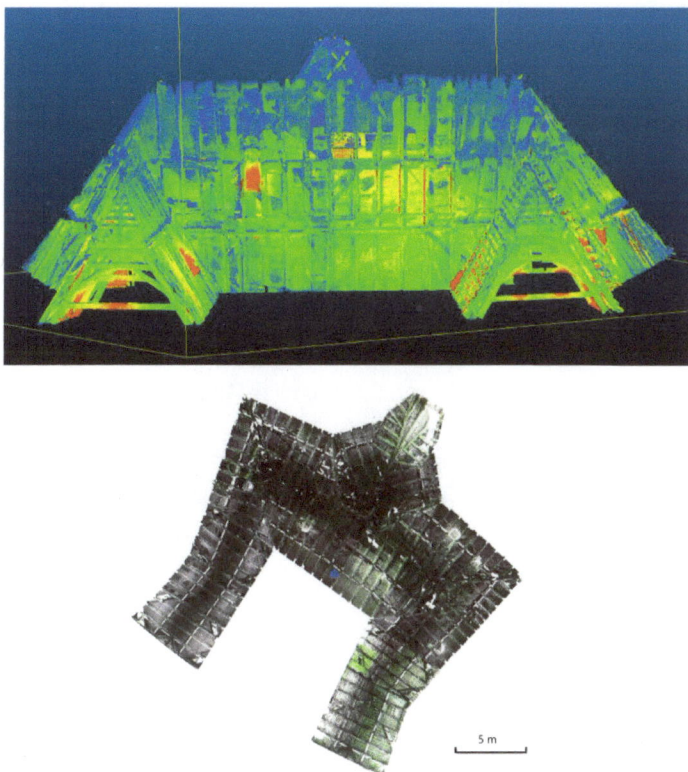

Figure 7. Terrestrial laser scanning data (left) serves as spatial reference in *Revit* (right).

3. Manual Parametric Modelling

BIM has been widely employed to obtain the parametric solid model by manual element creation. Based on the different data sources made of total station points and terrestrial laser scanning point clouds, two parametric models are created.

3.1. Wireframe Model from Total Station Points

When compared to datasets that are based on point clouds acquired by terrestrial laser scanning, the number of points that are recorded by total station is very low but each point is significant. Total stations deliver highly accurate single-points, often used as 3D surface reconstruction reference points or control points for other techniques. Although a limited number of measured points do not allow for a detailed study of the structure, it is still easy to rebuild wireframe rectangular beams.

315

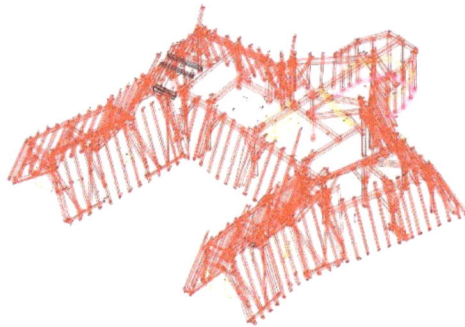

Figure 8. Wireframe model in *AutoCAD*.

Despite the increasing supporting of reality-based as-built data, *Revit* cannot directly support the ASCII format file with total station points. Thus, *AutoCAD* is employed to build the parametric model by manually connecting the beam edges. This modelling was very time consuming. About 400 beams were reconstructed (Figure 9). The obtained beam system is a linear wire-frame model (Figure 8). The wireframe model cannot define relationship and variable parameters that are describing the beam elements. The obtained wireframe model is supported for further BIM application and structural analysis in *Revit* and computational software.

3.2. Manual Scan-To-BIM Process

Laser scanning and photogrammetry can rapidly capture the most accurate documentation of reality, serving as the data source of as-built BIM. Currently, the geometric model can be accurately created from the surface point cloud, even for the irregular built heritage. Yet, the problem is how to convert the geometric model to a semantic model on which additional attribute information can be attached. As-built BIM software provides such a platform to manually create this kind of parametric model with the reference of point cloud. The friendly support for 3D point clouds has currently made *Revit* the main reality-based parametric modelling environment.

The point cloud can be directly loaded in Revit environment and three reference levels are created to confirm the location (Figure 9). The final parametric model is created based on the *Revit* beam family on the reference of point cloud (Figure 10 and Figure 11). Compared to the wire-frame model (Figure 6), the BIM solid model provides not only a definite relationship and variable parameters describing the beam elements, but also more reality results in terms of geometry and structure behavior (Bassier, 2016).

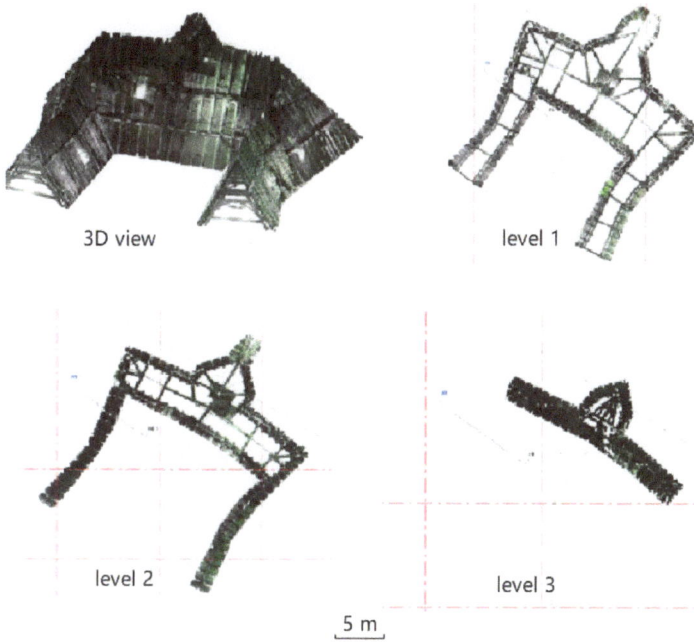

Figure 9. Different reference levels to model the beams.

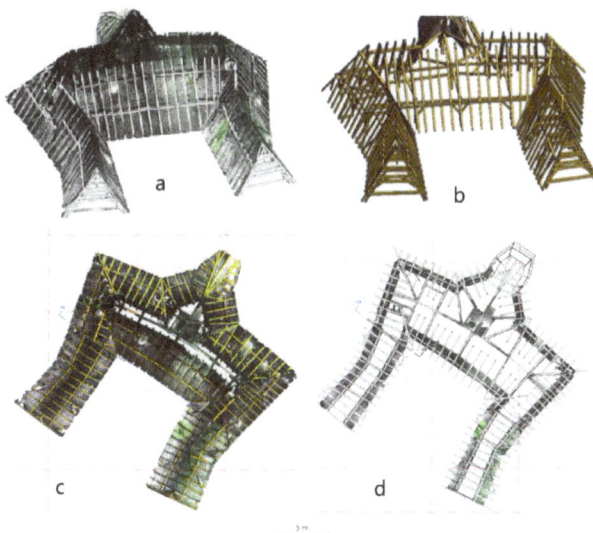

Figure 10. Parametric beam frame model from point cloud. (**a**) 3D viewer with reference point cloud (**b**) 3D frame (**c**) two-dimensional (2D) viewer with reference point cloud (**d**) 2D frame.

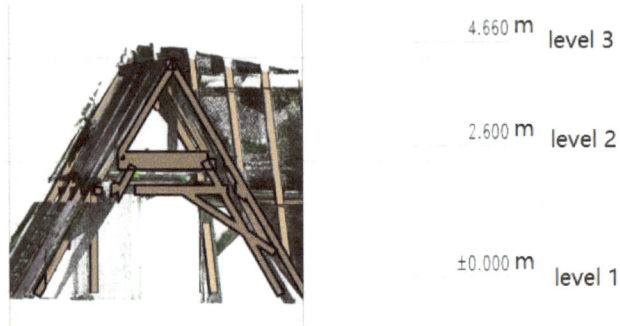

Figure 11. Local detailed information.

The complexity of the historic timber beam frame lies in the sloped and crossing distribution, instead of the geometry that tends to be rectangular based regular shape. However, most BIM software (including *Revit*) is not a 3D-centric and free-form geometry modeller. They mostly create the 3D model based on the stretch of prior defined position in 2D plane. It is feasible for the modern steel truss system, the beams are basically in horizontal and vertical distribution, and areperpendicularly connected. As for the sloped timber beam frame, the parametric model is created by alternating 2D plane positioning and 3D space drawing with prior defined angle value.

4. Revit API Implementation

Currently, Scan-to-BIM is widely accepted for constructing the parametric model directly, thus avoiding the transfer from geometry model and element segmentation to parametric and semantic model. However, the process is mostly finished manually in time-consuming ways, as described in Section 3.

In our project, the "Beam Frame Modelling" package is setup in order to characterize the geometry of timber roof structures from total station surveying and TLS-based point cloud. This plugin consists in a *Microsoft Windows* .Net 4.5 Dynamic Link Library (DLL), developed in C# by means of *Revit API* 2017. It is a *Revit API* plugin that can run automatically in *Revit* interface and conducts further processing with *Revit* functions. The plugin consists of two parts, dealing with the two kinds of datasets from total station points and laser scanning point cloud. The obtained model is defined and managed in the unique BIM environment with the framework of geometry, attribute and spatial relationship knowledge.

The current package focuses on the total station data processing and the next version, including point cloud processing parts is in preparation. The current version realizes the total station related functions: displaying total station data, cleaning point data, parameter and corner calculation, and parametric beam generation. If a timber roof structure is studied, then the plugin allows for the

318

geometry reconstruction of the beam frame from ASCII text files, and outputs the IFC format data.

4.1. Rectangular Beam Geometry Construction

Although the field work that was based on the total station points was time consuming, the beam construction process could be finished fast with a limited number of accurate points. It is expected to transfer the disordered points to parameters describing the cuboid beams. A cuboid beam can be described by a central point, three directions and extensions (Figure 12d), which is the basis to create the beam element in *Revit*. As the points are not always the corners, two parameters of the beam are not totally confirmed: one is the circle point and another is the height.

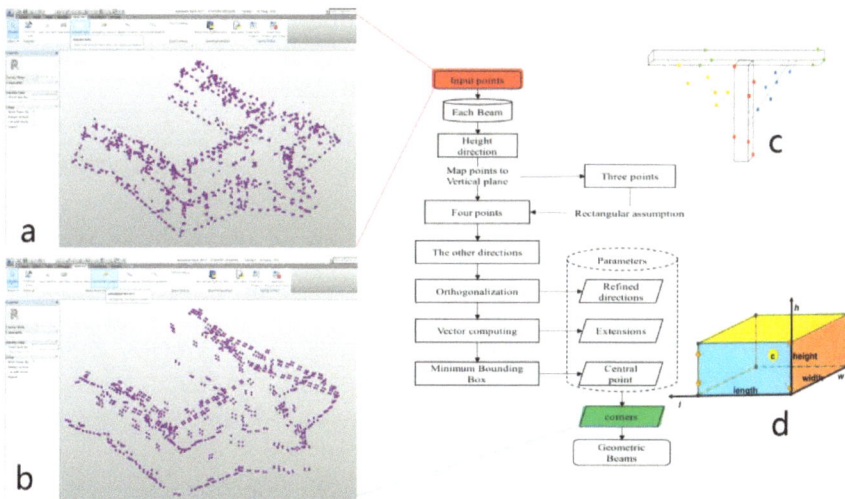

Figure 12. Beam reconstruction workflow (**a**) Initial total station points displayed in Revit after data cleaning; (**b**) Beam corners calculated by the algorithm in workflow; (**c**) Simplified description of the distribution of the initial points; and, (**d**) Rectangular beam description by central point, extensions, and directions.

A *Python* program has been developed to calculate the beam parameters based on the following algorithm workflow (Figure 12). Firstly, we can obtain the direction of height (*h*), considering that collected total station points are located in a set of parallel edges. Then, mapping the points to the vertical plane along with the height direction, we can obtain three (if the points are located in the three out of four edges) or four corners (if the points are situated in the whole four edges). In the former case, we can calculate the fourth point under the rectangular assumption of the beam shape. The other two directions (*l* & *w*) are thus confirmed. The three obtained directions are further refined because they may be not vertical to each other, owing to the error of total station points. Finally, the

319

extension can be calculated by vector computing. The circle point is confirmed by the minimum bounding box of the total station points. After constructing the blocks, the relationship and additional information need to be added and managed, which is the advantage of BIM technique.

Displaying total station data. This plugin is able to conveniently parse ASCII text files containing *XYZ* coordinates derived from real beam frame total station data capture, translating them into native reference points Revit's mass modelling environment.

Cleaning point data. Some total station data are redundant and mistaken, which then need to be eliminated before the beam parameter calculating. Inputting total station points is shown in Figure 12a, which are disordered and difficult to transfer to Revit beam structure directly.

Parameter and corner calculation. The parameters and corners are calculated by the proposed workflow in Figure 12. The algorithm is also a built-in function in the developed plugin. It can automatically transfer all of the disordered points to the corners of beam structures (Figure 12b).

Parametric beam generation. A Revit "Beam family" is created firstly, which is in the regular rectangular shape and basic wooden materials. Then, the family instances can be created in the central position (based on the calculated parameters) along the direction confirmed by the corners. The beam elements (Figure 13) are in standard BIM parametric type, which can be modified and exchanged either by the API or users.

The results show the potential of automating the parametric modelling by interactive API development in BIM environment. It also integrates the separate data processing and different platforms into the uniform *Revit* software. *Revit* BIM environment provides the attributional material and construction information management and structural analysis based on the obtained model.

The parametric model from total station is the pseudo solid model from wireframe model. Although the total station points provide the highest single point positioning accuracy, it is difficult to accurately describe the surface and geometry of the beam element. It can hardly monitor the dynamic change of the beam frame when considering the huge time cost.

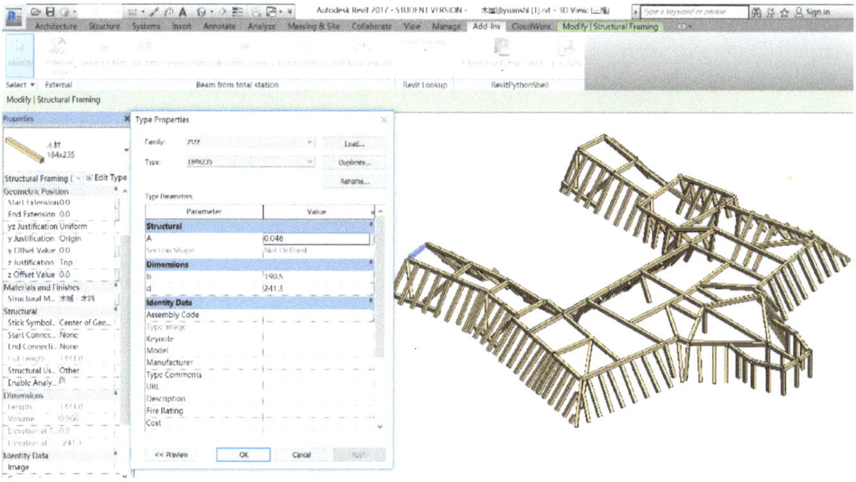

Figure 13. Display of the Parametric Beam frame system in Revit.

4.2. Scan-To-BIM Plugin

Revit 3D modelling tools focus primarily on delivering 2D documents with the added 3D capability. When creating complicated geometry and handling large amounts of data in 3D space directly, the user's freeform design is limited. This dramatically reduces project productivity and the accuracy of arbitrary tip-tilted distribution of beam frame.

The manual elements (Section 3.2) segmentation and parameterization process is time-consuming, especially for the historic timber truss system with sloped and crossing beams. The scan-to-BIM plugin for timber beam frame parametric modelling from point cloud is expected and is currently in preparation. The *API* development really maintains the need of beam construction in two ways: the first one is obviously the automation, and the other is the possible improvement of accuracy. The *API* directly calculate the parameters of parametric beam and then manipulate the Revit beam family with obtained parameters, which can return more accurate spatial location than manual drawing. Moreover, the regular geometric shape of the beam element makes the automatic segmentation feasible.

Some point cloud segmentation algorithms have been proposed for regular steel beam frame (Laefer et al., 2017) and columns (Díaz-Vilariño et al., 2015). Our developing plugin is expected to realize the semantic segmentation and parametric modelling in the BIM environment for sloped timber beam frame. Currently, we are trying to obtain the total station like-wise wire-framed model by feature points choosing from dense point cloud. The developed plugin addressed total station points, which are the sparse point cloud with specific spatial position. That is, the plugin can be utilized for dense point cloud if a certain number of points can be selected and define the edges. The workflow consists of following steps: (1) straight line detection by 3D Hough transformation to obtain the candidate edges

321

in point cloud space; (2) nearest neighbourhood matching to merge the wire-frame semantic model and point cloud together; and, (3) beam edge selection among the candidate straight lines based on the distance judgment between the straight lines and the obtained model. (4) the developed plugin can be utilized once the three or four edges for each beam have been selected.

5. Conclusions

The main objective of the project presented in this chapter is to explore the potential of a parametric modelling tool in as-built BIM environment for a timber roof structure whose elements are leaning and crossing beam frame. The tool does not only automate the semantic and parametric modelling, but also integrates together the traditionally separate geometric modelling, parametric elements management, and structural analysis.

BIM software, typically as *Autodesk Revit*, is increasingly supportive of reality-based data, normally called *Scan-to-BIM* process. Aiming at the timber beam frame, different parametric models are built from traditional total station points and terrestrial laser scanning data. The current element parametric procedure is generally manual, as the obtained model in Section 3.

The manual process ignores the existing semantic modelling and point cloud processing algorithms and libraries. The *Revit API* provides an interactive environment to utilize *Revit* functions and to develop specific algorithms simultaneously. The API development can manipulate the geometric/semantic modelling and structural analysis directly in BIM environment as well. For example, beam elements are traditionally constructed first, and their joint and interconnected relationship need to be judged and analyzed further, while the relationship is the core for BIM technique and the related functions are available within API development. Thus, the integrated environment not only avoids the data exchange, but also reduces some inefficient and repetitive work.

In conclusion, current reality-based modelling and structural analysis techniques and platforms are important for verifying the actual load-bearing capacity and structural safety of historic timber roof structures. The BIM provides the possibility to combine the separate processing, the software incompatibility, and data homogeneity to the uniform platform. The BIM based parametric modelling of a timber roof structure can provide: (i) an accurate complete survey on the geometry aspect; (ii) attribute, material, and relationship information of the sub-elements; (iii) possible deformations and changes over time; and, (iv) load-bearing and structural analysis. Generally, it provides the conservation professionals decision support with spatial, temporal, and multi-criteria analysis.

The Plugin is still in preparation. The current version is just supportive of total station data. Ongoing development focuses on the laser scanning data to obtain accurate modelling by combining *Revit API* with *PCL* functions.

The automatic path from a point cloud to a BIM structure is certainly an interesting goal, but remains very complex. Nevertheless, the implementation of an automatic process allowing for the use of a reduced set of significant points (from topographic measures) allows the user to foresee an intermediate stage using the 3D TLS point clouds. By using the TLS point cloud as reference recording, a digitization following the principles of the codification developed for the total station survey allows the user to extract the reduced significant point cloud, which will afterwards allow an automatic reconstruction of the beams and the beam structure in a BIM environment.

The intersection and crossing relationship among the beam frame are there still being addressed. The final frame consists of parametric beam elements, which can be applied to structural analysis.

Acknowledgments: This research is supported by the China Scholarship Council (Grant No.201504490008). We would like to thank Katia MIRANDE and Marie-Anaïs DHONT, Benjamin CHEREL and Pierre ARNOLD, Master Students of INSA Strasbourg for the organisation of the collection of the total station points (Figures 4 to 6) and the drawing of the beam structure in *AutoCAD* (Figure 8).

References

1. Arias, P.; Carlos Caamaño, J.; Lorenzo, H.; Armesto, J. 3D modeling and section properties of ancient irregular timber structures by means of digital photogrammetry. *Comput. Aided Civ. Infrastruct. Eng.* **2007**, *22*, 597–611.
2. Armesto, J.; Lubowiecka, I.; Ordóñez, C.; Rial, F.I. FEM modeling of structures based on close range digital photogrammetry. *Autom. Constr.* **2009**, *18*, 559–569.
3. Barazzetti, L.; Banfi, F.; Brumana, R.; Gusmeroli, G.; Previtali, M.; Schiantarelli, G. Cloud-to-BIM-to-FEM: Structural simulation with accurate historic BIM from laser scans. *Simul. Model. Pract. Theory* **2015**, *57*, 71–87.
4. Bassier, M.; Hadjidemetriou, G.; Vergauwen, M.; Van Roy, N.; Verstrynge, E. Implementation of Scan-to-BIM and FEM for the documentation and analysis of heritage timber roof structures. In *Euro-Mediterranean Conference*; Springer: Cham, Switzerland, 2016; pp. 79–90.
5. Bertolini-Cestari, C.; Invernizzi, S.; Marzi, T.; Spano, A. Numerical survey, analysis and assessment of past interventions on historical timber structures: The roof of valentino castle. *Wiad. Konserw.* **2016**, *45*, 87–97.
6. Cabaleiro, M.; Riveiro, B.; Arias, P.; Caamaño, J.C.; Vilán, J.A. Automatic 3D modelling of metal frame connections from LiDAR data for structural engineering purposes. *ISPRS J. Photogramm. Remote Sens.* **2014**, *96*, 47–56.
7. Chapman, M.J.; Norton, B.; Taylor, J.M.A.; Lavery, D.J. The reduction in errors associated with ultrasonic non-destructive testing of timber arising from differential pressure on and movement of transducers. *Constr. Build. Mater.* **2006**, *20*, 841–848.

8. Chi, H.L.; Wang, X.; Jiao, Y. BIM-enabled structural design: impacts and future developments in structural modelling, analysis and optimisation processes. *Arch. Comput. Methods Eng.* **2015**, *22*, 135–151.

9. Díaz-Vilariño, L.; Conde, B.; Lagüela, S.; Lorenzo, H. Automatic detection and segmentation of columns in as-built buildings from point clouds. *Remote Sens.* **2015**, *7*, 15651–15667.

10. Garagnani, S.; Manferdini, A.M. Parametric accuracy: Building Information Modeling process applied to the cultural heritage preservation. *Int. Arch. Photogramm. Remote Sens. Spat. Inf. Sci.* **2013**, *XL-5/W1*, 87–92.

11. ICOMOS, 2003. Recommendations for the Analysis, Conservation and Structural Restoration of Architectural Heritage. 37 Pages. Available online: http://www.icomos.org/en/about-the-centre/179-articles-en-francais/ressources/charters-and-standards/165-icomos-charter-principles-for-the-analysis-conservation-and-structural-restoration-of-architectural-heritage (accessed on 22 January 2017).

12. Klein, L.; Li, N.; Becerik-Gerber, B. Imaged-based verification of as-built documentation of operational buildings. *Autom. Constr.* **2012**, *21*, 161–171.

13. Koehl, M.; Viale, A.; Reeb, S. A Historical timber frame model for diagnosis and documentation before building restoration. *Int. J. 3-D Inf. Model.* **2015**, *4*, 34–63.

14. Laefer, D.F.; Truong-Hong, L. Toward automatic generation of 3D steel structures for building information modelling. *Autom. Constr.* **2017**, *74*, 66–77.

15. Leonov, A.V.; Anikushkin, M.N.; Ivanov, A.V.; Ovcharov, S.V.; Bobkov, A.E.; Baturin, Y.M. Laser scanning and 3D modeling of the Shukhov hyperboloid tower in Moscow. *J. Cult. Herit.* **2015**, *16*, 551–559.

16. Murphy, M.; McGovern, E.; Pavia, S. Historic Building Information Modelling–Adding intelligence to laser and image based surveys of European classical architecture. *ISPRS J. Photogramm. Remote Sens.* **2013**, *76*, 89–102.

17. Park, J. BIM-based parametric design methodology for modernized Korean traditional buildings. *J. Asian Archit. Build. Eng.* **2011**, *10*, 327–334.

18. Quattrini, R.; Malinverni, E.S.; Clini, P.; Nespeca, R.; Orlietti, E. From TLS to HBIM. High quality semantically-aware 3D modeling of complex architecture. *Int. Arch. Photogramm. Remote Sens. Spat. Inf. Sci.* **2015**, *40*, 367–274.

19. Sánchez-Aparicio, L.J.; Ramos, L.F.; Sena-Cruz, J.; Barros, J.O.; Riveiro, B. Experimental and numerical approaches for structural assessment in new footbridge designs (SFRSCC-GFPR hybrid structure). *Compos. Struct.* **2015**, *134*, 95–105.

20. Sánchezaparicio, L.; Villarino, A.; Garcíagago, J.; Gonzálezaguilera, D. Photogrammetric, geometrical, and numerical strategies to evaluate initial and current conditions in historical constructions: A test case in the church of San Lorenzo (Zamora, Spain). *Remote Sens.* **2016**, *8*, 60.

21. Saygi, G.; Remondino, F. Management of Architectural Heritage Information in BIM and GIS: State-of-the-art and Future Perspectives. *Int. J. Herit. Digit. Era* **2013**, *2*, 695–713.

22. Semplici, M.; Tampone, G. Timber Structures and Architectures in Seismic Prone Areas in the UNESCO World Heritage List (Progress report). In Proceedings of the 15th Symposium of the IIWC, Istanbul, Turkey, 20 September 2006; 10p.

23. Tampone, G.; Ruggieri, N. State-of-the-art technology on conservation of ancient roofs with timber structure. *J. Cult. Herit.* **2016**, *22*, 1019–1027.

24. Yang, X.; Koehl, M.; Grussenmeyer, P. Parametric modelling of as-built beam framed structure in BIM environment. *Int. Arch. Photogramm. Remote Sens. Spat. Inf. Sci.* **2017**, *XLII-2/W3*, 651–657.

25. Yang, X.; Koehl, M.; Grussenmeyer, P.; Macher, H. Complementarity of Historic Building Information Modelling and Geographic Information Systems. *Int. Arch. Photogramm. Remote Sens. Spat. Inf. Sci.* **2016**, *XLI-B5*, 437–443.

Yang, X.; Koehl, M.; Grussenmeyer, P. Automating Parametric Modelling From Reality-Based Data by Revit Api Development. In *Latest Developments in Reality-Based 3D Surveying and Modelling*; Remondino, F., Georgopoulos, A., González-Aguilera, D., Agrafiotis, P., Eds.; MDPI: Basel, Switzerland, 2018; pp. 307–325.

Development of an Open Source Spatial DBMS for a FOSS BIM

Sotiris Logothetis, Elena Valari, Eleni Karachaliou and
Efstratios Stylianidis

Faculty of Engineering, School of Spatial Planning & Development, Aristotle University,
Thessaloniki 54124, Greece; slogothet@auth.gr; evalarig@auth.gr; ekaracha@auth.gr;
sstyl@auth.gr

Abstract: Building information modelling (BIM) is currently a widely known and
used technology in many application areas, including architecture, civil
engineering, electrical and mechanical engineering, facility management, cultural
heritage management, etc. Nevertheless, in the times that the development of free
and open source software (FOSS) has been rapidly growing and their use tends to
be consolidated, there is a lack of integrated open source platforms able to
support historic building information modelling (HBIM) processes. The present
research aims to use a FOSS computer-aided design (CAD) environment in order
to develop an ecosystem of BIM plugins that will facilitate a comprehensive and
integrated BIM model analysis through a FOSS solution. The first step towards
this project concerns the development of an open source spatial database
management system (DBMS) to support data management.

Keywords: BIM; FOSS; DBMS; CAD; HBIM

1. Introduction

1.1. From CAD to BIM

In the early 1960s, information technology penetrated the field of AEC
(Architecture, Engineering, Construction), and introduced computer-aided design
(CAD) systems to the design process. The typical approach to representing a
model on a paper by hand was replaced by the use of digital and advanced CAD
tools, which reinforced the creation, modification, analysis, or optimisation of
design models (Garagnani and Manferdini, 2013). The world's first CAD software
(Sketchpad) was developed in 1963 by Ivan Sutherland (Turing Award in 1988) in
the course of his PhD (Sutherland, 2003). The developments that followed in the
CAD software industry allowed professionals and scientists to use specialized
software in order to represent any object with graphics, create databases for storing
the models, easily generate accurate graphical models of the object, manage

complex design analysis in a short time, store and recall the model, and modify it (Bilalis, 2000).

For many years, CAD systems used in AEC were implementing only 2D modelling, and they were able to process entity objects as graphic symbols, representing only the geometrical properties of each element (Ibrahim and Krawczyk, 2003). Building modelling based on 3D solid modelling was first developed in the late 1970s and early 1980s. It was then that the available CAD systems of that period started to develop their capabilities, and the concept of object-based parametric modelling came into light by representing objects by parameters and rules which determined the geometry along with non-geometric properties and features (Eastman et al., 2008). Nowadays, the parametric modelling in the field of AEC is mostly being represented by the so-called "BIM models". Building information modelling (BIM) models face the object entity not only as a graphical symbol with fixed geometry and properties, but as an integrated entity allowing different levels of information (geometric and non-geometric) to be included, as well as pointing to the relations among all the items of the model (Czmocha and Pekala, 2014).

BIM often is considered as the next generation of the CAD systems—an advanced approach which extends the capability of the traditional design methodology by applying and defining intelligent relationships between the elements in the designed model (Garagnani and Manferdini, 2013).

1.2. The BIM Market

According to MarketsandMarkets (Web 1), the BIM market is expected to grow from USD 3.16 Billion in 2016 to USD 7.64 Billion by 2022, at a compound annual growth rate (CAGR) of 16.51% during the forecast period. The communication and coordination through the asset lifecycle management process and government directives for the use of BIM are two of the major drivers for the BIM market. Nevertheless, high costs of BIM (commercial) software and long training are the key factors that are limiting the BIM market growth.

Amongst the main BIM market applications, the building applications are the first. The broad usage of BIM in buildings' construction is driving the market. Between 2016 and 2022, the industrial applications' sector is expected to reach the second-largest share in the BIM market. With respect to the geographic regions, the MarketsandMarkets report underlines that the market in the Asia-Pacific region is expected to grow at the highest CAGR during the chosen period.

The report ranks the following as major players in the BIM market: Autodesk Inc. (U.S.), Nemetschek SE (Germany), Trimble Navigation Limited (U.S.), Bentley System, Inc. (U.S.), Asite Ltd. (U.K.), AVEVA (U.S.), RIB Software AG (Germany), Dassault Systèmes (France), Archidata Inc. (Canada), Intergraph Corporation

(U.S.), Beck Technology, Ltd. (U.S.), Computers and Structures, Inc. (U.S.), Robert McNeel & Associates (U.S.), and Cadsoft (U.S.).

2. BIM Systems and Applications

2.1. BIM Description

The "BIM" concept has been expressed through several definitions in the literature, and the concept of modelling building information has been related to widely used terms (i.e., Asset Lifecycle Information System, Building Product Models, Integrated Design Systems, Integrated Project Delivery, etc.) (Succar, 2009).

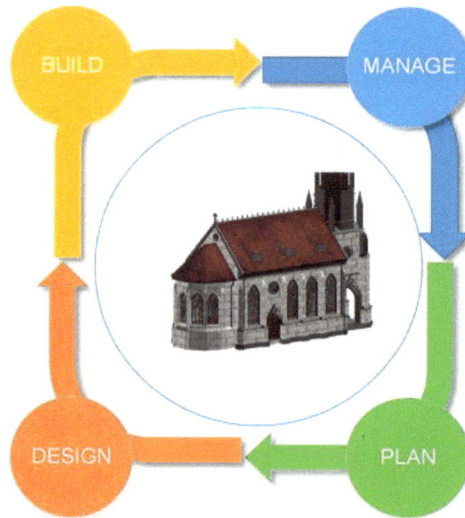

Figure 1. Building information modelling (BIM) concept.

From a technical point of view, BIM can be characterized as a five-dimensions (5D) digital representation of a building, structure, or environment with its physical and functional characteristics. It consists of intelligent building components or characteristics of environments which include parametric rules and data attributes for each object (Hergunsel, 2011). The five dimensions are: the three (3D) primary spatial dimensions (width, height, and depth), the fourth dimension (4D) is time, and the fifth (5D) is cost.

On the other hand, BIM can also be defined as "a methodology to manage the essential building design and project data in digital format throughout the building's life-cycle" (Penttilä, 2006), as it is illustrated in Figure 1.

2.2. BIM Systems' Characteristics and Features

According to Popov et al. (2006), the most important capabilities of the BIM technology are the following:

- to ensure integration management of graphical and informational data flows;
- to combine the graphical interface with the information flows and process descriptions;
- to develop the strategy of building project design, to construction and maintenance management;
- to perform life cycle operations of a construction project faster, more effectively, and with lower cost;
- to transform individual executors into teams and decentralize tools into complex solutions; this leads to individual tasks being implemented as complex processes.

BIM characteristics can be divided into two categories. The first category refers to the data modelling that is used in order to describe information regarding the building (i.e., CAD object, parametric object modelling, etc.) while the second refers to the way that the data model is being shared or exchanged (Vanlande et al., 2008). In the case of the data exchange model, a software system holds a master copy of the data and exports data snapshots to be used by other users (i.e., create a physical file on CD/DVD). The sharing model corresponds to a centralized control of ownership, where data sharing could be succeeded through an application programming interface (API) or a standard data access interface (SDAI), providing the way users access the BIM physical file according to their rights (Vanlande et al, 2008):

- a central database, meaning that multiple applications can access the products' data and make use of the database features, such as query processing and business object creation;
- federated databases, which allow a single unified view of multiple distributed databases;
- web services, either by giving access to the central project database where the BIM is stored, or by giving access to an API.

The main features of a BIM system are:

- its ability to manage, store, share, and exchange 3D geometries;
- its capacity to carry all the information related to each object and element (geometric, descriptive, functional);
- it covers all stages of a project lifecycle, from the design to the implementation, allowing and facilitating the cooperation of the various stakeholders involved in the project;

- the provision of a reliable basis for timely and rapid decisions, assisting the involved staff to increase quality and productivity in the design but also during project implementation.

The available BIM platforms can now be distinguished in the different sets of tools, as illustrated in Table 1 (Logothetis et al., 2015).

Table 1. Types of BIM platforms.

Type of tools	Role
Tools for designing 3D models (3D modellers)	Represent the real BIM tools, working with solid parametric objects in sufficient features to virtually construct the building.
Tools for projection and visualization of models (Viewers and surface modellers)	Aid to realise how the building will "look", a surface modeller or a viewer (all shapes are empty).
Tools for calculating models' analysis	Third-party software that communicates to the main BIM tool, e.g. analyse data from the 3D modeller to determine the model's energy efficiency or day lighting (Web 2).

2.3. IFC Format and Standard

In BIM systems, the Industry Foundation Classes (IFC) data model is used as a collaboration format, intended to describe building and construction industry data (Eynon, 2016). The IFC is an ISO (International Organization for Standardization) standard that defines and describes all the elements of a building in a project. At the same time, this standard allows interoperability among stakeholders so as to share information encoded in a unique and common way (Figure 2). IFC defines how information should be provided/stored in all stages of a building project lifecycle, being able to maintain data for geometry, cost, time management, etc., for many different professions (Web 3).

The IFC model is an entity-relationship model which uses the international standard data definition language known as "EXPRESS". It consists of a set of schemas, each of which corresponds to one IFC layer. Every IFC layer includes several classes that are all subtypes of the IfcRoot class, unless they are resource classes. The three basic and fundamental classes in an IFC model are all derived from IfcRoot (Web 3):

- object classes that record object characteristics and types (IfcObject);
- relations classes which define the relationship among the objects (IfcRelationship);
- properties classes that refer to the different types of property that follow an object (IfcPropertyDefinition).

IFC is an open file format, and its data model is developed by buildingSMART: a worldwide authority driving transformation of the built environment through creation and adoption of open international standards. It was founded in 1995, and has evolved to meet the demands of the building and infrastructure sectors (Web 4).

```
1   ISO-10303-21;
2   HEADER;
3   FILE_DESCRIPTION(('ViewDefinition [CoordinationView]'),'2;1');
4   FILE_NAME('C:/Users/user/Desktop/BOOK_CHAPTER/ifc/wall.ifc','2017-05-10T18:42:19',('',
5   FILE_SCHEMA(('IFC2X3'));
6   ENDSEC;
7   DATA;
8   #1=IFCPERSON($,$,'',$,$,$,$,$);
9   #2=IFCORGANIZATION($,'',$,$,$);
10  #3=IFCPERSONANDORGANIZATION(#1,#2,$);
11  #4=IFCAPPLICATION(#2,'0.16 build 6706 (Git)','FreeCAD','118df2cf_ed21_438e_a41');
12  #5=IFCOWNERHISTORY(#3,#4,$,.ADDED.,$,#3,#4,1494441739);
13  #6=IFCDIRECTION((1.,0.,0.));
14  #7=IFCDIRECTION((0.,0.,1.));
15  #8=IFCCARTESIANPOINT((0.,0.,0.));
16  #9=IFCAXIS2PLACEMENT3D(#8,#7,#6);
17  #10=IFCDIRECTION((0.,1.,0.));
18  #11=IFCGEOMETRICREPRESENTATIONCONTEXT('Plan','Model',3,1.E-05,#9,#10);
19  #12=IFCDIMENSIONALEXPONENTS(0,0,0,0,0,0,0);
20  #13=IFCSIUNIT(*,.LENGTHUNIT.,$,.METRE.);
21  #14=IFCSIUNIT(*,.AREAUNIT.,$,.SQUARE_METRE.);
22  #15=IFCSIUNIT(*,.VOLUMEUNIT.,$,.CUBIC_METRE.);
23  #16=IFCSIUNIT(*,.PLANEANGLEUNIT.,$,.RADIAN.);
24  #17=IFCMEASUREWITHUNIT(IFCPLANEANGLEMEASURE(0.017453292519943295),#16);
25  #18=IFCCONVERSIONBASEDUNIT(#12,.PLANEANGLEUNIT.,'DEGREE',#17);
26  #19=IFCUNITASSIGNMENT((#13,#14,#15,#18));
27  #20=IFCPROJECT('ce9e2352_6796_4cf8_99d',#5,'Unnamed1',$,$,$,$,(#11),#19);
28  #21=IFCCARTESIANPOINT((-0.11,0.04));
```

Figure 2. Example of an Industry Foundation Classes (IFC) data model.

2.4. BIM Application Areas

BIM models find application predominantly in the AEC industry. They provide many benefits regarding: the 3D visualization; the fabrication; the cost estimation; the code reviews; the construction sequencing; the conflict, interference, and collision detection; the forensic analysis; the facilities management; etc. BIM models are practically useful for any kind of AEC activity and lifecycle management of a building or other structure (Azhar, 2011).

Moreover, as Succar (2009) records, BIM fields of activity can be distinguished into technology, process, and policy fields, each of them including various "players" and "deliverables". The technology field refers to all organizations that generate software, hardware and network development aiming to advance BIM applicability in AEC. BIM process field concerns all the stakeholders (architects, engineers, contractors, facility managers, etc.) that use BIM software solutions in order to procure, design, construct, manufacture, use, and manage various structures and projects. The policy field concerns all the players that do not participate in the construction but are involved in the decision-making domain (Succar 2009). The three fields interact each other.

Moreover, BIM models could be developed to support the building maintenance and deconstruction, rather than only facilitating the new construction of a building. The latest trends/challenges in the field of BIM for existing buildings follow topics such as the automation of data collection and BIM development (without a previously developed BIM model), the update and maintenance of information in BIM, and the management and modeling of unknown data, objects, and relations in existing buildings (Volk et al., 2014).

2.5. HBIM for Cultural Heritage and Challenges

As BIM technology was initially developed in order to serve the needs of the construction industry, BIM environments were structured to generate and manage from zero parametric geometries and create objects using existing libraries of shapes and conditions. Nevertheless, in the past decade, a remarkable development and use of BIM technology in the field of cultural heritage has occurred. In general, tangible or intangible cultural heritage has cultural value, and it is worth preservation in order to retain its cultural significance while ensuring its accessibility to present and future generations (Kalamarova et al., 2015). In the case of cultural heritage documentation, the objects consist of components and materials whose geometry and characteristics are not represented by libraries of typical software. Thus, the Historic BIM (HBIM) approach was developed, referring to the process by which the architectural elements are collected; e.g. using a terrestrial laser scanner, and produced photogrammetric survey data are converted into parametric objects (Dore and Murphy, 2012). In this way, it is possible to document and manage any size or complex form of a cultural heritage object (Singh et al., 2011). Moreover, the BIM approach in the cultural heritage field allows the generation of a parametric model with specific standards and protocols that can be exchanged and processed from different professionals in order to be used for various use-cases (Osello and Rinaudo, 2016).

However, cultural heritage elements and management differ from the projects that are usually being generated with BIM tools and procedures. BIM is commonly used for a building before it is built, and follows its lifecycle. On the other hand, a cultural heritage BIM refers to an existing geometry (most of the times with undefined shapes or with missing parts), along with its characteristics and a lifecycle that includes many and various phases of uses and conservation. However, the BIM industry does not have shared libraries for built heritage elements (Oreni et al., 2013).

Fai and Sydor (2013) remarked on two main challenges in the use of BIM for the documentation of cultural heritage structures. The first challenge refers to the difficulty in adapting the generic templates and clusters that already exist in the available BIM software to the specific characteristics and the particular construction methods that characterize cultural heritage architecture. This is scenario which also does not economically benefit the stakeholders of the industrialized building systems field, since by developing BIM libraries for their products they are able to use the same templates and elements multiple times without changes, whilst the heritage entities are formed by materials and components that are unique. The second challenge concerns the optimal representation of the very particular form and geometry that characterize the heritage buildings, which is now achieved to a good extent with the use of digital recording tools and procedures. Thus, cultural heritage documentation requires a detailed record of the geometry and a BIM system which is able to store this kind of dataset (Fai and Sydor, 2013).

Osselo and Rinaudo (2016), also outline that in the field of HBIM there is a need to optimize the data management process rather than to develop new tools.

The complexity and the diversity of HBIM has led to the development of specific software that offer several possibilities to their users, allowing them to deal with various categories of historical structures' information; however, the majority of these tools are commercial software, and in most cases are quite expensive and inaccessible. Therefore, the challenge in the HBIM field is the development of BIM systems that will be able to respond to the data management needs of a cultural heritage structure while being low-cost or free to be useful and beneficial in a sector which is less rich than the construction and engineering industry.

3. The FOSS BIM Concept

3.1. Foss CAD & Foss BIM Software

The term "free and open source software" (FOSS) is used to define that the source code (i.e., any group of computer instructions written using a human-readable computer language) is freely accessible to everyone interested in it, who can freely use, copy, improve, change, and distribute it for any purpose (Statskontoret, 2003). Without any permission required, the software can be freely used, modified, and redistributed (Ambar et al., 2010). As Steiniger and Hunter (2012) underline, the benefits of FOSS are its freedom to be run for any purpose, to be adapted to the users' own needs, to be improved upon, and to have such improvements released to the public.

Regarding CAD software, very interesting free and open source CAD environments exist, some of which allow the users to exploit the programming environment inside the platform and extend their possibilities. Among the available FOSS CAD and 3D graphics software is FreeCAD, Blender, SketchUp, and B-Processor, which are widely known and used by designer, engineer, and modeller communities. Each of them has specific characteristics that differentiate it from the rest, whilst all of them offer a variety of functionalities.

As shown in Table 2, some of them are open source, while others are freeware; this means that they are free but it is not possible for the user to access the source code and extend its functionalities. Some differences are also observed in their API, though the majority of them can be run both in Windows and Mac environments. A deeper search in some web forums which are active on these specific fields (such as "alternativeto.net", "cad.softwareinsider.com") shows that the users are more interested in reviewing FreeCAD and SketchUp capabilities and improvement possibilities.

Table 2. Free and open source software (FOSS) 3D graphics, computer-aided design (CAD), and BIM characteristics (Logothetis et al., 2017). API: application programming interface.

	FOSS 3D Graphics, CAD, and BIM Software			
	FreeCAD	Blender	SketchUp	B-Processor
3D	✓	✓	✓	✓
License	open source	open source	FREEWARE	open source
Available extensions	✓	✓	✓	-
Programming API	python	python	Ruby	Java
Use	CAD	3D MODEL	CAD	BIM
Up-to-date	Ver. 0.16, Rel. 7 Apr 2016	Ver. 2.78c, Rel. 28 Feb 2017	Ver. 17.2.2555, Rel. 17 Nov 2016	Ver. M11, Rel. 15 Nov 2015
Runs on Unix	✓	✓	-	-
Runs on Windows	✓	✓	✓	✓
Runs on Mac OS	✓	✓	✓	-
Forum Support	✓	✓	✓	-

Amongst the BIM software that have been developed so far, there are no open source integrated platforms able to cover all stages of a BIM process. The users prefer commercial BIM software instead of FOSS, because there is no credible and comprehensive platform that can be used for the overall 5D digital representation of a building. There are only a few accessible and free BIM viewers to visualise the final produced model (Logothetis and Stylianidis, 2016).

The vision of this research is to use an open source software environment with CAD functionalities in order to extend its capabilities and develop a FOSS BIM plugins' ecosystem (Figure 3). To meet this objective, the following criteria have been set:

1. A FOSS that will be open source and not freeware, so as to have access to the source code and extend its functionalities.
2. The software should be able to accept extensions, with the preferable programming API to be in Python programming language.

3. It should also be available for all types of operating systems (UNIX, Windows, Mac) and commonly used and accepted by the users.

Within this concept, the FOSS parametric FreeCAD software meets all the requirements set and was selected in order to develop the BIM plugins ecosystem which will facilitate a comprehensive and integrated BIM model analysis.

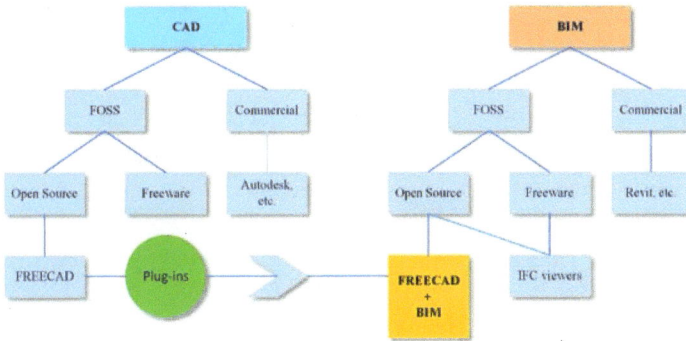

Figure 3. Concept of plug-ins development (Logothetis et al., 2017).

3.2. FreeCAD General Features and Use

FreeCAD is a 3D CAD modeller which meets all the requirements set in order to be used as the primary infrastructure for the FOSS BIM development. It is an open source platform and is extremely customisable, scriptable, and extensible. New functions and open source libraries can be added without modifying the main core system. It includes an embedded Python interpreter, and thus someone can use Python code in the software.

In the field of 3D modelling, the FreeCAD system has served a number of times as the subject of research work. Carl Schultz et al. (2015) used FreeCAD to demonstrate the way in which a parametric model may be constrained by the spatial aspects of conceptual design specifications and higher-level semantic design requirements, involving a combination of topological, visibility, and movement constraints. Guimaraes et al. (2016) used FreeCAD for 3D modelling geometry and assembling objects, and Horneber et al. (2014) used FreeCAD to build ".stl" files of different 3D structures.

3.3. Towards Extending FreeCAD into BIM

An integrated BIM model is able to display 3D entities which are directly linked with additional layers of information. In this way, it is possible to perform an advanced analysis of the model with extra information that permits different functionalities such as cost analysis, calculation of energy consumption, structural resistance, estimation of construction time, etc.

335

FreeCAD includes an "Arch module" option, which implements a series of tools and allows BIM workflow; however, it does not include the same tools and level of completion as the commercial alternatives (i.e., Autodesk Revit). Moreover, it is still under development (Web 5) and it lacks a complete add-on that implements 4D (time scheduling) and 5D (cost estimating) model analysis. For our approach, the absence of a database management system (DBMS) to support a BIM concept is vital, and is our first priority.

The current work aims to develop a plugins' ecosystem that will serve a BIM model analysis through a FOSS BIM solution. The high-level design (HLD) illustrated in Figure 4 elucidates the architecture that would be used for developing this specific software solution. The architecture diagram in the following figure (Figure 4) provides an overview of the entire ecosystem around FreeCAD, identifying the main components to be developed.

The first step towards the development of the FOSS BIM into the FreeCAD environment is the DBMS. It is so important because it concerns data management resourcefully and permits the users to perform several tasks with ease. A DBMS can store, organize, and manage an enormous amount of information within a single application. Such a system provides a very efficient method for handling multiple types of data. It is very important that data can be categorized and structured to meet the needs of the users to whom it is addressed.

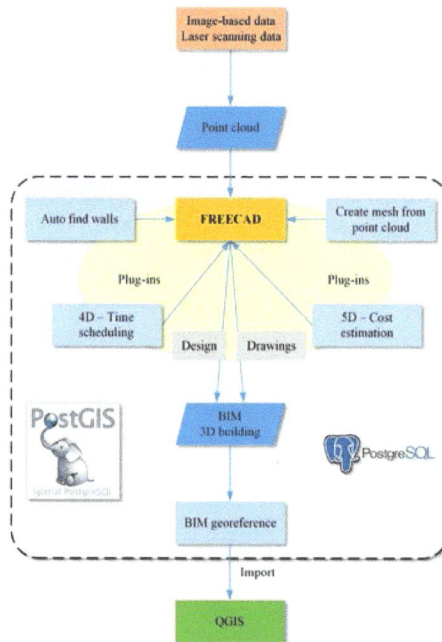

Figure 4. Workflow of the plugins ecosystem under development (Logothetis et al., 2017).

4. An Open Source Spatial DBMS for FreeCAD as a BIM

A DBMS is not supported by the FreeCAD software as in the case of other quite similar software. FreeCAD uses files in order to handle and store the data; in particular there is a FreeCAD standard file format called "FCStd". It is a standard zip file that holds files in a certain structure: (i) a document.xml file which stores all geometric and parametric objects definitions, (ii) a GuiDocument.xml which holds the visual representation details of objects, and (iii) some other files include brep-files for objects and drawing thumbnails (Web 3). The FreeCAD software can also support some well-known file formats such as IFC, DXF (Drawing Interchange Format), the SVG (Scalable Vector Graphics), etc.

File management quite a time-consuming procedure and has many disadvantages, especially with large files as in FreeCAD. An alternative and more efficient way to handle data is to use a DBMS. A DBMS is a software system that allows the users to define, create, and maintain a database, and provides controlled access to data. A DBMS is basically a collection of programs that enables users to store, modify, and extract information from a database as per the requirements. A DBMS is an intermediate layer between programs and data. In practice, the programs access the DBMS, which then accesses the data (Web 6).

4.1. Database Management System vs. File System

In the file system approach, each application has its own private files. These files cannot be shared between multiple applications. Such an approach may reflect extensive redundancy, which leads to storage space waste. By having all the data stored in a centralized database, most of this can be avoided, while it is not possible that all redundancy should be eliminated. A database has the ability to handle and control multiple copies of the same data, occasionally very important for technical reasons. This is very difficult to be considered in a file-based system approach. Moreover, having the data stored in a database always leads to accurate data. Incorrect information cannot be stored in a database. For example, in order to maintain data integrity, some (integrity) constraints are enforced on the database. A DBMS should offer capabilities for defining and enforcing such constraints (Ramakrishnan, 2002).

Using a DBMS provides many advantages. One of the most important advantages is that data inconsistency can be avoided. When we have duplication of data and changes are performed in one place which is not propagated to the other, this creates inconsistency, and the two entries concerning the same data will disagree. This is called data inconsistency. In a DBMS, backup and recovery are subsystems. In the case where we run a complex update program and the computer system fails, the recovery subsystem is responsible for securing that the database will be restored to the level prior to the update program execution. Using a DBMS rather than a file system provides additional advantages in data sharing and data integration. In a DBMS, data can be shared by authorized users. Usually, the database administrator (DBA) is responsible for data management and also for providing rights to the users to access data. It is also possible for many users to be authorized to access the same set of information at the same time. The remote users can share the same data, and data of the same database can be shared between different application programs. Concerning data integration, a single database contains multiple tables, and relationships can be created between tables (or associated data entities). This makes it easy to retrieve and update data (Silberschatz et al., 2011).

4.2. DBMS Overall Architecture

A representative DBMS system has a layered architecture. Figure 5 illustrates one of the potential architectures; each system could have its own variation. Figure 5 presents a seven-layered structure which contains the following layers:

Disc space management: The main component of a Minibase (a DBMS) is the disk space management system, which is responsible for allocating and deallocating the pages into a database. The disk manager carries out reads and writes of pages, to and from the disk, and transfers the block or page requested by the file manager. It supplies a "logical" file layer into a DBMS.

Buffer management: The primary purpose of a database system is to store and retrieve data; as a result, the main characteristic of a DBMS is an "exhaustive" disk input/output (I/O). The disk I/O operations are time and resources consuming, so the DBMS focuses on making I/O very efficient by using the buffer manager. The buffer management component consists of two mechanisms: the buffer manager to access and update database pages, and the buffer in order to reduce database file I/O (Sacco and Schkolnick, 1986).

Files and access methods: There are many data access methods that can enhance performance, and warns of situations to avoid. Figure 6 gives all the available database access methods.

Relation operations: In a database, it is possible to define some very important relational operators which are responsible to serve as a basis for formulating retrieval and update requests. The first and most important relational operators are: (i) SELECT (originally called RESTRICT), (ii) PROJECT, (iii) JOIN, (iv) PRODUCT, (v) UNION, (vi) INTERSECT, (vii) DIFFERENCE, and (viii) DIVIDE (Date, 2013).

Query optimization and execution: The query optimization layer is not available directly to the users. After the queries' submission to a database, these queries are passed to the query optimizer, where optimization occurs. In some database systems, the user can guide the query optimizer by giving "advices". The queries are executed based on a query execution plan or just query plan. Every DBMS system has at least one query execution plan in order to define a set of ordered steps which have to be followed to access data.

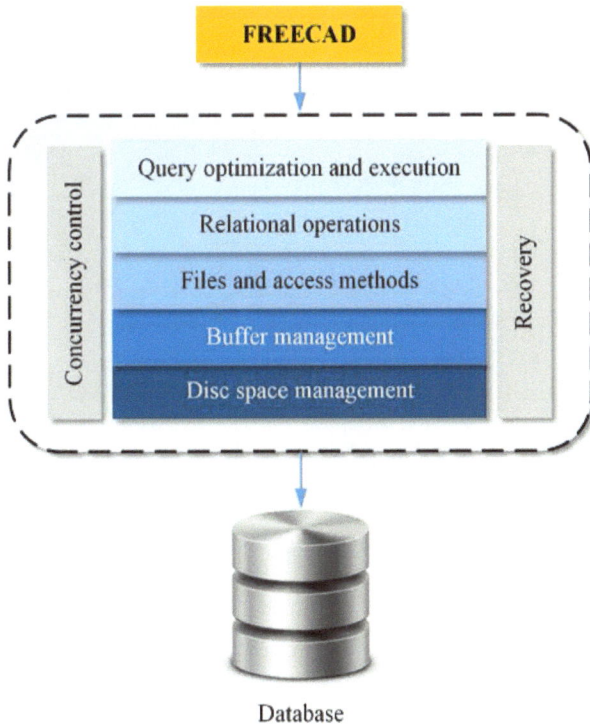

Figure 5. Database management system (DBMS) and the layered architecture.

Method	Scope	Comments
EMBOSS	•	Uses a B+ tree index from the programs dbxflat (used for "flat" files, i.e. files in their native database format) or dbxfasta (FASTA-format files)
EMBLCD	•	Uses an EMBLCD index from the programs dbiflat (used for "flat" files, i.e. files in their native database format) or dbifasta (FASTA-format files).
SRS	•	This calls getz locally, using the -e switch to return whole entries in original format. Query fields supported are 'id' 'acc' 'gi' 'sv' 'des' 'org' and 'key'.
SRSFASTA	•	As for SRS, but uses getz -d -sf fasta to read the sequence in FASTA format. Query fields supported are 'id' 'acc' 'gi' 'sv' 'des' 'org' and 'key'.
SRSWWW	single entry	Uses a defined SRS WWW server to read a single entry. Query fields supported are 'id' 'acc' 'gi' 'sv' 'des' 'org' and 'key'.
MRS	•	This uses a defined MRS server to read a single entry. Query fields supported are 'id' and 'acc'
Entrez	•	This uses Entrez at NCBI to read data. Query fields supported are 'id' 'acc' 'gi' 'sv' 'des' 'org' and 'key'. Data is returned in the original format (e.g. Genbank)
Dbfetch	•	This uses Dbfetch REST acces at EBI to read data. Query fields supported are 'id' and 'acc'
BLAST	•	Uses an EMBLCD index from the program dbiblast.
EMBOSSGCG	•	Uses a B+ tree index from the program dbxgcg to access a database reformatted for GCG 8, 9 or 10 by GCG programs such as embltogcg
GCG	•	Uses an EMBLCD index from the program dbigcg to access a database reformatted for GCG 8, 9 or 10 by GCG programs such as embltogcg
DIRECT	all	Opens the database file(s) and returns each entry sequentially. Query fields supported are 'id' 'acc' 'gi' 'sv' 'des' 'org' and 'key'
URL	single entry	Uses any other Web server (for example the EBI's emblfetch or swissfetch queries) to return an entry.
APP EXTERNAL	•	Run an external application or a simple script which returns one/more/all entries.

Figure 6. Database access methods (Source: Web 7).

Concurrency control: Concurrency control is part of a DBMS system which is used to address conflicts with the concurrent accessing or altering of data that can occur in a database. The concurrency control layer is responsible for coordinating simultaneous transactions while maintaining data integrity (Rob and Coronel, 2004).

Recovery: The recovery control is used to restore data that has been lost, corrupted, or unintentionally deleted. Once the data is inserted into the database, it should never be deleted except as a result of some explicit user request. The DBMS systems support many recovery techniques; some of the most common are: (i) salvation program, (ii) incremental dumping, (iii) audit trail, (iv) differential files, (v) backup/current version, (vi) multiple copies, and (vii) careful replacement (Date, 2013).

4.3. A Proposed DBMS for FreeCAD

After an extensive research, the DBMS which will be used in order to be connected with the FreeCAD is the PostgreSQL. This choice was based on the following reasons (Web 8):

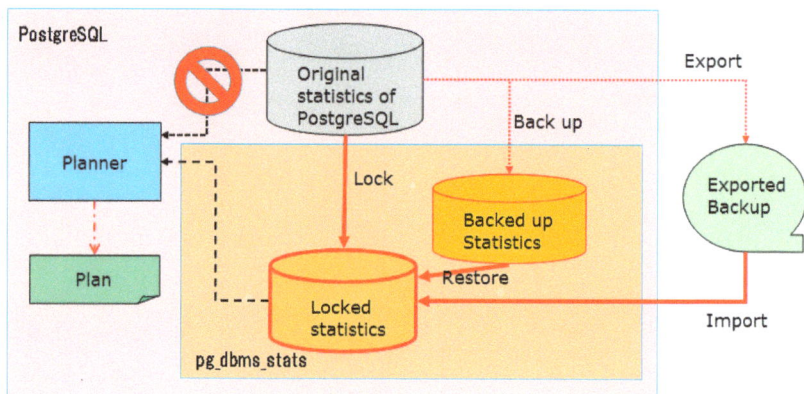

Figure 7. PostgreSQL overall architecture (Source: Web 9).

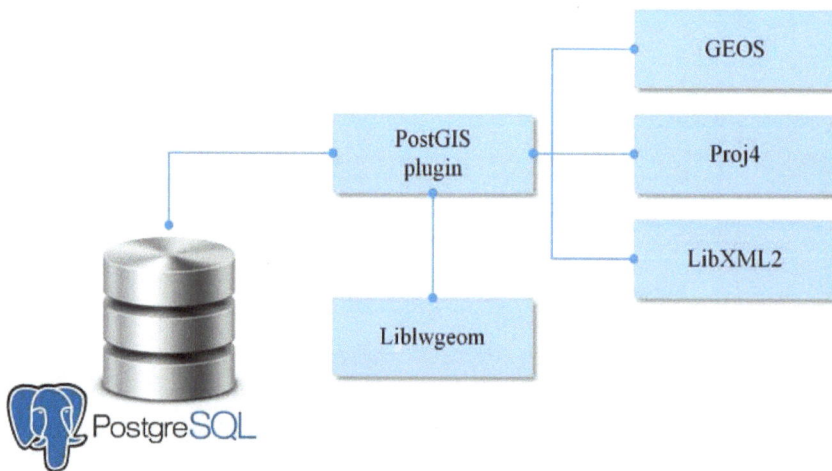

Figure 8. PostgreSQL and PostGIS connection.

340

- It is a FOSS. PostgreSQL is released under the PostgreSQL license—a liberal open source license, similar to the BSD or MIT licenses. FreeCAD is also open source software; as a result, this approach will be totally free and open source as well.
- PostgreSQL is widely used in large systems where read and write speeds are crucial and data needs to be validated. In addition, it supports a variety of performance optimizations that are available only in commercial solutions such as Geospatial data support, concurrency without read locks, etc. (e.g., Oracle, SQL Server). PostgreSQL performance is utilized best in systems requiring the execution of complex queries. The objects which are produced from the FreeCAD software are usually very large and complex objects.
- PostgreSQL can be extended in an easy and efficient way in order to support spatial information management. PostGIS is an open source software program that adds support for geographic objects to the PostgreSQL object-relational database. Our ultimate goal is to connect the FreeCAD software with a DBMS system which can support spatial information management.

Figure 7 and Figure 8 show the overall architecture of PostgreSQL as well as the connection of the PostgreSQL with PostGIS plugin.

4.4. How to Connect FreeCAD and Postgres DBMS

As a first step, Postgres DBMS was imported into FreeCAD through a Python script, as illustrated in Figure 9. As it was mentioned in the previous paragraphs, FreeCAD can export files in various well-known file types. One of these types is the ".ifc" data type. Therefore, our first approach is to use the IFC standard in order to store the BIM data into the database.

Due to the high complexity of the BIM models and the IFC standards, this work focuses on simple objects (e.g., 3D walls, etc.). As a result, and based on the file characteristics of a simple object (i.e., a 3D wall) that was exported in .ifc file format, we started the development of a Postgres database, following the IFC standards for BIM data (Figure 10).

The proposed method is established considering the following steps:

- Step 1: The .ifc file is exported following an algorithm like the one presented in Figure 9. In practise, the attributes data are retrieved from the IFC schema.
- Step 2: Read the .ifc file and create a table in the database with columns agreeing with the IFC parameter and insert the appropriate database again on the file (Figure 10).
- Step 3: The inserted table can be shown to the user via the FreeCAD software. The user can preview and also edit the data of the table (Figure 11).

Algorithm 1 GET ATTRIBUTES DATA FROM IFC SCHEMA

Input: IFC shema
Output: Database schema

```
1:  def getAttributes(self, name):
2:      # Get all attributes f an entity, including supertypes
3:      ent = self.entities[name]
4:      attrs = []
5:      while ent != None:
6:          this_ent_attrs = copy.copy(ent["attributes"])
7:          this_ent_attrs.reverse()
8:          attrs.extend(this_ent_attrs)
9:          ent = self.entities.get(ent["supertype"], None)
10:     end while
11:     attrs.reverse()
12: return;
```

Figure 9. Get attributes data from Industry Foundation Classes (IFC) schema.

Algorithm 2 EXTRACT ATTRIBUTES FROM ENTITIES

Input: IFC entities
Output: Entities' attributes

```
    # Search for a phrase and return the content of the line
    # being displayed
1:  def getAttributes(self, name):
2:      with open(file_name) as f :
3:      for i, line in enumerate(f, 1) :
4:          if phrase in line:
5:              return line

    # Processing text into parentheses
6:      get_line_number ('IFCEXTRUDEDAREASOLID',)
        'wall.ifc'
7:      n = s[s.find(" (") + 1 : s.find(")")]
8:      print n

    # Splits words using comma as separator
9:      mylist = n.split(',')
10:     print mylist
```

Algorithm 3 SAVE IFC ENTITY TO THE DATABASE

Input: IFC entity
Output: Create an entity table in the database

```
1:  conn = psycopg2.connect("dbname =' wall_db2',
    user =' postgres', password =' 123'")
2:      cur = conn.cursor()
3:      query = INSERT INTO
        "IFCEXTRUDEDAREASOLID"("SweptArea","Position",
        "ExtrudedDirection","Depth")
        VALUES(%s,%s,%s,%s);
4:          data = (mylist[0], mylist[1], mylist[2], mylist[3])
5:          this_ent_attrs.reverse()
6:          cur.execute(query, data)
7:  conn.commit()
```

Figure 10. Extract attributes from entities and save IFC entity to the database.

Algorithm 4 RETRIEVING OBJECT'S DATA FROM THE DATABASE

Input Cursor object
Output Object's data

```
     # psycopg2 is a Python module which is
     used to work with the PostgreSQL database.
1:   import psycopg2
2:   import sys
3:   con = None
     # Connection to the database.
4:   try:
     # The connect() method creates a new database
     # session and returns a connection object.
5:       con = psycopg2.connect(database =' wall_db', user =
         'postgres', host =' localhost', password =' 123')
     # From the connection, we get the cursor object.
     # The cursor is used to traverse the records
     # from the result set. We call the execute() method
     # of the cursor and execute the SQL statement.
6:       cur = con.cursor()
7:       cur.execute('SELECT * FROM "Application"')
     # We access the data from the while loop.
     # When we read the last row, the loop is terminated.
8:       while True:
     # fetchone() method returns the next row from
     # the table. If there is no more data left,
     # it returns None. In this case we break the loop.
9:           row = cur.fetchone()
10:          if row == None:
11:              break
     # The data is returned in the form of a tuple.
12:          print row[0], row[1], row[2], row[3]
     # In case of an exception, we print an error message
     # and exit the script with an error code 1.
13: except psycopg2.DatabaseError, e:
14:     print 'Error %s' % e
15:     sys.exit(1)
     # In the final step, we release the resources.
16: finally:
17:     if con:
18:         con.close()
```

Figure 11. Retrieving object's data from the database.

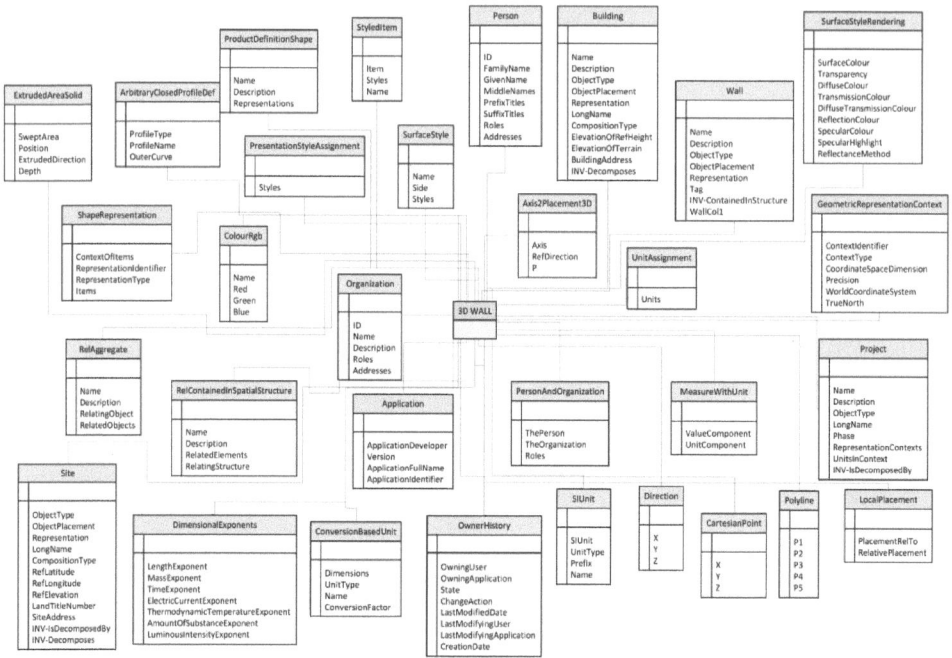

Figure 12. Database schema.

5. Conclusions and Future Work

BIM technology has gained ground in the building, construction, and architecture industry, but the high cost of software licenses, hardware upgrades, and the setup of the libraries is still a considerable obstacle to their market uptake and use. Especially in the field of HBIM, the challenges are many. The most important issue in heritage buildings is how to bring the existing structure into the 3D BIM process, while the requirement to record and preserve many architectural and structural features is an another important challenge.

Based on the idea to use a FOSS CAD software in order to extend its capabilities and transform it gradually into a FOSS BIM platform, the current work presented a first approach of a spatial DBMS development. The spatial DBMS is established to support the proposed FOSS BIM concept.

FreeCAD software met the specific criteria set, such as the programming API, the possibility to intervene, modify, and extend the source code, the operating systems, and the users' preference. These are the reasons for selecting this open source CAD to be integrated with a set of plugins that will reinforce the software with advanced BIM functionalities.

However, FreeCAD does not support a DBMS for data management. The development of a spatial DBMS in the FreeCAD environment was done. This system is responsible for reading the .ifc file and creating a table in the database with the same IFC standard's parameters and then inserting the appropriate database again in the file. At the same time, the system allows the user to preview and edit the data in the table.

The next steps in our research will focus on storing issues, more complex objects—and thus .ifc files—as well as on handling the data directly to the database without need to read the file, and finally on supporting spatial information storage.

References

1. Ambar, K.; Debashish, S.; Sitanath, B. Application of Free and Open source software and its Impact on society. *Int. J. Comput. Sci. Inf. Technol.* **2010**, *1*, 226–229.
2. Azhar, S. Building Information Modeling (BIM): Trends, Benefits, Risks, and Challenges for the AEC Industry. *Leadersh. Manag. Eng.* **2011**, *11*, 241–252.
3. Bilalis, N. Computer Aided Design CAD. In *INNOREGIO: Dissemination of Innovation and Knowledge Management Techniques*; Technical University of Crete: Chania, Greece, 2000.
4. Czmocha, I.; Pekala, A. Traditional Design versus BIM Based Design. *Procedia Eng.* **2014**, *91*, 210–215.
5. Date, C. Basic Database Concepts. In *Relational Theory for Computer Professionals*; O'Reilly Media, Inc.: Newton, MA, USA, 2013.
6. Dore, C.; Murphy, M. Integration of historic building information modeling and 3D GIS for recording and managing cultural heritage sites. In Proceedings of the 18th International Conference on Virtual Systems and Multimedia: "Virtual Systems in the Information Society", Milan, Italy, 2–5 September 2012; pp. 369–376.
7. Eastman, C.; Teicholz, P.; Sacks, R.; Liston, K. BIM Tools and Parametric Modeling. In *BIM Handbook: A Guide to Building Information Modeling for Owners, Managers, Designers, Engineers, and Contractors*; John Wiley & Sons, Inc.: Hoboken, NJ, USA, 2008; pp. 25–65.
8. Eynon, J. *Construction Manager's BIM Handbook*; John Wiley & Sons, Inc.: Hoboken, NJ, USA, 2016.
9. Fai, S.; Sydor, M. Building Information Modelling and the Documentation of Architectural Heritage: Between the 'Typical' and the 'Specific'. In Proceedings of the IEEE 2013 Digital Heritage International Congress (DigitalHeritage), Marseille, France, 28 October–1 November 2013.
10. Garagnani, S.; Manferdini, A. Parametric Accuracy: Building Information Modeling Process Applied to the cultural heritage preservation. *Int. Arch. Photogramm. Remote Sens. Spat. Inf. Sci.* **2013**, *XL-5/W1*, 87–92.
11. Guimaraes, A.V.; Brasileiro, P.C.; Giovanni, G.C.; Costa, L.R.O.; Araujo, L.S. Failure analysis of a half-shaft of a formula SAE racing car. *Case Stud. Eng. Fail. Anal.* **2016**, *7*, 17–23.
12. Hergunsel, M.F. Benefits of Building Information Modeling for Construction Managers and BIM-Based Scheduling. Master's Thesis, Worcester Polytechnic Institute, Worcester, MA, USA, 2011.

13. Horneber, T.; Rauh, C.; Delgado, A. Numerical simulations of fluid dynamics in carrier structures for catalysis: Characterization and need for optimization. *Chem. Eng. Sci.* **2014**, *117*, 229–238.

14. Ibrahim, M.; Krawczyk, R. The Level of Knowledge of CAD Objects within the Building Information Model. In Proceedings of the 2003 Annual Conference of the Association for Computer Aided Design In Architecture, Indianapolis, IN, USA, 24–27 October 2003.

15. Kalamarova, M.; Loucanova, E.; Parobek, J.; Supin, M. The support of the cultural heritage utilization in historical town reserves. *Procedia Econ. Financ.* **2015**, *26*, 914–919.

16. Logothetis, S.; Delinasiou, A.; Stylianidis, E. Building Information Modelling for Cultural Heritage: A review. *ISPRS Ann. Photogramm. Remote Sens. Spat. Inf. Sci.* **2015**, *II-5/W3*, 177–183.

17. Logothetis, S.; Karachaliou, E.; Stylianidis, S. From OSS CAD to Bim for Cultural Heritage Digital Representation. *Int. Arch. Photogramm. Remote Sens. Spat. Inf. Sci.* **2017**, *XLII-2/W3*, 439–445.

18. Logothetis, S.; Stylianidis, E. BIM Open Source Software (OSS) for the documentation of Cultural Heritage. In Proceedings fo the 8th International Congress on Archaeology, Computer Graphics, Cultural Heritage and Innovation 'ARQUEOLÓGICA 2.0', Valencia, Spain, 5–7 September 2016; pp. 28–35.

19. Oreni, D.; Brumana, R.; Georgopoulos, A.; Cuca, B. HBIM for conservation and management of built heritage: Towards a library of vaults and wooden bean floors. *ISPRS Ann. Photogramm. Remote Sens. Spat. Inf. Sci.* **2013**, *II-5/W1*, 215–221.

20. Osello, A.; Rinaudo, F. Cultural Heritage Management Tools: The Role of GIS and BIM. In *3D Recording, Documentation and Management of Cultural Heritage*; Whittles Publlishing: Dunbeath, UK, 2016.

21. Penttilä, H. Describing the changes in architectural information technology to understand design complexity and free-form architectural expression. *ITCON* **2006**, *11*, 395–408.

22. Popov, V.; Mikalauskas, S.; Migilinskas, D.; Vainiunas, P. Complex Usage of 4D Information Modeling Concept for Building Design, Estimation, Scheduling, and Determination of Effective Variant. *Technol. Econ. Dev. Econ.* **2006**, *12*, 91–98.

23. Ramakrishnan, R.; Gehrke, J. *Database Management Systems*; McGraw-Hill: New York, NY, USA, 2002.

24. Rob, P.; Coronel, C. *Database Systems: Design, Implementation, and Management*; Thomson/Course Technology: Cambridge, MA, USA, 2004.

25. Sacco, G.M.; Schkolnick, M. Buffer Management in Relational Database Systems. *J. ACM Trans. Database Syst. TODS* **1986**, *11*, 473–498.

26. Schultz, C.; Bhatt, M.; Borrmann, A. Bridging qualitative spatial constraints and feature-based parametric modelling: Expressing visibility and movement constraints. *Adv. Eng. Inf.* **2015**, *31*, 2–17.

27. Silberschatz, A.; Korth, H.; Sudarshan, S. *Database System Concepts*; McGraw-Hill: New York, NY, USA, 2011.

28. Singh, V.; Gu, N.; Wang, X. A theoretical framework of a BIM-based multi-disciplinary collaboration platform. *Autom. Constr.* **2011**, *20*, 134–144.

29. Statskontoret. *Free and Open Source Software—A Feasibility Study*; Statskontoret: Stockholm, Sweden, 2003.

30. Steiniger, S.; Hunter, A.J. The 2012 free and open source GIS software map—A guide to facilitate research, development, and adoption. *Comput. Environ. Urban Syst.* **2012**, *39*, 136–150.

31. Succar, B. Building information modelling framework: A research and delivery foundation for industry stakeholders. *Autom. Constr.* **2009**, *18*, 357–375.

32. Sutherland, I. Sketchpad: A man-machine graphical communication system. Technical Report No 574; University of Cambrigde Computer Laboratory: Cambrigde, UK, 2003

33. Vanlande, R.; Nicolle, C.; Cruz, C. IFC and building lifecycle management. *Autom. Constr.* **2008**, *18*, 70–78.

34. Volk, R.; Stengel, J.; Schultmann, F. Building Information Modeling (BIM) for existing buildings—Literature review and future needs. *Autom. Constr.* **2014**, *38*, 109–127.

35. Markets and Markets. Available online: http://www.marketsandmarkets.com.

36. AWCI. Available online: http://www.awci.org/pdf/bim.pdf.

37. Available online: http://www.it.civil.aau.dk/it/education/reports/building_smart/WS3_IDM_WhatIsTheIFCModel.pdf.

38. BuildingSMART. Available online: http://buildingsmart.org.

39. FreeCAD. Available online: http://www.freecadweb.org.

40. Thakur, D. What is DBMS? Advantages and Disadvantages of DBMS. Available online: http://ecomputernotes.com/fundamental/what-is-a-database/advantages-and-disadvantages-of-dbms.

41. Database Access Methods. Available online: http://emboss.openbio.org/html/adm/ch04s03.html.

42. PostgreSQL vs MySQL. Available online: http://www.2ndquadrant.com/en/postgresql/postgresql-vs-mysql.

43. Pg_dbms_stats. Available online: http://pgdbmsstats.osdn.jp/pg_dbms_stats-en.html.

Logothetis, S.; Valari, E.; Karachaliou, E.; Stylianidis, E. Development of an Open Source Spatial DBMS for a FOSS BIM. In *Latest Developments in Reality-Based 3D Surveying and Modelling*; Remondino, F.; Georgopoulos, A.; González-Aguilera, D.; Agrafiotis, P.; Eds.; MDPI: Basel, Switzerland, 2018; pp. 326–347.

BIM Application for the Basilica of San Marco in Venice: Procedures and Methodologies for the Study of Complex Architectures

Luigi Fregonese, Laura Taffurelli and Andrea Adami

Politecnico di Milano, Dept. ABC, Mantova Campus, Piazza d'Arco, 3, Mantova, Italy;
luigi.fregonese@polimi.it; laura.taffurelli@polimi.it; andrea.adami@polimi.it

Abstract: The BIM (Building Information Model) of the Basilica of San Marco contains the solutions to the many problems encountered during its acquisition and modelling stage. The complexity of the church and the variety of its materials (golden mosaics, capitals of different styles and origins, statues and decorations in many different marble types), the large and continuous stream of visitors, and the request for high-resolution models and orthophotos forced us to devise a strategy for the digitization process: a multiscale photogrammetric approach allowed us to acquire all materials and decorations of the basilica and, according to the use of a reference topographic network, we could split the whole work into smaller parts. Later, in the modelling stage, the decision to use a non-commercial BIM software allowed us to use NURBS (non-uniform rational B-spline) for a more accurate restitution of architectural elements and decorations and to integrate high-resolution orthophotos for the description of all surfaces (both marbles and golden mosaics). The established workflow started with the initial acquisition of images and resulted in both final models and high-quality orthophotos, so we were able to obtain different outcomes to answer the specific needs of the church, its managers, and its users.

Keywords: HBIM; modelling; cultural heritage; complex architecture

---

1. Introduction

The Basilica of San Marco is certainly one of the most visited monuments in Italy, probably all over the world, and for many justified reasons: the eventful history of its constructions, the provenance of its artistic elements, the richness of the decoration, and the opulence of materials, not to forget the uniqueness of the position in San Marco square in central Venice. All these characteristics combined make this church one of the most important masterpieces of Cultural Heritage.

However, the Basilica of San Marco also has a special feature which relates to its liveliness. In the cultural heritage field, many monuments became museums of themselves, independently of their function, and their architecture seems to be frozen, suspended in an ideal state. This behavior, which has the aim of preserving the monument from damage, also involves the risk of changing the nature of the place and of the building. The condition is very different in San Marco, because the church hosts many activities every day. It is open for both simple visitors and worshippers who either visit the church or follow the religious activities. It is also a high-value location for the very famous evening concerts which take place occasionally inside the Basilica. The liveliness of this architectural complex involves not only the aforementioned public activities, but also the ones carried out behind closed doors. The church itself is a large and animated construction site: a team of construction workers, carpenters, electricians, marble workers, and mosaicists work every day to maintain the building in the best possible conditions. Even in this sense, this cultural heritage artifact shows a peculiarity of its own: although it underwent some major restoration works during the centuries, the maintenance carried out daily by this team of different specialists is what makes San Marco a unique example of intervention in the conservation scene.

Such a complex management effort requires the Procuratoria (the institution which oversees all the activities) to be provided all the instruments and tools that can optimize this massive work. The knowledge of the architecture and its state of preservation cannot be demanded of only one responsible (the "proto della fabbrica", the architect who has the responsibility of the architecture of San Marco), but must be managed and shared between all the participants in the process. This approach can be helped drastically by providing a BIM (Building Information Model) which allows each single intervention on the Basilica to be observed, managed, and evaluated together. For this reason, in 2013 the Procuratoria of San Marco asked the Politecnico of Milano to produce a 3D volumetric model to be used as the base for a subsequent BIM.

This work is only part of the collaborative activities carried out between the Politecnico di Milano and the Procuratoria of San Marco which led, over the years, for the monitoring of the Basilica, the survey of the entire mosaic floor, up to the detailed survey and modelling of the volumes between the vaults and the roofs.

1.1. A BIM for the Basilica

Although the theme of the BIM is starting to be of great relevance in the Italian heritage scene, the transition from the theoretical approach to practical operations is always fairly complex.

The possible uses of BIMs are a new trendy topic for professionals who deal with Cultural Heritage, because its advantages are relevant and noticeable. Firstly, it ensures the coordination between the various interventions in the building and the different experts involved during the processes. Moreover, a model designed according to BIM criteria allows you to collect different types of data in one single database: drawings and historical maps, archival documentation concerning the evolution of the artifact, its material, the state of preservation, the previous interventions which regarded the building, and CAD drawings related to more or less recent surveys.

The model is also an excellent knowledge tool and starting point for further investigation and tests to be made by experts in the field. In addition, it is potentially interoperable with different software systems; for example, for structural analysis. Therefore, the insertion of the model in structural calculation software provides information on the state of conservation of the building, and it allows projections to be made of its behavior in the short- and long-term, with or without the prevision of consolidating interventions.

It is not wrong to think that BIMs could be used for educational purposes. A BIM can provide a 3D image that gives a general idea of the architectural complex you are visiting, or more specifically, it can illustrate—for each element of molded material—construction characteristics and period of construction, to give an overall picture of the building's evolution over time.

Moreover, from the point of view of the administration, a BIM-oriented model can facilitate the management of the building, ensuring the ability to extract from the model itself documents related to security, such as evacuation plans, management previsions, and maintenance cost summaries.

Unfortunately, all these benefits collide with some issues which cannot be completely solved when the aim of the BIM project is to build a correct 3D model to be used for restoration and conservation purposes (Fai et al., 2001; Murphy, McGovern, 2009). There are many possibilities and different strategies to arrive at the aforementioned 3D model, but it is very difficult—quite impossible—to find only one ultimate solution.

Before delving into all the possible options to reach a choice, it is necessary to understand and define exactly which are the main objectives of the BIM. As the use of BIM in Italy is still limited and the procedure for BIM adoption—if it even exists—is not applied and intended for private institutions, the identification of the requirements from the customer and the definition of possible uses and future perspectives are a necessary first step to designing an efficient BIM.

The layout of the Basilica of San Marco is quite unique; consequently, its BIM cannot be considered in the same way as other case-study buildings. In the Basilica, the most urgent necessity was the knowledge of the geometry of the building. It could have been very difficult to set some single sections and plans to survey because, as we said, the conservation process is continuous and repeatedly treats each section of the church. The possibility of having a 3D model of the entire volume allows to put off the choice: the BIM system in fact is intended for the extraction of infinite drawings simply by defining the position of the cutting plane.

The efforts of survey and modelling are concentrated in one single stage, whilst the outcomes can be computed whenever necessary, theoretically until the building suffers some new changes.

However, the interest of the Procuratoria was not only focused on the geometrical layout: they also wanted detailed information about the decorated surfaces, covered with mosaic and stone. About ten years ago, the Politecnico di Milano realized the complete survey and modelling of the pavement of the Basilica to extract an orthophoto at a scale of approximately 1:1. The intention was to give the restorers a document which they could use for their activity of mosaic conservation and renovation: they could use the printed 1:1 scaled orthophoto as a reference to reassemble the tesserae of the mosaics (Fregonese et al., 2006). Likewise, the Procuratoria asked for a very similar outcome for all the surfaces of the church, be them mosaic or marble tiles.

These requirements which we had to meet demonstrate that the BIM for the Basilica of San Marco can be considered exceptional not only for the richness, importance, and complexity of the building itself, but also for the necessity of managing high-resolution images (orthophotos). Therefore, the main discussion about the design of the BIM regarded the choice of the best solution capable of guaranteeing these results. Initially, the choice was between the use of commercial BIM software or a self-implemented one. For a review of both systems, see Logothetis et al., (2015).

On one side, the approach through commercial BIM authoring software such as Revit, Archicad, or Allplan guaranteed a good level of integration with many other software and a high level of spread. This meant that the management of the model could have been quite easy and widespread, and that it would have been possible to count on a large community of users. Some doubts were connected with the geometric complexity of the church. As that kind of software operates on the concept of standardization, through the concept of parametric families, it was very difficult to translate each element of the Basilica in a dictionary of elements, each one precisely defined in all its parts. This behavioral model was fighting not only with the concept of Cultural Heritage, where each element should be considered unique (and modelled as such), but especially with the history of the Basilica itself. The architectural complex we see today is the result of the addition of many different pieces, all coming from different places of the Venetian supremacy and therefore with different characteristics, and it was unconceivable to reduce all this complexity into defined classes. Another problem was concerned with the management of high-resolution orthophotos. Even if it is given that each BIM software allows different kinds of information (including images) to be linked, the problem was the quality of the connections. Because high-resolution orthophotos reach very large dimensions (in terms of number of pixels) and memory demand (megabytes or even gigabytes each), the commercial software could barely manage those attributes. Moreover, it was required to develop a more user-friendly system to allow the employees of the Procuratoria to deal with it easily.

On the other hand, a self-implemented software required many efforts assemble all the necessary aspects which characterize a BIM environment. From modelling and data visualization to content sharing and database connection, all these elements needed to be merged in a single system. The risk connected with this self-implemented approach was the complexity of use: many open source systems, even if very efficient in terms of calculations and outcomes, are disadvantaged because of their interface.

The solution to this dilemma came with the experience of the 3D Survey Group of Politecnico di Milano on the Main Spire of Milano Cathedral (Fassi et al., 2015). They implemented a specific solution for the Cathedral of Milano which allowed the realization of a highly customizable system, able to easily adapt itself to the different cases and to work with different types of objects, different representation scales and resolutions, also showing different types of information. The system can visualize huge dimensions and high numbers of different 3D models which can be realized in any external software (in their case, Rhinoceros software of Robert McNeel and Associates). The advantages of this systems are on several levels. The system can in fact be implemented to manage high-resolution images and, in a second phase, to project the database according to the requirements and needs of the Procuratoria. Another positive element is that the modelling stage is disconnected from the BIM environment: this means that we could decide which is the best software to realize the model, only after importing the model in the BIM system. This is convenient not only during the first steps of modelling, when the model is created by specialists in the field of geomatics and architecture, but also in a following phase when the 3D is to be managed by the architects of the Basilica. Lastly, the database of the whole system can be installed on a remote server or on the cloud, in order to be accessible by more workstations. However, if needed, the database can be both local or placed within a local network.

This research discusses the construction of the BIM for the Basilica of San Marco, considering different aspects. The entire activity is split into survey stage and modelling stage, although this subdivision is quite fleeting and it is very difficult to exactly set the domain of survey and the one of modelling. We include in the survey the stage of data acquisition (both photogrammetric and topographic), the registration of all data in a single reference system, and the data processing to obtain the point-clouds. Together with the description of the techniques used, there is also a discussion about the logistic and technical problems which characterized the entire work.

The following step—3D modelling—regards the modelling workflow from point-cloud to the final 3D model. In particular, it concerns the approaches used for modelling different items which characterize the church (walls, arches, vaults) and the enriching decorative elements such as moldings, capitals, and statues.

The last step of the process discusses the method used to obtain the orthophoto.

2. The Basilica of San Marco in Venice

2.1. Characteristics and Complexities

The current structure is the result of more than 12 centuries of transformations, and it is evident in its structure and its decoration. The Basilica has a floor plan in the shape of a Greek cross, with a main dome over the crossing and another dome on each of the four arms. The nave and the transept have a central aisle and two side aisles divided by an arcade. The nave of the upper arm corresponds to the presbytery, raised by a few steps, below which there is the crypt. To the sides of the presbytery there are two chapels. The area of the presbytery is separated from the rest of the Basilica by the iconostasis, which itself is divided into three distinct sections: the central one, right in front of the presbytery, and the two lateral parts, in front of the chapels. The upper inner perimeter of the entire Basilica, except for the apse, is crossed by the women's galleries. A narthex wrapped around the west end disguises the cross shape, but creates a wider surface for the main facade.

This architectural structure of the church and its ornaments are byzantine, but we can also find Gothic and Romanesque styles in the decorations. These are indeed the result of a continuous addition and re-use of ornaments, sculpture, and precious marbles acquired during travels and exhibited as a symbol of power. A certain repetition of similar shapes and dimensions can commonly be noted in complex architectures such as this, but in San Marco it practically never happens.

One of the main distinctive features of the Basilica is definitely its ornamentation; composed of marble slabs, sculpture, and golden mosaics, it completely hides the sight—from both inside and outside—of the main brick structure of the church.

Inside, due to the impressive gilded mosaics cover, the upper structure, composed of vaults and domes, looks like a continuous surface without edges, while the lower part of the building is completely enclosed by marble slabs and columns, in the same way as the exteriors. This is contrary to common architecture, where the break lines are the most important elements to be represented.

Moreover, the entire building presents some huge irregularity in shape, because of the constant addiction of architectural structure and decorations over the centuries, such as the heavy wooden and lead domes overlapping the original ones or the gothic spires of the façade.

The complexity of the shape is accompanied by the degradation which the church is subjected to daily due to the continuous presence of visitors and the salty air, considering its location (Figure 1). The high water and its salinity have been deteriorating surfaces and decorations for a long time now and, today again, in the narthex, the high water due to floods is very frequent and it continues to cause damages. Together with these surface degradations, we also found geometric deformations due, for example, to the instability of the soil: the effects are particularly evident in the pavement of the central nave, where the deformations have great values and can be perceived by the naked eye.

All these characteristics—even if they define the absolute value of the Basilica of San Marco—caused many challenges for both the survey and modelling stages. Apart from the difficulties in surveying particularly reflective materials, such as marble and mosaics, many problems were connected to the logistical aspects.

The Basilica was open to the public every day, some sections were just for praying and others purely for touristic purpose; many liturgical functions, even special ones, were being carried on and this made the church constantly busy with people during the day. We also had to consider that we could not enter at all in some of the construction sites inside the church and the narthex for the survey activities. All these aspects made it difficult to completely isolate an area and work freely. Additionally, night work sessions were not allowed because of safety and economic reasons.

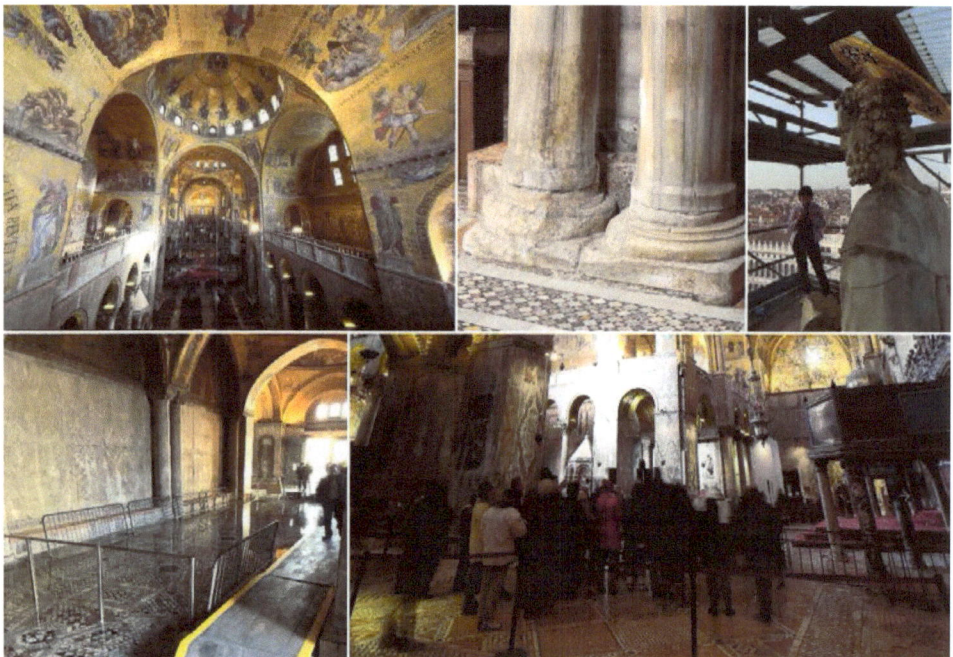

Figure 1. The interior of the Basilica with many of the difficulties that were encountered: tourists, high water, light, degradation.

354

Apart from these aspects, we also had to face some problems connected with the lighting inside the church. Of course, one of the main interests of the church's administration was to guarantee the best possible visitor experience, so they set up a system of lights of different intensity and color to create a certain "atmosphere". Together with the sunlight coming in through the rose-windows of the main and south façade, these lights made the acquisition operations very difficult, being almost impossible to achieve the diffuse illumination which is more suitable for the survey.

2.2. Previous Works and Data Reuse

During the last century, several survey campaigns have been conducted because of the necessity of monitoring the Basilica's structural situation, given its particular location. All this information has been georeferenced in the same reference system, such that a huge database for the Basilica has been created.

The decennial campaign of 1983–1993, entrusted to a private company, led to the complete geometrical representation of the interior of the Basilica, with plants, sections, and facades, and external parts (narthex, lateral side, and facades). The drawings produced on different scales (1:20, 1:50, 1:100) made it possible to carry out the following diagnostic surveys of the 1990s, and to create a mathematical model of the structure.

In fact, the plans have been used for floor-to-floor chemistry, static studies, and plant design, as well as having been the graphic support for the performance of monitoring results.

Since 2000, the collaboration with the Politecnico di Milano has allowed the implementation and improvement of the quality of the Basilica surveys needed for a correct approach to the planned conservation of the building. In 2001–2002, a laser scanner approach was tested on the internal side of the dome of the "Ascensione", a part of the mosaic floor (including some of the narthex) and for the detection of the external facade of the Basilica.

Since 2003, a collaboration project with the San Marco Procuratoria has been signed and we have acquired about 1,700 photogrammetric shots of the entire floor of the Basilica for a total of 2,600 m², with the aim of obtaining an overall geometric model of the pavement and an orthophoto representing each part of the mosaic. The final product has been a representation that provides information on various levels, all important for both direct application and accurate study of the mosaic, its degradation, iconography, and more specific geometric and material information (Fregonese et al., 2006).

Subsequently, in the years 2005–2008, the necessity of placing the roof systems under fire-protection has led to the survey of the crawl spaces of the Basilica. This survey also had the important aim of understanding the complexity of the structure and the irregularity of its morphology. In fact, it highlighted the profile and thickness of the masonry to the level of the cover and defined the actual conformation of the vaults.

During the same years, the geometric information regarding the dome of Pentecoste and of the Profeti was surveyed (Fregonese et al., 2012). During 2010, the areas of the Sacristy and the Chapel of St. Leonard were added to the large database of the whole Basilica.

3. Survey Stage

3.1. The Architectural Survey

The last few years have not seen any major innovation in the architectural survey scene, particularly in terms of data acquisition instruments and techniques. After the apparent prevalence of laser scanners on photogrammetry, and vice versa, we are now heading towards a more mature phase, where these two main methods are studied and compared in terms of efforts, data processing, achievable results, costs, and processing times.

The most recent innovations in laser scanner techniques concern, above all, the development of new instruments customizable to the necessities of the customers (with increasing acquisition speed) and optimized to allow an easier and more efficient acquisition process, paying particular attention to the costs. Another issue is the one concerning the automation in the acquisition workflow, whose aim is to avoid the use of a topographic network (surveyed by a total station), in favor of an automatic algorithm for registration in order to minimize the acquisition time (Vosselman, Maas, 2010). The topic of automation also deals with the research of methods for point-clouds segmentations, in order to provide 3D data with meaningful attributes (Grilli et al., 2017).

The last applications of digital photogrammetry—especially the ones about dense image matching, based on many algorithms (global or semi-global matching) (Szeliski, 2010) according to the different stages of the photogrammetric process—are geared to process optimization and software validation (Remondino et al., 2014). The attention is focused on new algorithms that can refine the results, from the image processing to a better dense matching (Gaiani et al., 2016) and to the use of particular sets of images such as the oblique images (Remondino et al., 2014) or wide-angle images (Perfetti et al., 2017, Barazzetti et al., 2017).

Although each of the two techniques have specific features that allow you to prevail in different areas, it has been clear that there is no real winner in this competition. In the practice of architectural surveying, for some years now, we have been witnesses of the stabilization of laser scanners and photogrammetry, and the integration of the two methods. When possible, the integration of a laser scanner (to acquire geometry) and photogrammetry (to acquire appearance) seems to be the most effective solution.

In the case of the Basilica of San Marco, the integrated approach cannot be used, as the materials of the church—especially its mosaics made of gilded glass tessera and marble—cause some geometric errors in laser scanner point-clouds. As described in (Godin et al., 2001), there is a cause–effect relationship between the laser spot diameter and the estimated noise levels in the geometric measurements. A bias in the depth measurement can be observed, and it is supposed to be the result of scattering on the surface of small crystals near the surface.

Therefore, the entire acquisition necessary to the model (both the general architecture and the details) has been conducted by using only the photogrammetric approach, integrated with the traditional topographic survey.

Moreover, complete automation in the field of architectural surveying is very difficult to achieve. The latest pieces of software try to avoid the specialist's intervention in favor of an automatic point-cloud registration (as Autodesk Recap does when importing new point-clouds) or by using black-box solutions which can make serious mistakes. As we will see, mostly in the next chapter, the automation is completely defeated in the further step, where the intervention of the architect— as the person who knows extensively about the architecture and its construction— is fundamental.

3.2. Survey Activity

As described before, the survey of the Basilica of San Marco was conducted using the photogrammetric method based on dense image matching. While on one hand it was clear from the beginning which method to use for that, we cannot say the same about the process of data acquisition.

Because of the above-mentioned logistical issues, the most relevant problem was the impossibility of conducting the survey in a single and continuous work session. Because of the touristic and religious activities, the presence of some construction sites in some areas of the church, and the special events which were scheduled in the Basilica's calendar, the survey had to be broken down into smaller campaigns. Moreover, the sequence of areas to be surveyed did not allow a unique linear development to be followed, but it had to be studied, discussed, and agreed upon day-by-day with the technical office. These constraints not only influenced the schedule of the work, but its overall methodology as well: we developed a program based on the photogrammetry and the topographic survey which answered to all these requirements. We worked with a multiscale approach in order to obtain different levels of detail according to the characteristics, dimensions, and complexity of the object. We used three high-resolution cameras (2 Canon EOS 5D mark III and 1 Canon EOS 5DS R), which enabled simultaneous work in different areas, and we used different lenses (24 mm – 35 mm – 85 mm – 200 mm) as shown in Figure 2. The choice of the lens was done according to the elements to be acquired, the distance from the object to survey, and the pixel-size needed for the processing. The result was a multiscale project in which the same elements were surveyed with different density of points. Usually, we realized several photogrammetric models for each area where we worked. The main model—the one obtained with the 24 or 35 mm lenses—was used to render the overall dimensions of the space, its geometry, and its shape. It was noted later that the architectural details and decorations could not be correctly extracted from this main model because of the point density—not accurate enough to describe these smaller details. In order to avoid this problem, some other models were computed for the same area. With the 85 mm lens, we obtained denser point-clouds of the details such as capitals and decorations. The upper mosaic surfaces needed a very high-resolution image acquisition for the creation of the final orthophotos, given the high distance from there (the dome's height reaches about 28 m above the ground floor). In those cases, we went so far as to use tele-lenses (up to 200 mm) in order to acquire the most distant mosaics. All images were acquired with the help of the tripod and according to the poor lighting conditions.

Additionally, we applied two different strategies in georeferencing the photogrammetric models (Figure 2). The main models—made with the 24 or 35 mm lenses—were georeferenced in the reference system we materialized, and adjusted for the survey of the pavement which occurred in 2004–2010 (Fregonese, 2004). This choice had many advantages: besides the re-use of the topographic network, after the necessary validation, each new point-cloud was integrated into the previous survey by using a single reference system. The result is an upgradable database over time, georeferenced to the same general topographic network. This allowed, for example, to avoid the survey of the pavement and to use (both for modelling and texturing) the previously acquired high-resolution images. The general photogrammetric models were geo-referenced thanks to a topographic survey by a total station (Leica TS30) and an automatic recognition of code targets (to minimize the selection errors). The use of the topographic network has also allowed the control of the georeferencing to be maintained, avoiding the propagation of any errors in the images' alignment.

Figure 2. Images of two models at different resolutions, according to the needs of modelling and orthophoto. From the left: model of the whole area with 35 mm, model of a capital with 85 mm, model of the dome with 200 mm for orthophoto.

The second strategy relates to the detailed models made with the tele-lens (from 85 to 200 mm). Obviously, we could not use coded targets for different reasons: the detailed models would have required too many targets. Moreover, the models often incorporated inaccessible parts (domes, vaults) where it was not physically possible to put some targets. In these cases, we decided to georeference the models using the Iterative Closest Point (ICP) algorithm (Zhang, 1994) which is implemented in the software we used for the photogrammetric alignment and dense image matching (Agisoft Photoscan).

The automatic multi-image photogrammetry required us to take images with large overlap between consecutive shots and to use a homogeneous illumination to acquire realistic photos without shadow areas or flash effects. This was surely quite a big issue, given that the lights used inside the Basilica came from various sources and we were not allowed to work at night. To minimize the effects of artificial lighting (and also of the sun), we used some large lighting balloons equipped with a discharge lamp that generated neutral light, and some adjustable LED spotlights which were more manageable than the former (Figure 3). As a general rule, the balloons were used in larger spaces, where it was easier to manage them and possible to illuminate high surfaces; as a matter of fact, they were provided by telescopic tripods that permitted regulation of the height up to 8 m. On the other hand, the smaller LED spotlights were used in complex areas within small spaces (e.g., the galleries' level), and where it was necessary to correct the illumination coming from different directions to obtain a more homogeneous condition. On the contrary, the problem of contrasting the natural light coming from the rose windows was not solved in any other way than organizing the acquisition according to the hours of the day in order to avoid the naturally sun-lit areas.

All the methods described in this paragraph were validated beforehand in a test area. The Baptistery of the Basilica represented a very suitable test bed to try the complete photogrammetric approach, and there we set the survey workflow to use for the entire church. This area was optimal for test acquisition: the Baptistery is independent of the rest of the church, but concurrently it is constituted by the same architectural elements that are present in the whole building: domes, arches, vaults, columns, capitals, sculptures, etc., so that the problems faced in this test area were going to substantially be the same in the rest of the church. Technically, in the other spaces we had to adjust the survey process and the instruments according to the specific case, but the sequence of actions for surveying (and modelling) was the same. More precisely, the dimension of the main church is greater than the one of the Baptistery, and this required different lenses for the acquisition phases to obtain the most suitable resolution, but the method was exactly the same.

After the validation of the workflow through practice, the same method was applied to the whole church. The entire complex was split into smaller spaces considered independently (even if they were not so). For each of them, we realized photogrammetric models at different resolutions, but included in the same reference system, according to the increasing necessity of information for details.

Figure 3. Images of the interior of the Basilica with the large balloons and LED lights to set the best lighting conditions.

The data processing which followed the acquisition step is well known. The images were oriented and georeferenced in Photoscan by using the previously described methods. Then, we calculated the point-clouds that were merged in a single point-cloud model of the interior, as in Figure 4.

4. Modelling Stage

4.1. BIM Modelling of Cultural Heritage: State of the Art

3D modelling in the field of Cultural Heritage is currently a topic of great relevance. In a sense, we could affirm that the research has moved its interest towards data modelling rather than data acquisition, and there are several reasons for that. Firstly, there has been a greater integration of research areas that, until a few years ago, worked autonomously: an example is the combined effort of geomatics with computer sciences such as computer graphics. Another reason should be searched in the outbreak of low-cost methods for obtaining 3D point-clouds, both using laser scanner and, mainly, photogrammetry. However, after the acquisition stage, the primary question remains how to use the amount of data, and how it can be applied for the valorization of Cultural Heritage. The first answer has been, of course, representation: to describe, to imagine, to virtually reconstruct a cultural asset. The figures who tried to give first-hand answers have come from the computer science community, as highlighted before.

However, now that 3D models started being used for other purposes, the situation has grown increasingly complex, and the case of BIM modelling is an example in this sense. In fact, many problems arise when the goal is to be very accurate and fitting to reality, to attach many kinds of information to a 3D object, and to deal with different software (to be used by both experts and non-experts) to follow the changes over time. The BIM approach has not caused these problems, but it has enhanced their appearance.

Practically, the difficulties connected to 3D modelling can be grouped into different levels. The first challenge is the choice of the most suitable approach for modelling. As described in Tommasi et al. (2016), different approaches can be used from direct modelling to parametric object-oriented modelling. The main method of modelling inside the BIM environment should be the parametric one, which allows objects to be modified with the use of simple parameters. This approach applies very well to new building projects, but its use is more complex in Cultural Heritage because it requires the use of previously-built libraries (i.e., Revit, ArchiCAD, etc.) or to work with a complex series of commands (i.e., Grasshopper).

For the Basilica of San Marco, we chose the direct modelling approach, which allowed us to be very accurate in fitting the point-cloud model. This method is quite time consuming, but it allows to the exceptional elements (variations in the moldings, relevant damages, etc.) to be modeled whenever necessary.

Figure 4. The result after merging all the point-clouds of the interior.

Particularly, we adopted non-uniform rational B-spline (NURBS) modelling with Rhinoceros software to describe the complex elements which could not have been represented with other methods. We considered surface modelling as the most flexible and effective method to describe the complexity of the Basilica.

Moreover, this choice allowed us to complete and make available blocks of the model without necessarily having the whole survey of the adjacent area of the building. This peculiarity allowed us to test the realization process from the survey to the texturing of the modelled object, individuating the guide-lines for the successive areas, like in the case of the baptistery. This method also responded to the necessity of integrating the model in a complex system for data management; as a matter of fact, the plugin to import a Rhino model into BIM3DGS already existed.

The accuracy of the model was a relevant and complex task. The literature and the regulations already gave some instructions, mainly related to the concepts of level of development and level of detail (the same acronym — LOD — for different concepts, as explained in Ciribini, 2013). However, it is not equally clear how these suggestions can deal with the traditional — and diffusely adopted — concept of scale of representation. In the modelling of the architecture of San Marco, the principles we applied derive from traditional architectural drawing, where the scale of representation gives information not only about the relationship between the object and its representation, but also about the amount of detail, and define the smallest element that can be represented. In this case, at a scale of 1:50, the smallest element to be modelled had to be larger than 1 cm.

The last level of complexity refers to the possibility of being able to manage the model. The large amount of survey data — even if processed and strongly reduced — could produce models which are quite large (in terms of memory occupation), which cannot be easily used in navigation and exploration. Parametric modelling produces very efficient models, but it has the same limitations we described before. On the contrary, the NURBS modelling allows an optimal file dimension, as well as a very accurate description of cultural heritage assets.

4.2. Modelling of the Basilica of San Marco

The amount of data related to the interior of the Basilica is enormous and, of course, it was not possible to use all those points together. As we did for the survey stage, we subdivided the work into smaller areas to be modelled independently. This subdivision was very suitable for the acquisition stage, but it showed some problems in the modelling stage, as each part could not be considered as completely independent. In fact, to model the borders of the area, it was necessary to use the next model as a reference. The main pillars in the church, which support the main dome, were a typical example of this problem: they are located at the conjunction point of the central and the lateral nave with the transept, and each one of them belongs to at least three or four of the areas, and the number even grows if we consider the union with the second level of the Basilica with the matroneum (the women's gallery on the first floor). Therefore, the work went from modelling each single area to refining the borders and the conjunctions between areas in a second step. This approach has highlighted another technical problem. In fact, to refine the borders it was necessary to open at least two models (in some cases more than two models, as in the case of the connection between the pavement and the wall) in the same file, with obvious problems of calculation and of shared work. We solved this problem by using a feature of Rhinoceros (the software selected for modelling) called "work session". It allows the work to be shared between all participants in the working group: it is a base file which manages a series of 3D models and allows multiple users to navigate and work on them like it were a unique one, avoiding that the dimension (byte) of the original files make the process more demanding.

Apart from the technical problems of managing such a huge amount of data, the modelling stage presented some other difficulties. As we saw before, the variety of elements and their characteristics did not permit the creation of "standard" objects, but some of them with similar features could be grouped, and a standard modelling methodology was decided for each group. The method was based on the fundamental concept of the extraction and generation of sections of the object from the photogrammetric point-cloud. To do this, the point-clouds extracted from the photogrammetric models were converted into Pointools file format (.pod) and then imported into Rhinoceros. The Pointools plugin allowed the extraction of new views and slices from the point-clouds directly in the Rhinoceros environment, to build the generating curves.

Figure 5. Example of modelling of a complex architectural element (gothic window) from a point-cloud to a non-uniform rational B-spline (NURBS) model.

For the main structure of the walls, the aim was to obtain a simple surface which could describe the complex shape of the object, by the extraction of some control curves in strategic points such as the junction of the marble slabs or the connection with cornices, columns, and other decorations. The same approach was used for the construction of smaller objects consisting of simple surfaces or objects with constant geometric profiles (Figure 5). We included the marble cornices in this category, even if they have a refined sculptural decoration. This choice helped both

363

to avoid a large use of meshes and to respect the representation scale chosen for the 3D model (1:50). In fact, at this scale, some of the tiny decorations (smaller than 0.01 m) could be omitted from the model. This information is given by the applied textures and the orthophotos, rather than from the aspect of the object. Additionally, the large use of mesh model did not allow simple navigation of the model, being too heavy to manage.

Therefore, for each of the aforementioned objects we extracted a geometric construction section, which was modified in terms of dimension, rotation, and distortion to fit the point-cloud (Figure 5). The modifications were applied with a frequency based on the extension of the object, its deformations, and the presence of key points like corners, fractures, and junctions. As for most of the operations of the modelling stage, an automatic approach was very difficult to use, as each element had to be considered as unique. So, we could not set, for example, a predefined number of sections to be used to model a molding: it depended on the length, the direction, the presence of discontinuities, and the state of preservation.

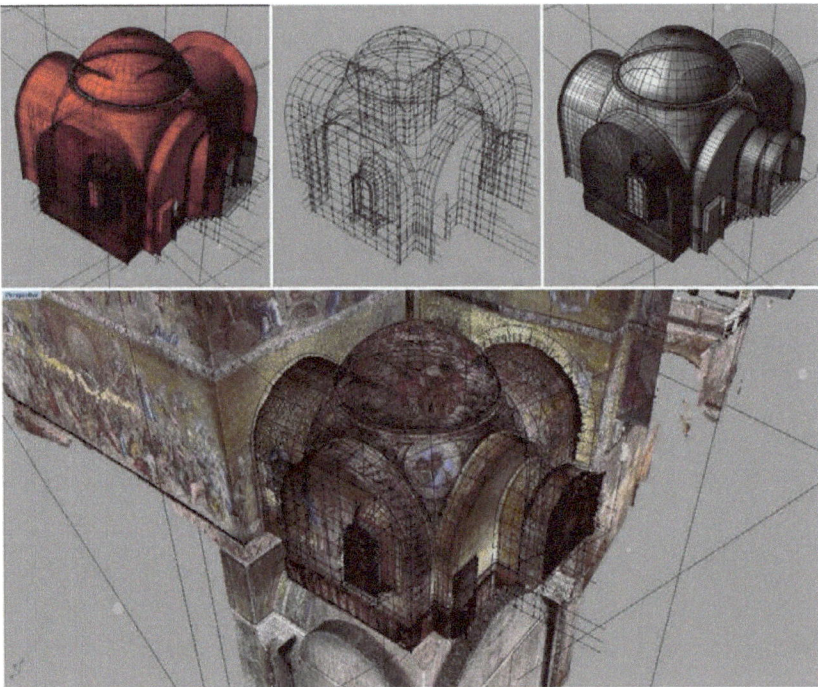

Figure 6. Different steps in the constructions of domes.

With the same method, we also modelled the many vaults, domes, and arches in the Basilica. The complex surface of the vaults was obtained by an interpolation of sections extracted from the data using a reference grid (Figure 6). This method was selected after several attempts. Firstly, we tried to obtain the dome directly from the point-cloud as a mesh model. Of course, the 3D model was actually fitting

364

the real surface, but it was very heavy for usage. We also tried to convert the mesh into a NURBS curve, with an automatic algorithm allowed by Rapidform and Geomagic, but the problem was the presence of some missing areas (because of the shadows due to the eaves of the moldings) and the difficulty in filling them. Instead, the method of interpolation of sections has allowed us to be very accurate and, at the same time, to close some of the holes in the vaults.

The outline and the pace of the grid depended on the dimension, geometrical features, and complexity of the shape that had to be obtained. For example, a simple and quite smooth barrel vault required a regular square grid with a 0.5 m step, whereas the small dome within the pillar (each pillar has a structure like the nave's, with a small dome and four smaller pillars) with its pendentive, required a radial pattern grid in a single direction, and a close series of horizontal sections that can reach a 0.2 m or smaller distance.

A completely different approach was employed for sculptural objects like capitals and statues, which cannot be represented by simple geometric shapes (Figure 7). In these cases, we went slightly beyond the limits that the scale of representation imposed on the rest of the model, both because these kinds of objects needed a particular attention and for the extreme complexity of their shape, which did not allow a geometric simplification. Those were the cases where the NURBS model gave way to a mesh model directly extracted from the dense point-cloud and triangulated and processed in Rapidform. If this kind of model was really fitting the surface and very much accurate, its problem resided in its memory occupation. Then, we noticed a second problem with this mesh model: the impossibility of using it in Rhinoceros for the automatic extraction of sections. One of the advantages of the geometric NURBS modelling of the Basilica was the possibility of extracting infinite sections from the model—a function present in all types of commercial BIM, and not a difficult one to use (compared with the mesh model, which made it impossible). Right now, we are defining the best workflow for converting the mesh models of statues and capitals into NURBS: of course, the problem of this conversion is the level of accuracy.

As the extraction of the model was finished, the 3D model obtained needed to be divided into singular elements, which had to be classified with a code and assigned to a layer to be used in the BIM3DSURVEY system. The classification was made according to the typology of architectural components; for example: vault, arch, dome, wall, cornice, column, floor, and windows.

Figure 7. Two capitals modelled in different ways according to their complexity. Above, a simpler capital modelled with NURBS, below the complex one, modelled with mesh.

5. Orthophotos

5.1. The Orthophotos for Architecture

In recent years, orthophotos have increased in terms of diffusion and use. The reason is quite evident, and it is related to the latest innovations in digital photogrammetry—Structure-from-Motion in particular (Chiabrando et al., 2015). The availability of systems that similarly, in a quasi-automatic way, allow a 3D model to be built, solved one of the problems of the process necessary to obtain orthophotos. This workflow is well-known, and it requires three elements: a DTM (digital terrain model) as detailed as possible, according to the scale of the orthophoto), the images oriented in the same reference system, and a reference plane where to project the object. Today, digital photogrammetry allows us to orient images and to obtain 3D models quite easily, and that has increased the diffusion of orthophoto usage.

In architecture, orthophotos represent a very important and valid tool, even if they are sometimes still not completely understood. The possibility of simultaneously acquiring measurements and qualitative information about an object is very advantageous in terms of time—not only in the survey stage, but also, for example, in the evaluation of the state of preservation.

366

5.2. The Orthophoto of San Marco Decorated Surfaces

As described before, in the Basilica of San Marco, the use of orthophotos was intended above all for the monitoring and the eventual restoration or reconstruction of mosaics. With this purpose, the orthophoto had to be of very high-resolution: for instance, the orthophoto should present not only the single tessera but also the grout lines. In practical terms, this means that we should also be able to recognize the elements with a dimension of about 1 mm.

This requirement influenced the acquisition stage: in fact, we created some photogrammetric models dedicated specifically to orthophotos: we used the 200 mm tele-lens in order to acquire these high-resolution images.

Regarding the process for orthophoto construction, after the dense point-cloud calculation, we decided to use (as 3D model for the orthophoto) the model we realized in Rhinoceros. This choice avoided the construction of high-resolution mesh models and the use of the NURBS models.

To achieve this, the NURBS objects were converted into a polygonal model and were used in the photogrammetric software Agisoft Photoscan.

The so-obtained polygonal object was imported in the original photogrammetric model and used. The texture of the object was given by the projection of the original oriented images used for the construction of the dense point-cloud on the object surface.

This passage was not essential to obtain the orthophoto, but it permitted the mesh to be colored enough to obtain a textured model for navigation purposes. In fact, in this procedure (which was initially planned for representation), we have no metric control of texture resolution on images, and it is only possible to set the final dimension of the texture atlas (typically in power of two to better interact in computer graphic application). If the object to be represented is not planar, the software constructs an atlas of pieces of the orthophoto, and it is difficult to deal with it for any photographic correction.

For the construction of the real orthophoto, after the choice of the projection plane, we had to select the best images to use. Not all cameras were necessary to realize the final orthophoto. So, we made a first selection by checking the quality of the images, to keep the best ones and to control the predetermined pixel-size, according to the analysis of Photoscan in terms of image quality and a direct review by the operator. Once the best-quality images were selected, we decided to use only the most nadir-positioned cameras in order to avoid an excessive stretching of pixels. Finally, the pictures that showed inhomogeneous light conditions were discarded to achieve the best possible radiometric information, which affected the general quality of the orthophoto.

Following this workflow, the orthophotos were produced with a pixel-size of 0.0005 m, which allowed the border of the tiniest mosaic tiles to be seen clearly, permitting use for restoration or analysis (Figure 8).

Figure 8. Unwrapped orthophoto of one of the main arches inner mosaic (**top**). Detail view of the same area, where we can recognize each single tessera and grout line (**bottom**).

6. Results for BIM Application

At the end of the process, we obtained three different results that could be used in the BIM environment with different purposes. In particular:

(a) a geometric NURBS model for the subsequent insertion into the BIM environment, with the possibility of extracting two-dimensional drawings such as plans and sections;

(b) a mesh model with low-resolution textures for online navigation;

(c) a high-resolution orthophoto.

These three results represent the overall answer to the requirements of the customer.

The NURBS model will be used especially in the BIM application to link the database (about history, restoration, planned conservation). It guarantees a good accuracy of the model, but at the same time, it is not so difficult to manage in terms of memory occupation and file size. In Rhinoceros, it is possible—as in other BIM software—to extract plans and elevations directly from the 3D model. This is a very relevant function for the management offices of the Basilica, as it allows innumerable sections and plans of each part and element of the architecture. The whole model can be uploaded in different parts according to the necessity of working on a particular area, as implemented in BIM3DSURVEY software.

The mesh model can be considered as a by-product of the workflow. To obtain the orthophoto, it was necessary to use a mesh model in Agisoft. This allows a realistic navigation inside the church, and can be used to see the mosaics in their real position in the 3D of the church.

Finally, the orthophoto represents the other request of the Procuratoria of San Marco. The high-resolution allows recognition of each individual tessera of the mosaic, but they are not easy to manage because of their file size.

To handle these three kinds of results (Figure 9), strictly connected but independent from each other, we selected the BIM system BIM3DSURVEY, developed by Politecnico di Milano for the main spire of the Duomo of Milano (Rechichi et al., 2016). It allows easy management of large-dimension 3D models, so it was particularly suitable for the architecture of San Marco. It maintains a strict relationship with the original Rhinoceros model, so it is possible to update the final model or add new parts not involved in this work. Furthermore, it allows the uploaded model to be linked to many other pieces of information. For this application, the system has been implemented with a function which allows the use of the model as a 3D spatial index for the orthophoto. The user can select an element and then it is possible to view and download the orthophoto of that element.

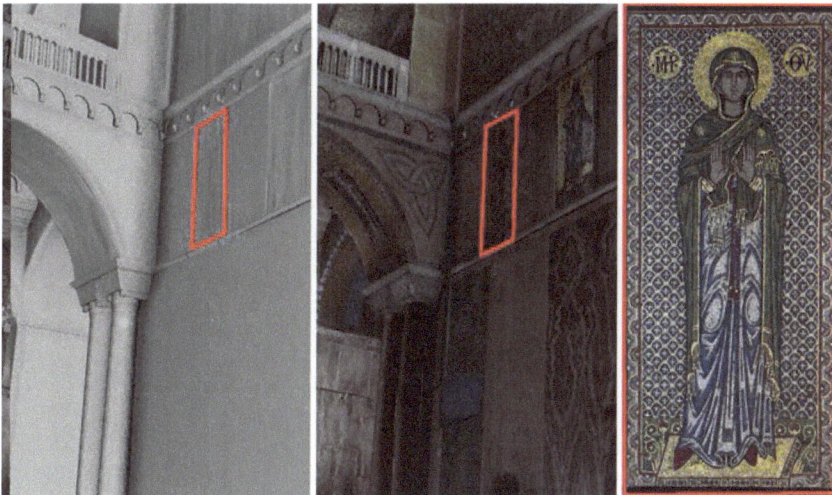

Figure 9. The three results of the previously described workflow which can be used in BIM environment. From left to right: 3D NURBS model, texturized mesh model, detail of orthophoto.

7. Conclusions

This research concerns the modelling of the Basilica of San Marco in Venice for a BIM environment. The characteristics of the church itself and the expected future use of the BIM make this case study a very particular one in the context of BIM applied to Cultural Heritage (HBIM). The BIM will be used to extract geometric and dimensional information and as an index for the orthophoto of mosaic and marble surfaces.

Both the acquisition and modelling stages were influenced not only by the material and geometrical features of the church, but also by its liveliness (considering the countless streams of visitors and continuous conservation activities that were being carried on). With regards to the survey, the experience of digital acquisition of the Basilica showed that digital photogrammetry—especially dense-image matching—is a very effective and flexible tool for architectural surveyors, especially if the project is referenced to a well-defined topographic network. The possibility of having photogrammetric models with different levels of point-density according to the complexity of the church was a very relevant issue, and allowed organization of the entire work and management of all data non-linearly. It also satisfied all the requirements of the customer and, at the same time, did not stop the entire working process.

The 3D model extracted from the point-clouds confirmed that, in the context of Cultural Heritage, we are quite far away from automation. Actually, automation in the field of dense point-cloud production requires an important human intervention, necessary to understand each architectural element, its role, and how it can be represented. NURBS modelling is a very effective tool, but only when used by specialists—experts not only in the field of survey, but also with a good knowledge of the architecture.

Some problems were not solved at all, but only faced to find a temporary solution. This is the case of the lighting set-up in the church for image acquisition: the spatial configuration of the Basilica with large rose windows and the system of light for architecture enhancement and touristic visit made it very difficult to delete all the reflections and color dominants that affect the images.

References

1. Adami, A.; Scala, B.; Spezzoni, A. Modelling and accuracy in a BIM environment for planned conservation: The apartment of Troia of Giulio Romano. *Int. Arch. Photogramm. Remote Sens. Spat. Inf. Sci.* **2017**, XLII-2/W3, 17–23, doi:10.5194/isprs-archives-XLII-2-W3-17-2017.
2. Barazzetti, L.; Previtali, M.; Roncoroni, F. Fisheye lenses for 3D modeling: Evaluations and considerations. *Int. Arch. Photogramm. Remote Sens. Spat. Inf. Sci.* **2017**, XLII-2/W3, 79–84, doi:10.5194/isprs-archives-XLII-2-W3-79-2017.

3. Chiabrando, F.; Donadio, E.; Rinaudo, F. SFM for orthophoto generation: A winning approach for cultural heritage knowledge. *Int. Arch. Photogram. Remote Sens. Spat. Inf. Sci.* **2015**, *XL-5/W7*, 91–98.

4. Ciribini, A. Level of Detail e Level of Development: I processi di committenza e l'information modelling. *Techne* **2013**, *6*, 90–99.

5. El Hakim, S.; Beraldin, J.A.; Picard, M.; Godin, G. Detailed 3D Reconstruction of Large-Scale Heritage Sites with Integrated Techniques. *IEEE Comput. Graph. Appl.* **2004**, *24*, 21–29.

6. Fai, S.; Graham, K.; Duckworth, T.; Wood, N.; Attar, R. Building Information Modeling and Heritage Documentation. In Proceedings of the 23rd International CIPA Symposium, Prague, Czech Republic, 12–16 September 2011.

7. Fassi, F.; Achille, C.; Gaudio, F.; Fregonese, L. Integrated strategies for the modelling of very large, complex architecture. *Int. Arch. Photogramm. Remote Sens. Spat. Inf. Sci.* **2011**, *XXXVIII-5*, 105–112

8. Fassi, F.; Achille, C.; Fregonese, L. Surveying and modelling the Main Spire of Milan Cathedral using multiple data sources. *Photogramm. Rec.* **2011**, *26*, 462–487.

9. Fassi, F.; Achille, C.; Mandelli, A.; Rechichi, F.; Parri, S. A new idea of BIM system for visualization, sharing and using huge complex 3D models for facility management. *Int. Arch. Photogramm. Remote Sens. Spat. Inf. Sci.* **2015**, *XL-5/W4*, 359–366.

10. Fregonese, L.; Taffurelli, L. Il pavimento musivo della Basilica di San Marco a Venezia: Ortofoto digitale 3D a grande scala a supporto dell'attività di tutela, di progetto e di cantiere. In *E-Arcom 2004—Tecnologie per Comunicare L'architettura*; CLUA Edizioni: Ancona, Italy, 2004:

11. Fregonese, L.; Monti, C.; Monti, G.; Morandi, S.; Taffurelli, L.; Vio, E. Il pavimento della Basilica di San Marco. La realizzazione dell'ortofoto 3D in digitale alla scala reale 1:1. In Proceedings of the XXII Convegno Scienza e Beni Culturali, Pavimentazioni storiche: Uso e conservazione, Bressanone, Italy, 11–14 July 2006; Arcadia Ricerche: Venice, Italy, 2006; pp. 99–108.

12. Fregonese, L.; Monti, C.; Monti, G.; Taffurelli, L. The St. Mark's Basilica pavement. The digital ortophoto 3D realisation to the real scale 1:1 for the modelling and the conservative restoration. In Proceedings of the Innovations in 3D Geo Information Systems, First International Workshop on 3D Geoinformation, Kuala Lumpur, Malaysia, 7–8 August 2006.

13. Fregonese, L.; Achille, C.; Taffurelli, L.; Fassi, F.; Monti, C.; Vio, E. 3D database of the knowledge of material data: Analysis of the complex structure of the Pentecoste dome in St. Mark's Basilica in Venice. In Proceedings of the International Congress Domes in the World, Florence, Italy, 19–23 March 2012.

14. Fregonese, L.; Taffurelli, L.; Adami, A.; Chiarini, S.; Cremonesi, S.; Helder, J.; Spezzoni, A. Survey and modelling for the BIM of Basilica of San Marco in Venice. *Int. Arch. Photogramm. Remote Sens. Spat. Inf. Sci.* **2017**, *XLII-2/W3*, 303–310, doi:10.5194/isprs-archives-XLII-2-W3-303-2017.

15. Gaiani, M.; Remondino, F.; Apollonio, F.; Ballabeni, A. An advanced pre-processing pipeline to improve automated photogrammetric reconstructions of architectural scenes. *Remote Sens.* **2016**, *8*, 178, doi:10.3390/rs8030178.

16. Godin, G.; Rioux, M.; Levoy, M.; Cournoyer, L.; Blais, F. *Marble Surfaces*; National Research Council of Canada: Ottawa, ON, Canada, 2001.

17. Grilli, E.; Menna, F.; Remondino, F. A review of point-clouds segmentation and classification algorithms. *Int. Arch. Photogramm. Remote Sens. Spat. Inf. Sci.* **2017**, *XLII-2/W3*, 339–344, doi:10.5194/isprs-archives-XLII-2-W3-339-2017.

18. Logothetis, S.; Delinasiou, A.; Stylianidis, E. Building Information Modelling for cultural heritage: A review. In Proceedings of the 2015 25th International CIPA Symposium, Taipei, Taiwan, 31 August–4 September 2015; Volume II-5/W3, pp. 177–183.

19. Murphy, M.; McGovern E.; Pavia, S. Historic building information modelling (HBIM). *Struct. Surv.* **2009**, *27*, 311–327.

20. Murphy, M.; McGovern, E.; Pavia, S. Historic Building Information modelling—Adding intelligence to laser and image based surveys. In Proceedings of the ISPRS Trento 2011 Workshop, Trento, Italy, 2–4 March 2011; Volume XXXVIII-5/W16.

21. Perfetti, L.; Polari, C.; Fassi, F. Fisheye photogrammetry: Tests and methodologies for the survey of narrow spaces. *Int. Arch. Photogramm. Remote Sens. Spat. Inf. Sci.* **2017**, *XLII-2/W3*, 573–580, doi:10.5194/isprs-archives-XLII-2-W3-573-2017.

22. Rechichi, F.; Mandelli, A.; Achille, C.; Fassi, F. Sharing high-resolution models and information on web: The web module of BIM3DSG system. *Int. Arch. Photogramm. Remote Sens. Spat. Inf. Sci.* **2016**, *XLI-B5*, 703–710, doi:10.5194/isprs-archives-XLI-B5-703-2016.

23. Remondino, F.; Spera, M.G.; Nocerino, E.; Menna, F.; Nex. F. State of the art in high density image matching. *Photogramm. Rec.* **2014**, *29*, 144–166.

24. Remondino, F.; Rupnik, E.; Nex, F. Automated processing of oblique imagery. *GIM Int.* **2014**, *28*, 16–19.

25. Simeone, D.; Cursi, S.; Toldo, I.; Carrara, G. B(H)IM—Built Heritage Information Modelling. Extending BIM to historical and archeological heritage representation. In *Proceedings of the 32nd International Conference on Education and research in Computer Aided Architectural Design in Europe*; Northumbria University: Newcastle upon Tyne, UK, 2014; Volume 1, pp. 613–622.

26. Szeliski, R. *Computer Vision: Algorithms and Applications*; Springer Science & Business Media: New York, NY, USA, 2010.

27. Tommasi, C.; Achille, C.; Fassi, F. From point-cloud to bim: A modelling challenge in the cultural heritage field. *Int. Arch. Photogramm. Remote Sens. Spat. Inf. Sci.*2016, *XLI-B5*, 429–436, doi:10.5194/isprs-archives-XLI-B5-429-2016.

28. Volk, R.; Stengel, J.; Schultmann, F. Building Information Modelling (BIM) for existing buildings—Literature review and future needs. *Autom. Constr.* **2014**, *38*, 109–127.

29. Vosselman, G.; Maas, H.G. (Eds.) *Airborne and Terrestrial Laser Scanning*; Whittles: Dunbeath, UK, 2010; Volume 318.

30. Zhang Z. Iterative point matching for registration of free-form curves and surfaces. *Int. J. Comput. Vis.* **1994**, *13*, 119–152.

MDPI AG
St. Alban-Anlage 66
4052 Basel, Switzerland
Tel. +41 61 683 77 34
Fax +41 61 302 89 18
http://www.mdpi.com